Dr. DAVID BUDESCU
DEPT. OF PSYCHOLOGY
UNIVERSITY OF HAIFA
HAIFA 31999 ISRAEL

FUZZY SETS IN PSYCHOLOGY

ADVANCES IN PSYCHOLOGY

56

Editors:

G. E. STELMACH

P. A. VROON

NORTH-HOLLAND
AMSTERDAM · NEW YORK · OXFORD · TOKYO

FUZZY SETS IN PSYCHOLOGY

Edited by

Tamás ZÉTÉNYI
*Eötvös L. University
Budapest, Hungary*

1988

NORTH-HOLLAND
AMSTERDAM · NEW YORK · OXFORD · TOKYO

© ELSEVIER SCIENCE PUBLISHERS B.V., 1988

All rights reserved. No part of this publication may be reproduced, stored in a
retrieval system, or transmitted, in any form or by any means, electronic, mechanical,
photocopying, recording or otherwise, without the prior permission of the copyright owner.

ISBN: 0 444 70504 X

Publishers:
ELSEVIER SCIENCE PUBLISHERS B.V.
P.O. Box 1991
1000 BZ Amsterdam
The Netherlands

Sole distributors for the U.S.A. and Canada:
ELSEVIER SCIENCE PUBLISHING COMPANY, INC.
52 Vanderbilt Avenue
New York, N.Y. 10017
U.S.A.

PRINTED IN THE NETHERLANDS

PREFACE

This volume provides an up-to-date picture of the current status of theoretical and empirical developments in the application of fuzzy sets in psychology; and will be of interest both to the advanced student in the field as well as to anyone possessing a basic scientific background.

There is little doubt that psychology requires formal, even mathematical frameworks for handling graded categories with blurried boundaries. It is believed that not only are the phenomena of human thought and behavior inherently fuzzy, but researchers in this field must also use concepts and theoretical schemata which themselves are fuzzy. Most verbally expressed hypotheses in psychology contain fuzzy nouns, predicates, modifiers and comparatives. The problem of fuzziness in general is of a fundamental nature, and suggests that we should treat human(istic) systems in terms of fuzzy logic based linguistics calculus rather than using discrete numerical language.

Lofti A. Zadeh (1965) was the first to formalize the idea by defining characteristic functions for sets which map onto the closed interval zero and one, rather than onto the conventional set of values zero and one. Since then, there have been many contributions to fuzzy set theories of various kinds, and some of them enjoy wide application in engineering and applied science. Zadeh attempted to develop a linguistic calculus "fuzzy calculus" that should be applicable to variuos sorts of systems. This present volume is concerned with many facets of fuzzy sets in psychological systems in this genuine sense.

Fuzzy set theory could benefit researchers in at least two ways: first, as a metaphor or model for ordinary thought, and secondly as an aid to data analysis and theory construction. One can find examples for both kinds in the volume.

The dialogue between fuzzy set theory and psychology has just begun to address these possibilities, but the area has developed so rapidly during recent years that I felt there was a need to put together a volume which covered the newest contemporary issues. This work will serve as a resource for scholars who have been interested in fuzzy logic and to provide them with an up-to-date survey of contemporary work in the field.

I would like to thank the authors who have contributed their time and effort, for without their help the completion of the volume would not have been possible. I am indebted to George E. Stelmach for his guidance and support. I wish to express my gratitude to William Barratt who took upon himself the laborious work of checking the English of some of the texts. Very special thanks to K. Michielsen and T. Kraaij of North-Holland Publishing House.

T.Z.

CONTENTS

    Preface   v
    Content   vii

1. Possibility Theory, Fuzzy Logic, and Psychological Explanation   1
   Michael Smithson

    Has psychology inherited the wrong mathematics? 2
    The nature of possibility 9
    Measuring freedom and constraint 19
    Weak entailment and prediction 33
    Possibilism in psychological theory and research 40
    References 47

2. Quantifiers As Fuzzy Concepts   51
   Stephen E. Newstead

    Quantifiers and syllogistic reasoning 52
    Quantifiers and rating scales 57
    Psycholinguistic studies of quantifiers 61
    Overview and synthesis 63
    Summary and conclusions 68
    References 69

3. A Common Framework For Colloquial Quantifiers And Probability Terms   73
   A.C. Zimmer

    Introduction 74
    Fuzzy numbers and qualifiers 78
    Fuzzy arithmetic as a model for reasoning 81
    Context specificity of qualifiers 84
    Conclusions 87
    References 87

4. An Empirical Study Of The Integration Of Linguistic Probabilities   91
   R. Zwick, D.V. Budescu and T.S. Wallsten

   Fuzzy set aggregation connectives 92
   Previous empirical results 97
   Method 102
   Results 106
   Discussion 120
   References 121

5. A Fuzzy Propositional Account Of Contextual Effects On Word   127
   Recognition
   Jay Rueckl

   Contextual effects on word identification 129
   An alternative paradigm 131
   Experiment 1. 134
   Experiment 2. 139
   General discussion 145
   Conclusion 150
   References 152

6. M-Fuzziness In Brain/Mind Modelling   155
   György Fuhrmann

   Introduction 156
   The conception of m-fuzziness 157
   Possible implementation in the brain 175
   On m-fuzzy relations 189
   Concluding remarks 195
   References 197

7. The Weighted Fuzzy Expected Value As An Activation Function   203
   For The Parallel Distributed Processing Models
   David Kuncicky and Abraham Kandel

   Introduction 204
   Activation functions 208
   The fuzzy expected value as activation function 210
   The weighted fuzzy expected value as activation function 217
   Density and orthogonality assumption 219
   Comparison of weighted mean and WFEV 225
   Conclusion 227
   References 231

8. Fuzzy Logic With Linguistic Quantifiers: A Tool For Better   233
   Modeling Of Human Evidence Aggregation Processes?
   Janus Kacprzyk

   Introduction 234
   On some fuzzy logic based calculi of linguistically quantified
   propositions 237
   Multiobjective decision making models based on fuzzy logic
   with linguistic quantifiers 241
   Multistage decision making (control) models based on
   fuzzy logic with linguistic quantifiers 246
   Group decision making models with a fuzzy majority
   based on fuzzy logic with linguistic quantifiers 250
   Concluding remarks 259
   References 260

9. Origins, Structure, and Function Of Fuzzy Beliefs   265
   Jósef Chwedorowicz

   Formation of fuzzy beliefs 269
   Language of fuzzy beliefs 273
   System of fuzzy beliefs 276
   Structure of fuzzy beliefs 285
   Conclusion 286
   References 292

10. Brain Activity and Fuzzy Belief 297
    G. Greco and A.F. Rocha

    Introduction 298
    The brain activity 300
    Recalling the text 304
    Methods 306
    Results 308
    Discussion 312
    References 315

11. Acquisition Of Membership Functions In Mental Workload 321
    Experiments
    I.B. Turksen, N. Moray and E. Krushelnycky

    Introduction 322
    Structure of the experiment 324
    Scale properties of data 327
    Significance tests 335
    Linguistic interpretation 339
    Conclusions 341
    References 342

12. A Fuzzy Set Model Of Learning Disability: Identification 345
    From Clinical Data
    J.M. Horvath

    Difficulty with traditional mathematics 348
    The search for models 350
    Learning disability - a fuzzy set 368
    The process of quantification 369
    Conclusions 379
    References 380

13. Towards A Fuzzy Theory Of Behaviour Management   383
    Vladimir B. Cervin and Joan C. Cervin

    Introduction 384
    Reference case 389
    Practitioner's operations 391
    Expected subject events 392
    Symbolic statement of fuzzy algorithms 406
    Concluding remarks 417
    References 421

14. Practical Applications And Psychometric Evaluation   425
    Of A Computerized Fuzzy Graphic Rating
    Beryl Hesketh, Robert Pryor, Melanie Gleitzman and Tim Hesketh

    The need for fuzzy measurment in psychology 426
    Fuzzy graphic rating scale 432
    Evaluation of fuzzy graphic rating scale 438
    Results 440
    Potential applications 445
    References 450

Author Index   455
Subject Index   463

POSSIBILITY THEORY, FUZZY LOGIC, AND PSYCHOLOGICAL EXPLANATION

Michael SMITHSON

*Behavioural Sciences Department
James Cook University,
Queensland 4811 Australia*

Psychologist have reiled almost exclusively on statistical models and methods for the quantitative analysis of human behavior. Because they invoke stochastic determinism, such models are incapable of incorporating human intentionality, purposive choice, or agency along with constraints and influences on behavior. However, many psychological accounts and theories mix voluntarism with constraint, and those theories are not adequately translated by conventional statistical models. Fuzzy logic and possibility theory offer an alternative framework which is compatibel with psychological explanations that permit choice under partial and uncertain constraints.

This paper outlines that framework and demonstrates its application to research and theory construction in psychology. The main components of the framework are "weak" prediction and entailment models provided by fuzzy logic, and the modeling of domains of choice via possibility theory. Accordingly, the conceptual foundations of fuzzy logic and possibility theory are outlined from the standpoint of psychological theory and explanation.

Applications include models of preference and freedom of choice, a calculus of freedom, fuzzy predictive models, and possibilistic theories of human behavior.

Has Psychology Inherited the Wrong Mathematics?

This paper proposes possibility theory as a basis for an alternative to the conventional statistical paradigm employed by researchers in psychology. Before a case may be made for possibility theory, we must have a motivating argument for alternatives in the first place. Accordingly, this section presents several criticisms of the statistical tradition in psychology which are more radical than the usual corrections of technical misuse and abuse. The principal claim here is that both psychological theory and data analysis have been needlessly constrained and distorted by the widely held assumption that a peculiar dialect of mathematics (the Neyman-Pearson-Fisher statistical framework, hereafter denoted by NPF) is the only one suited to quantitative data analysis in the human sciences.

The mathematical legacy bequeathed to psychology is dominated by the NPF, which in turn bears recognizable traces of its origins in military strategic and decision analysis, industrial quality control, and agricultural experimentation. The NPF was designed to solve the problems posed in those fields, not the human sciences, and yet it has been adopted almost without modification by psychologists. So widespread is its influence that often debates over quantitative versus qualitative research styles, critiques of positivism, and the development of innovative approaches to psychological research have rested on the assumption that mathematical or quantitative analysis is synonymous with the NPF. Furthermore, the NPF has pervaded the very design of psychological studies as well as theory construction itself. After all, the nature of researchability is linked to the descriptive, expressive, and analytical capabilities of the mathematical language within which "data analysis" is couched. On a more mundane level, it is highly likely that many fledgling researchers have made their choices for and against whole schools of research methods partly on the basis of this assumption.

That assumption is mistaken in principle, of course, since mathematics encompasses far more than the NPF and even probability theory boasts several schools of thought and practice. But does it matter? If exclusive reliance on the NPF prohibits mathematization or systematic investigation of key theoretical concepts in psychology, then the answer to this question must be yes. To assess whether this is the case, we must analyze how the NPF translates psychological theories into statistical models and tests. Without much loss of generality I shall take the general linear model (GLM) as the primary object of discussion. Not only does the GLM include the most prized data analytic techniques in psychology (the varieties of ANOVA, linearizable regression models, log-linear analysis, discriminant analysis, factor analysis, and their progeny), but it also embodies the properties most sought-after in most nonparametric techniques.

The GLM cleaves variation in human behavior into two components: That predicted by the instrumental variable(s), and the unpredicted component which is treated as arising from either unobserved influences or random processes. The predictions allowed in the GLM are necessarily one-to-one by virtue of their restriction to a linearizable model. This constraint entails translating any psychological prediction into a one-to-one relation between states or values on the independent variables and those on the outcome variable. Human agents, therefore, are not permitted a range of alternative outcomes in these models. The sole source of unpredictability allowed these actors by the GLM lies in the random processes responsible for that part of the unpredicted component of variation not eventually explicated in terms of additional predictors. A literal reading of the GLM yields a stochastically deterministic view of human behavior. Insofar as people are not behaving in accordance with a one-to-one prediction, they are behaving randomly.

Modern currents in many fields of psychological research, however, are not confluent with the GLM translation of theories into its stochastically deterministic terms. Indeed, several prominent schools of thought have explicitly rejected that viewpoint while nevertheless continuing to use the GLM and NPF in research. Cognitively oriented psychologists long ago rejected simple S-R accounts, for instance, in favor of a view in which behavior is understood in relation to the actor's mental images of the world and his or her plans, intentions, or strategies (cf. Miller, Galanter, & Pribram, 1960). Cognitivism virtually requires concepts like intentionality, purposiveness, choice or decision, and agency, which have no place in a one-to-one predictive model with randomicity as the sole source of uncertainty.

Perhaps the most obvious examples of theories embodying this view come from social psychology. Classic examples include cognitive dissonance theory and attribution theory, both of whose original formulations portrayed people as mindful actors whose choices are only partly constrained by situational or structural influences and not even strictly determined by motivations. The subsequent development of those theoretical frameworks became increasingly tainted by the GLM requirement (often confused with an imperative of the experimental method itself) that an identical response be predicted for all subjects under the same experimental conditions, despite the fact that this was never necessitated by the theories themselves.

This trend has been echoed in several psychological research areas, and it directly opposes the paradigmatic shift in social psychology during the last 25 years which, as Ginsburg (1979, p. 2) points out, is towards the view that people are active agents, "capable of making plans and pursuing objectives, of acting as well as reacting, of doing things for reasons as well as having been forced to do them by causes". Rule-following or rule-using behavior, for instance, necessitates a model that allows for multiple possible responses to a situation. Likewise, a skills-oriented or competence and performance model of behavior finds explanations in "enabling conditions" rather than causes in the old sense. The focus on enabling or empowering has characterized a number of modern therapeutic schools in clinical psychology as well. Unfortunately, the GLM cannot tell the clinician whether the treatment was sufficient but not necessary, necessary but not sufficient, or weakly contributing to improvement in the clients. A "statistically significant" difference between the means of treatment and control groups could arise from any of those outcomes.

These shortcomings are not in any way tied to the experimental method itself. Survey researchers are no better off. Coleman's recent critique of survey methods pinpoints a key missing ingredient in statistical survey research as "an explicit purposive or intentional orientation" and remarks that "...the statistical association basis for inference in survey analysis seemed to have little natural affinity for the intentions or purposes of individuals" (1986, p. 1314). But Coleman strays wide of the mark when he recommends replacing statistical association with "meaningful connections" between variables. The difficulty lies not in the use of associations per se, but in the GLM assumption that it must be one-to-one. Weaker models of association might be more compatible with concepts like purpose, intention, or choice because they could permit a range of possible outcomes for actors under identical conditions.

A defender of the NPF or GLM might object that these problems can be handled by using confidence intervals to express the range of likely responses, and that the bounds of those intervals could be the subject of theoretical predictions. After all, what the GLM really amounts to is a "moving average" model that predicts the central tendency of actors' responses conditional on the values of various instrumental variables. A "dual" GLM approach that explicitly accounts for conditional central tendency and dispersion should capably represent the variability in actors' responses. I raise this argument in anticipation, since it comprises the heart of Cheeseman's (1985) defense of probability against nonprobabilistic alternatives for representing uncertainty in the artificial intelligence literature. It also highlights the most fundamental limitation of the NPF, and that is the assumption that probability is capable of representing any form of uncertainty in human thought or behavior.

The NPF (and, for that matter, any other statistical paradigm such as the Bayesian approach) utilizes probability to handle two kinds of uncertainty. One is directly attached to the model of behavior and constitutes uncertainty about what the actor(s) will do. The second kind concerns the state of our own or others' knowledge about what the actor(s) may do. Probability is incapable of entirely capturing the first kind of uncertainty because it deals only in likelihood or relative frequency of occurrence rather than in concepts such as potentiality, possibility, or necessity. That I can ride my bicycle to work (but may or may not) is not representable in probabilistic terms. Nor is the datum that 43% of the employed adults in my suburb could ride their bicycles to work. Nonprobabilistic kinds of uncertainty have a central role in many psychological theories, especially where counterfactuals form part of the basis for explaining behavior. In a skills model of behavior, for instance, Clarke (1983, p. 202-203) points out that what an actor does is limited mainly by what she or he can do or knows how to do. A similar argument could be advanced for a host of perspectives in psychology that are guided by concepts such as competence, resources, potential, decision making, strategic behavior, and reflexivity.

Incomplete knowledge is not satisfactorily captured by probability either. Both normative and psychological arguments support this claim. Given the simplest case of a binary outcome setup (A and B, say), probability fails to distinguish uncertainty about whether A will occur from ignorance about whether A will occur. The proposition that $P(A) = 1/2$ could mean that we know A and B are equally likely, or it could mean that we are utterly ignorant of the likelihood of A or B. This argument has recurred in several contexts - see Shafer (1976) for a philosophical account, and Cohen (1977) for a discussion of legal uncertainty. Likewise, probability is incapable of capturing any ambiguity, vagueness, or fuzziness in the nature of A and B - as Zadeh (1980) has forcefully argued.

Given the limitations of the NPF and its faulty translation of essential elements in modern psychological theories, it is no accident that many of the more sophisticated and innovative paradigms in recent psychology are fundamentally innumerate. Nor is it always the prospect of quantification itself that puts their adherents off statistical methods (cf. Collins, 1984). Clearly we require a formal framework that is capable of handling quantitative data without excluding agency and choice on the one hand, but which incorporates constraint, entailment, and prediction, on the other. This framework can be nothing other than formal or mathematical at least in part, since we are dealing with quantification. Before moving on to outline a candidate for such a framework, I should like to argue that it is possible to construct one without violating any known rules of evidence for imputing voluntarism, determinism, or randomicity. These arguments are well-known, but they bear repeating in this context.

First, it should be obvious that there is no incompatibility between predictability and voluntarism or agency. People may choose to behave predictably. Conversely, unpredictability is no evidence for indeterminacy. The algorithms for generating so-called nonrepeating decimals are deterministic, but their output not only is unpredictable to a remote observer but passes tests for randomicity as well. Likewise, there is no litmus test for randomicity. The empirical demonstration that events fit a particular distribution is not sufficient to imply an underlying random process. What is plausible for dice (since they have no preferences) is not so for people (who do have preferences and may exercise them).

The conclusion following from these arguments is that the decision to use statistical models for analyzing quantitative data rests on interpretive and theoretical, rather than logical or empirical, grounds. Figure 1 displays two scatterplots whose patterns are nearly identical. In both plots mere inspection reveals that high X is sufficient but not necessary to produce high Y. Are these causal relationships, or do they reflect choice under partial restriction? Is the variation in the upper triangle of each plot generated randomly, or does it reflect individual choices made under weak constraints? The uppermost plot is taken from a study of attitudes towards the use of violence for political ends (Muller, 1972). X refers to intention to engage in violence for political ends (IPV) and Y to approval of the use of violence for such ends (APV). The cases in this plot are individual responses. The second plot is adapted from a study of national breast cancer death-rates as a function of per-capita consumption of animal fat (Carroll, 1974). The cases in this plot are nations.

Most medical models would hold that the relationship between animal fat intake and risk of cancer is causal, while some social scientists might hesitate on whether approval of violence is causally related to intention to engage in it. Likewise, while it might make sense to most social scientists to conclude from Figure 1 that people who approve of violence may choose to engage in it or not, most medical researchers would not wish to claim that countries in which people do not eat much animal fat may choose whether to have high death-rates from breast cancer. The data in these plots say the same things logically and empirically. The only reasons for assigning causality or intentionality to one or the other stem from one's interpretive or theoretical perspective. Therefore, the framework proposed in this paper should not be construed as a replacement for statistical models in all circumstances, but instead as permitting the articulation and investigation of interpretations that cannot be handled by the statistical perspective.

	IPV					
APV		0.0	0.2	0.4	0.6	0.8	1.0
	1.0	2	2	2	4	4	5
	0.8	2	4	11	8	3	2
	0.6	9	17	8	13	1	0
	0.4	43	45	39	3	1	0
	0.2	89	75	5	0	0	0
	0.0	97	5	0	0	0	0

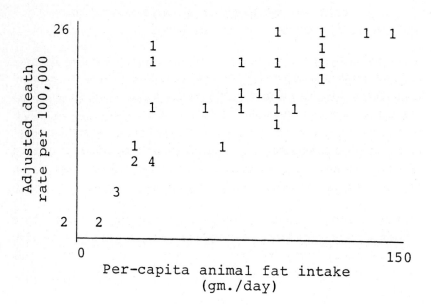

FIGURE 1
Two Scatterplots

Our framework involves the analysis of domains of possibility for action under constraints induced by "weak" models of entailement or prediction. One-to-many and other weak entailement models combine possibilistic and statistical concepts and, unlike their one-to-one counterparts, permit a range of options for human choice or intention. Why possibility? First, choices may be made only when possibilities exist. Secondly, structural or other constraints may be thought of as restrictions on the domain of conceivable possibilities. Third, within a range of possible actions, behavioral variation need not be dismissed as erroneous or random. Instead, it may be intepreted by motivational or other theories of choice: "An action is the outcome of a choice within constraints" (Elster, 1983 p. vii).

The Nature of Possibility

Although the concept of possibility often is implicitly referred to in theories of action, possibility itself is rarely explicated. Until recently there was no analytic framework for possibility on a par with the well-developed calculus of probability. I will introduce a framework based on developments by Zadeh (1978) and his colleagues which enables sensible definitions of graded possibility, as well as joint, conditional, and marginal possibility. First, however, we require some clarification of the concept of possibility.

Broadly speaking, an event is possible if it could happen. It need not have happened yet, nor need its occurrence be probable or inevitable. On the other hand, if an event is impossible then it would seem to have a probability of 0 as well. Beyond these general remarks, at least three questions need to be addressed:

(1) Are there distinct kinds of possibility?

(2) How is possibility related to probability or necessity?

(3) How may possibility be quantified and measured?

Several modern writers have decided that there are different kinds of possibility, and have made two kinds of distinction: semantic and ontological. Elster (1978) and Lukes (1978) both emphasize the difference between "real" and "abstract" possibility in social theory. The former refers to real situations, while the latter is essentially a logical device. Wilkinson (1981), influenced by the possibilistic psychology of G.Kelly (1955, 1969) as well as modal logic, proposes a three-level hierarchy which includes the linguistic theoretic sense of possibility as opposition, the more commonsensical notion of possibility as a range of alternatives, and the general concept from modal logic of "possible worlds".

Hacking (1975) and Zadeh (1978) have pointed out that there are at least two semantically distinct usages of the term in ordinary language. In Zadeh's terms, possibility may be thought of as either compatibility or feasibility. The statement "John is tall" may be viewed as a restriction on John's possible height, whereby in a Western context 6'2" is more compatible (possible) with the statement than 5'2". Feasibility or ease of accomplishment does not apply to intransitive concepts such as tallness, but it does to events or actions (e.g., "it is almost impossible to find a needle in a haystack"). A related sense is possibility as availability or accessibility (e.g., "it is possible for Lisa to have cake"). Hacking's major distinction is between "possible for" and "possible that". The former refers to feasibility or availability, while the latter denotes a state of knowledge (e.g., "it is possible that George may have changed his mind"). Hacking calls the first kind of possibility "de re" and the second "de dicto".

These latter distinctions have a history that can be traced directly to the origins of probability. As Hacking persuasively argues (1975, ch. 14), the definition of probability itself was initially couched in terms of "equipossibility" and "grades of possibility". Not only did 17th and 18th century mathematicians such as Leibniz, Jacques Bernoulli, and Laplace hold by this conception, but so did modern classicists such as Borel and von Mises. Events were recognized as being possible in degree, and von Mises makes the case in 1928 for grades of possibility using exactly the same intellectual weapons as Zadeh does 50 years later. Furthermore, several of the early probabilists distinguished between de re and de dicto possibility as a basis for a similar distinction between physical and epistemic probability. Of course, identifying probability with possibility is a conceptual error, and as mathematicians and philosophers found better foundations for probability theory possibility was abandoned and ignored.

The distinction between de re and de dicto possibility points to two arguments for possibility as a graded concept. Zadeh's (1978) proposal focuses on the de dicto when he argues on intuitive grounds that some actions may be more difficult, hence less possible, than others. A "frequentist" case may be made for degrees of de re possibility as well. If, say, 90% of the people in suburb A own cars but only 45% of the people in suburb B do, then private automotive transport is twice as possible for the suburbanites in A as for those in B. Both the subjectivist and frequentist cases support Zadeh's proposal that possibility be measured on a scale from 0 to 1 in a manner similar to probability, with 0 indicating entire impossibility and 1 entire possibility.

Clearly possibility is not a kind of probability, nor can probability represent possibilistic uncertainty. A phenomenological case for this rests on the observation that some kinds of uncertainty are nonprobabilistic; they do not involve references to likelihood or chance. That an option is possible says nothing about the likelihood that it will be chosen. A frequentist argument is that a grade of possibility puts an upper limit on probability. If 45% of a given population own cars, then at most 45% could possibly use them for transportation. The actual percentage of people using private automotive transport might well be less than 45% but it cannot be any greater. Likewise, the necessity of a given option places a lower bound on probability. If, say, 15% of the population have no other option than driving their own cars, then at least 15% must necessarily use private automotive transport. Formally, if we denote the possibility of the ith option by po_i, its probability by pr_i, and its necessity by ne_i, then

$$ne_i \leq pr_i \leq po_i. \tag{1}$$

The simple relationship in (1) immediately suggests two useful tools in a possibilistic calculus: the measurement of freedom of action, and the measurement of relative preference. Insofar as an action is possible but not necessary, then people are free to perform it or not. A natural definition of freedom of action for the ith option, then, is

$$F_i = po_i - ne_i. \tag{2}$$

"Preference" in many surveys is measured simply by the percentage of people who select a given option. But this practice assumes that all options have a possibility of 1 and a necessity of 0. True preference is indicated by the extent to which people select an option when they are free to do so, and a valid relative preference measure may be defined by

$$S_i = (pr_i - ne_i) / (po_i - ne_i). \tag{3}$$

This definition hints that theorists may well find it profitable to consider two sets of influences on human behavior: Those influencing the po_i and those influencing the ne_i.

By way of illustration, consider a survey (Smithson, 1983) of elderly citizens' preference patterns for transportation in a northern Australian community. For those elderly who could not drive a car, there were four conceivable transportation options: (1) being driven by a friend or relative, (2) taking a taxi, (3) taking a bus, and (4) self-locomotion (bicycling, walking, etc.). The elderly were asked questions that determined whether each of these options was possible, necessary, and which actually were used. Table 1 displays the portions of the sample for whom these options were possible, necessary, or used on various occasions. While 32% of the elderly used buses and 32% were driven by someone for shopping, we cannot conclude that these two options were equally preferred because they were not equally possible. From (3) we have $S_1 = (0.32-0.02) / (0.42-0.02) = 0.75$, while $S_3 = (0.32-0.12) / (0.72-0.12) = 0.43$. The elderly prefer being driven by a friend or relative almost twice as much as they do taking the bus, when it is possible.

TABLE 1
Transport Options for Elderly

Transport Mode	Shopping	Chosen for Possible	Necessary
Being driven	32%	42%	2%
Taxi	21%	70%	0%
Buses	32%	72%	12%
Other	20%	93%	8%

An intuitively plausible candidate for negation in possibility theory would be $1-po_i$, but before welcoming this as our definition we must distinguish between the "impossibility" of the ith option and the "possibility of not-i". Conventionally the possibility of not-i is called "internal" negation, while the term for the impossibility of i is "external" negation. Elster (1983) provides an example of this distinction from Zinoviev's account of the Soviet government's tendency to replace internal with external negation. Thus, while for a time it became possible not to quote Stalin the government responded by making it impossible to quote him. Here $1-po_i$ refers to the impossibility of the ith option, while $po_{i'}$, the possibility of not-i, is measured by the degree to which the ith option is not necessary:

$$1 - po_i = ne_{i'}, \text{ and} \qquad (4)$$

$$po_{i'} = 1 - ne_i. \qquad (5)$$

For instance, the portion of elderly who could take the bus is $po_i = 72$ so $1-po_i = 0.28$. On the other hand, since the portion who must take the bus is $ne_i = 0.12$, $po_{i'} = 0.88$.

Given measures for possibility, necessity, negation, and freedom, a calculus of possibility may be constructed along the lines of the standard theory of probability. I will do so here using a frequentist approach, delaying the discussion of a basis for subjective or epistemic possibility until the next section. This perspective differs somewhat from Zadeh (1978), whose framework does not often distinguish physical from epistemic possibility. The resulting definitions of joint and conditional possibility not only differ from their probabilistic counterparts, but they also have a rather different epistemological status.

Suppose that the portions of people in a given population for whom the ith and jth options are possible are po_i and po_j respectively. What is the joint possibility $po(i \text{ and } j)$ of having both the ith and jth options available? In frequentist terms, for what portion of people could options i and j both be possible? The smallest value $po(i \text{ and } j)$ could take is 0 if $po_i + po_j \leq 1$, but if this sum exceeds 1 then $po(i \text{ and } j)$ must be at least $po_i + po_j - 1$. The largest value for $po(i \text{ and } j)$ occurs when the portions po_i and po_j entirely overlap one another (i.e., when one includes the other). Hence, the range of values for joint possibility is

$$\max(0, po_i + po_j - 1) \leq po(i \text{ and } j) \leq \min(po_i, po_j). \tag{6}$$

In the absence of empirical information about $po(i \text{ and } j)$, there are two definitions one could adopt. One is a "pointwise" definition that stipulates a single value for $po(i \text{ and } j)$. Given that our definition of possibility connotes the most that could happen, our pointwise candidate would be $po(i \text{ and } j) = \min(po_i, po_j)$, which is Zadeh's definition. The second definition is "intervalic", and requires us to view $po(i \text{ and } j)$ as a "fuzzy number" by virtue of the fact that in the absence of further information no single value in the interval specified by (6) is any more plausible than any other value. Hence the definition of $po(i \text{ and } j)$ is expressed by (6) itself. For the time being, suffice it to say that there are occasions favoring each approach, and under some conditions the choice between them boils down to what the researcher means specifically by possibility. In the frequentist paradigm there is no intervalic definition for joint necessity, since the ith and jth options are assumed exclusive. Given that $ne_i + ne_j \leq 1$, $ne(i \text{ and } j) = 0$. There are, however, some subjectivist arguments for nonzero values of $ne(i \text{ and } j)$.

In the frequentist paradigm, the value of $po(i \text{ and } j)$ reflects the extent to which the availability of the ith and jth options are related to one another. If the possibility of i excludes the possibility of j, then $po(i \text{ and } j) = \max(0, po_i + po_j - 1)$. If everyone who has access to option i also has access to j (or vice versa), then $po(i \text{ and } j) = \min(po_i, po_j)$. Statistical independence between the availability of options i and j implies that $po(i \text{ and } j) = po_i po_j$.

By a similar argument to the one just given, we may establish that the relevant interval for $po(i \text{ or } j)$ is

$$\max(po_i, po_j) \leq po(i \text{ or } j) \leq \min(po_i + po_j, 1). \tag{7}$$

If the pointwise definition of $po(i \text{ and } j) = \min(po_i, po_j)$ is adopted, then we are forced to conclude $po(i \text{ or } j) = \max(po_i, po_j)$. Otherwise the intervalic expression in (7) becomes the definition. In the event that we know the value of $po(i \text{ and } j)$, we have

$$po(i \text{ or } j) = po_i + po_j - po(i \text{ and } j). \tag{8}$$

From the data in Table 1, for instance, we may conclude that the portion of elderly for whom both taking a taxi and being driven by someone are possible options is not greater than $\min(0.42, 0.70) = 0.42$ and not less than $\max(0, 0.42 + 0.70 - 1) = 0.12$. In this particular survey, crosstabulation revealed that the joint possibility is $po(1 \text{ and } 2) = 0.34$. Nearly all the elderly who can be driven by a friend or relative could also take a taxi. Likewise, from (8) we know that $po(1 \text{ or } 2) = 0.42 + 0.70 - 0.34 = 0.78$.

Turning now to the concept of conditional possibility, there are at least two senses in which we may speak of possibility being "conditioned". One is a very general notion of influence, whereby a grade of possibility may change with some event or circumstance. Thus, the possibility that Dr. Gorgonzola may obtain a position at a university in Carbuncle is 0 if no university exists there in the first place, but rises conditional on a university being built there, a suitable position becoming available, Gorgonzola applying for it, being offered the job, and so forth. Examples abound in everyday life of individuals and institutions altering possibility distributions, and in some ways the concept is at the center of many psychological theoretical concerns.

The second sense of a conditional possibility, however, is formally defined in terms of joint and simple possibilities:

$$po(j/i) = po(i \text{ and } j)/po_\iota, \qquad (9)$$

where po(j/i) denotes the possibility of j given i. In frequentist terms, po(j/i) is the portion of the people for whom option i is possible who also have access to option j. In set theoretic terms, it measures the degree to which option i is included by option j. Returning to the example above, given that 0.42 of the elderly can be driven by someone and that the portion who have that option and the taxi available is 0.34, the possibility of taking a taxi conditional on having a friend or relative to drive is po(2/1) = 0.34 / 0.42 = 0.81.

Definitions of joint and conditional possibility enable the multivariate analysis of possibility data. In conjunction with the definition of relative freedom of action given in (2), this basic calculus opens the way for a multivariate calculus of freedom and constraint which will be outlined in a section to come. First, however, I will discuss some critical issues in the measurement of epistemic and physical possibility that do not fall within the frequentist paradigm.

The distinction between objective and subjective possibility implies that at least two kinds of measurement foundations are required, somewhat analogous to those found in probability and statistics. Epistemic possibilities pose some of the same questions haunting subjective probability: What rating tasks are valid for human judges, how reliable are the judges, and what heuristics do (or should) people adopt in making possibility judgments? Research programs along the lines of Wallsten, Zwick, Budescu, and their colleagues (e.g., Wallsten et. al., 1986; Zwick & Wallsten, 1987) show considerable promise in providing useful answers for psychologists who wish to scale subjective judgments of possibilistic uncertainty.

To what extent can existing "possibilistic" instruments be analyzed in a readymade fashion using possibilistic techniques? Some practical suggestions are available in Smithson (1987). To begin with, negation, joint, and even conditional possibility may be defined on any bounded scale in an interval, say, [a,b]. The impossibility of the ith option becomes $a + b - po_i$, which amounts to "reverse coding" the scale. If the subjective scale has a 0-point corresponding to absolute impossibility and the scale points are numerical, then converting that interval to [0,1] is trivial. Likewise, the min-max pair of operators for joint possibilities po(i and j) and po(i or j) use only ordinal information, and so may be applied to even ordinal judgment scales.

Less intuitively, we may incorporate certain kinds of unbounded subjective scales into the possibility calculus described thus far. Under some conditions, possibility rating scales may not have apparent upper or lower bounds. Some concepts of "ease", "difficulty", "ability", and "power" are possibilistic but do not seem to have absolute anchoring points at either end of the scale. If, however, they have a single midpoint as an anchor, then they pose no difficulty for our calculus. An example is a judgment or rating that refers to a norm (e.g., "abnormally difficult", "unusual ability", or z-scores). Let q be the midpoint. Then the impossibility of the ith point or option is $2q-po_i$, and the min-max operators for joint possibility are again entirely appropriate.

Turning now to physical possibility, the same remarks that have been made regarding subjective rating scales apply to scales measuring difficulty, ease, costliness, ability, or availability from which possibility scales are derived by definitional arguments. Furthermore, the direct estimation of possibilities from either proportions of people for whom options are possible or various rating scales designed to measure possibility directly may be handled by standard measurement theory. A distinction must be made, however, between possibilities arising from restrictions at the individual level and those derived from group-level constraints. That is, $po_i = 0.73$ could arise because each of

27% of a group does not have the ith option available, or it could reflect a constraint that specifies only up to 73% of the group may exercise that option. Examples of the latter kind of possibility include the utilization of any limited resource, as in grazing cattle on a commons, withdrawing money from a bank, or distributing 73 cookies among 100 children. This distinction has consequences for the measurement of relative freedom or constraint.

The greatest difficulties for inferring degrees of possibility arise, however, when there is no basis for doing so other than sample observations of behavior or other events. Such observations yield point-estimates of probability, which in turn place s a lower bound on possibility. Usually, there is no way of obtaining a pointwise estimate of possibility under these conditions. Some investigators have gone as far as to propose explicit greatest lower bound (glb) estimates for poi as a function of sample estimates of pr_i (cf. Civanlar & Trussell, 1985). These glb candidates have the form

$$glb(po_i) = pr_i + f(s_i, c) \qquad (10)$$

where s_i is the sample standard deviation of pr_i and c is the confidence level desired.

These glb proposals lead, however, to some intepretive problems. First, we must ask exactly what is implied about possibility by a probability of any kind. The strict answer is only that the true possibility must equal or exceed that probability. Hence, glb's of the form shown in (10) assert, paradoxically, that the higher the confidence interval the higher the imputed value for $glb(po_i)$. The true state of affairs, unfortunately, is that while we might be 95% confident that pr_i lies between two bounds, we cannot be 95% confident that po_i lies outside those bounds. There is a loose sense in which one may claim that such data provides information about whether people are behaving "as if" certain options were possible, but the researcher is merely using a probability distribution to model possibility.

Measuring Freedom and Constraint

The characterization of uncertainty in possibilistic terms motivates a measure of how relatively free the actors in a system are to make choices, given the constraints of that system. Our initial definition in (2) serves as a starting point for a more general measure, capable of assessing the collective freedom of individuals on the basis of their relative access to the set of conceivable choices.

It is beyond the scope of a single article to fully elaborate a calculus of freedom and constraint. I shall outline the beginnings of one such calculus from a frequentist standpoint. Let N individuals each have possibility values po_{ij} and necessity values ne_{ij} (for the ith individual on the jth option) for each of r conceivable options. Let the values of po_{ij} and ne_{ij} be either 0 or 1. Obviously, $ne_{ik} = 1$ only if $po_{ij} = 0$ for all but the kth option. Then the relative amount of freedom enjoyed by the ith individual is r

$$F_i = \sum_{j=1}^{r} (po_{ij} - ne_{ij})/r. \tag{11}$$

We may recover the marginal possibility and necessity distributions over the r options by N

$$po_j = \sum_{i=1}^{N} po_{ij}/N, \text{ and}$$

$$ne_j = \sum_{i=1}^{N} ne_{ij}/N. \tag{12}$$

The relative freedom for (portion of people free to choose) the jth option is then measured by

$$F_j = po_j - ne_j, \qquad (13)$$

which is a recovery of the definition given by (2). Likewise, relative preference for the jth option, S_j, may be defined as in (3).

Now, if we move to the group level and wish to measure the freedom available to the entire collection of N individuals, there are three definitions that could be adopted. One is simply to measure the average F_i or F_j, which clearly may be done whether possibilities have been defined at the individual or only at the group level. But this definition ignores both partitions and permutations. The method of permutations leads us to define system freedom F^* by

$$F^* = \prod_i rF_i/rN,$$

$$= 0 \text{ only if all } F_i = 0, \qquad (14)$$

where only nonzero F_i are multiplied together. The Nth root of F^* is of course the geometric mean of the nonzero F_i. As in probability, the permutations approach to measuring freedom involves dividing the freedom accorded all permissible permutations by the freedom accorded all conceivable permutations.

The third definition ignores permutations and considers only partitions. There are several situations where this might be desirable. One is where the system or group distinguishes only among different partitions rather than among permutations (akin to the Bose-Einstein statistics for photons), so that each permutation is considered equipossible. Another such condition is when only marginal possibilities are known, as with the imposition of constraints at the group level. Group constraints are constraints on the margins (partitions) only, and do not yield deterministic estimates of simple individual possibility. I shall briefly outline the calculus of relative freedom using the method of partitions.

Given r alternatives with equipossible partitions, there are two kinds of constraint that must be taken into account when calculating relative group freedom. One is the constraints on possibility and necessity for each of the alternatives. The second constraint is that the partitions themselves must always add to 1, even though the po_j and ne_j need not. This latter restriction implies that r options may be represented in r-1 dimensions (or, in conventional statistical terms, "degrees of freedom"). Figure 2 shows the geometry of freedom domains for $r=2$ and 3. In the case where $r=2$ the domain is a segment on the line defined by $p_1 + p_2 = 1$ whose endpoints are determined by po_1, po_2, ne_1, and ne_2. This line has length $\sqrt{2}$. For $r=3$, the domain is an area inside the equilateral triangle defined by $p1 + p2 + p3 = 1$, whose edges have length $\sqrt{2}$. For $r=4$ the domain is a volume inside a regular tetrahedron, and so on.

Conceptually, the computation of the relative group freedom is simple. One computes the portion of length, area, volume, etc. that is left over once possibility and necessity constraints have been taken into account. The precise nature of these computations will be elaborated shortly. For now, suffice it to say that freedom is decremented in terms of $(1-po_j)^{r-1}$ and ne_j^{r-1}. Thus, in the first example provided in Figure 2, if $po_1 = 0.75$, $po_2 = 0.80$, $ne_1 = 0.10$, and $ne_2 = 0.15$, then the relative group freedom (FG) is

$$FG = [1 - (1-po1) - (1-po2)] \sqrt{2} / \sqrt{2}$$

$$= 1 - (1-po1) - (1-po2) = 0.55. \qquad (15)$$

In the second example, if $po_1 = 0.75$, $po_2 = 0.80$, $po_3 = 0.56$, $ne_1 = 0$, $ne_2 = 0$, and $ne_3 = 0.15$, then

$$FG = (1-ne_3)^2 - (1-po_3)^2 - (1-po_2-ne_3)^2 = 0.52 \qquad (16)$$

Of course, these examples have been made straightforward.

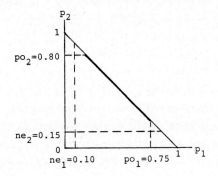

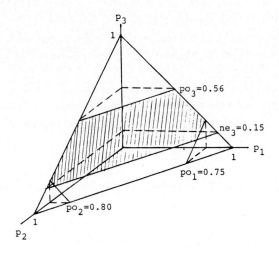

FIGURE 2
Relative Freedom for 2 and 3 Dimensions

Even simple examples, however, may be intuitively interesting. Consider a teacher who gives students some choice on how they weight three pieces of assessment. Which is the more flexible: (1) A rule that specifies no piece of assessment gets less than a 20% weight, or (2) a rule that no piece of assessment gets more than 60%? Under rule (1), FG = $(1 - 3(0.20))^2$ = 0.16, while under rule (2) FG = $1 - 3(1-0.60)^2$ = 0.52. Rule (2) therefore gives 3.25 times the relative freedom of rule (1). How low would the maximum weight restriction in rule (2) have to be before its FG equals that of rule (1)? The required equation is 0.16 = $1 - 3(1-po_i)^2$, and the solution is po_i = 0.47, or a 47% upper weight limit.

Let us move toward the general (r-option) case. Given equipossible partitions, and the constraint that the partitions must sum to 1, the space for r alternatives may always be represented as an r-1 dimensional hypertetrahedron. The relative freedom of a collectivity that has possibilistic constraints imposed on the availability of the r options is simply a hypersubvolume inside that space, and therefore FG may be defined as the ratio of that subvolume to the volume of the hypertetrahedron. The computation of FG consists entirely of subtracting terms in po_i and ne_i and adding back those portions that have been subtracted twice. These terms are always taken to the r-1 power. The result is a sum of groups of terms that alternate in their signs.

For the 2-option case,

$$FG = (1-ne_1-ne_2) - \max(0, 1-po_1-ne_2) - \max(0.1-po_2-ne_1). \qquad (17)$$

The 3-option model yields

$$FG = (1-ne_1-ne_2-ne_3)^2 - \max(0.1-po_1-ne_2-ne_3)^2$$

$$- \max(0, 1-po_2-ne_1-ne_3)^2 - \max(0, 1-po_3-ne_1-ne_2)^2$$

$$+ \max(0, 1-po_1-po_2-ne_3)^2 + \max(0, 1-po_1-po_3-ne_2)^2$$

$$+ \max(0, 1-po_2-po_3-ne_1)^2, \qquad (18)$$

and the 4-option model gives

$$FG = (1 - ne_1-ne_2-ne_3-ne_4)^3 - \max(0,1 - po_1-ne_2-ne_3-ne_4)^3$$

$$- \max(0,1 - po_2-ne_1-ne_3-ne_4)^3 - \max(0,1 - po_3-ne_1-ne_2-ne_4)^3$$

$$- \max(0,1 - po_4-ne_1-ne_2-ne_3)^3 + \max(0,1 - po_1-po_2-ne_3-ne_4)^3$$

$$+ \max(0,1 - po_1-po_3-ne_2-ne_4)^3 + \max(0,1 - po_1-po_4-ne_2-ne_3)^3$$

$$+ \max(0,1 - po_2-po_3-ne_1-ne_4)^3 + \max(0,1 - po_2-po_4-ne_1-ne_3)^3$$

$$+ \max(0,1 - po_3-po_4-ne_1-ne_2)^3 - \max(0,1 - po_1-po_2-po_3-ne_4)^3$$

$$- \max(0,1 - po_1-po_2-po_4-ne_3)^3 - \max(0,1 - po_1-po_3-po_4-ne_2)^3$$

$$- \max(0,1 - po_2-po_3-po_4-ne_1)^3. \qquad (19)$$

The general equation for the r-option model may be expressed by

$$FG = (1 - \sum_{i=1}^{r} ne_i)^{r-1} - \sum_{i=1}^{r} \max(0, 1 - po_i - \sum_{j \neq i} ne_j)^{r-1}$$

$$+ \sum_{i=1}^{r} \sum_{j > i} \max(0, 1 - po_i - po_j - \sum_{k \neq i,j} ne_k)^{r-1} - \ldots$$

$$\ldots + (-1)^{r-1} \sum_{i=1}^{r} \max(0, 1 - ne_i - \sum_{j \neq i} po_j)^{r-1}. \qquad (20)$$

Thus, returning once again to the survey on transportation for the elderly, we may compute the relative freedom of our sample from the data supplied in Table 1. There are 4 options, so from (19) we have

$$FG = (1-.02-.12-.08)^3 - (1-.42-.12-.08)^3$$
$$- (1-.70-.02-.12-.08)^3 - (1-.72-.02-.08)^3 = 0.41. \qquad (21)$$

These, however, are the most straightforward constraint models for categorical data. Cross-classified systems are considerably more complicated, and mixed systems with constraints on sums across several categories more so again. Nevertheless, they are not incomputable and both classical and Monte Carlo algorithms have been proposed for evaluating the volumes of the required polytopes (e.g., Cohen & Hickey, 1979). Within reason, it is feasible to compute FG for cross-classified systems of options under a variety of assumption about (non)independence.

We are now ready to move to the case where N individuals are assigned possibilities over a range of values on a continuous variable rather than over discrete options. For the most part, this is a simple generalization of the scheme discussed above. Let X be a continuous variable with range [d,u], and let X the ith individual be assigned a possibility $po_i(x) = 1$ for all x in some subrange $[a_i, b_i]$ and 0 for all y outside this subrange. Then F_i is defined as

$$F_i = (b_i - a_i)/(u-d) \qquad (22)$$

The average F_i and F^* are defined as above, with u-d replacing r in (14). More generally, let $po_i(x)$ and $ne_i(x)$ take values from the [0,1] interval. Then (22) is generalized to

$$F_i = \int_d^u (po_i(x) - ne_i(x))dx/(u-d) \qquad (23)$$

A somewhat different but widely useful case is where each state x of X is assigned a possibility and/or necesssity value: po(x) and ne(x). Most likely these possibilities would be epistemic, but the important issue here is that po(x) and ne(x) do not refer to individuals. They may refer, for instance, to the ease with which x is attainable. If the researcher wishes to compute the relative freedom for each individual, then in this instance F_i is identical for all i, and it is defined by (23) without the use of subscripts in the right-hand terms.

The reader may have noticed that I have been silent on the issue of evaluating FG in systems characterized by constraints on probability density functions over interval-level (continuous) variables. The reason for this is that there is, to my knowledge, no satisfactory framework available for defining FG in such systems. For practical purposes in some instances, we may approximate the continuous case via the standard ploy of dividing the appropriate range of the variable into r equal intervals and t reating it as an r-option system. However, the limiting behavior of (20) is unknown at present, and of course in the vast majority of practical problems (20) cannot be applied and we have no explicit expression for the general case. We are left with computational approaches along the lines suggested by Cohen and Hickey (1979).

Finally, there is a class of situations in which the density functions constrained by possibility and necessity need not accumulate to 1. Instead, 1 becomes merely a limit on their accumulation value. Consider a Commons-type setup in which M shareholders may use a resource up to a limit, L. Each shareholder's usage of the resource also is limited by his or her share of the holdings, h_i. Let H denote the sum of the holdings h_i, w_i the amount actually used by the ith shareholder, $b_i = w_i/H$, and B the sum of the b_i. Then $po(b_i) = min(h_i/H, L/H)$ and $po(B) = min(L/H, 1)$. Clearly if HL then the shareholders face a Commons dilemma, since they are less than perfectly free to use as much of their holdings as it is possible for them to do. There are meny Commons situations in the real world, but perhaps the one familiar to most people is a bank. If too many people withdraw their savings from a bank in a too short time, the bank fails. Furthermore, the relative freedom for any individual to withdraw money hinges on how much others have withdrawn recently.

In these problems, FG may be calculated as a subvolume of an M-dimensional rectrangular solid. The case for $M=2$ is shown in Figure 3 below. Were $L \geq H$ then shareholders would be free to maximize their b_i. Hence, FG is the proportion of the area $po(b_1)po(b_2)$ that is bounded by the line $b_1 + b_2 = L/H$:

$$FG = [1/2po(b_1)po(b_2)]$$

$$[(L/H)^2 - \max(0, L/H - po(b_1))^2 - \max(0, L/H - po(b_2))^2] \qquad (24)$$

For these general case, we have M "shareholders" each with a share h_i, which in turn sum to H, the total holdings. Each user may make a withdrawal w_i subject to the collective limitation that the sum, W, of the w_i may not exceed L. This scheme may be translated into possibilisitic terms by setting $b_i = w_i/H$ and B equal to the sum of the b_i. Then as before, $po(b_i) = h_i/H$, and $po(B) = \min(L/H, 1)$. FG is then evaluated as the portion of the hyper-rectangular solid defined by the $po(b_i)$ bounded by the constraint $po(B) = L/H$. For $M=2$, FG is defined as in (24). For $M=3$,

$$FG = [1/(6po(b_1)po(b_2)po(b_3))]$$

$$[(L/H)^3 - \max(0, L/H - po(b_1))^3 - \max(0, L/H - po(b_2))^3 -$$

$$\max(0, L/H - po(b_3))^3 - \max(0, L/H - po(b_1) - po(b_2))^3 -$$

$$\max(0, L/H - po(b_1) - po(b_3))^3 + \max(0, L/H - po(b_2) - po(b_3))^3] \qquad (25)$$

and for the general case $M = r$,

$$FG = [1/(r! \prod_{i=1}^{r} po(b_i))][(L/H)^r - \sum_{i=1}^{r} \max(0, L/H - po(b_i))^r$$

$$+ \sum_{i=1}^{r} \sum_{j>i} \max(0, L/H - po(b_i) - po(b_j))^r - \ldots$$

$$\ldots + (-1)^{r-1} \sum_{i=1}^{r} \max(0, L/H - \sum_{j \neq i} po(b_j))^r]. \tag{26}$$

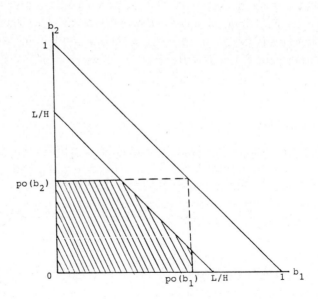

FIGURE 3
Two-dimensional Commons Setup

Possibility Theory, Fuzzy Logic, and Psychological Explanation

Before moving on to consequences of our definitions, several issues remain to be discussed. One of them is the relationship between the "freedom" measures proposed here and traditional notions such as "degrees of freedom". In both the strongly and weakly constrained r-option systems, the statistical concept of degrees of freedom corresponds with the dimensionality of the space in which the FG are computed. In this formalism, then, freedom has two components: dimensionality and magnitude. One consequence is that direct comparisons of freedom across systems having different dimensionality is not straightforward. Can FG = 0.4 be directly compared with FG = 0.3, if the former system has 3 options and the latter has 10?

This last consideration raises the possibility of norming FG with respect to its exponent. Consider a strictly sum-constrained M-option system, and let us add K more options onto it. If we assume that the sum of the po_i for $i = 1,...M$ exceeds 1 and that the addition of these options does not alter the po_i or ne_i, then for all $j > M$ we have the following limits:

$$po_j < 1 - \sum_{i=1}^{M} ne_i, \text{ and}$$

$$ne_j > 0. \qquad (27)$$

These new options could be said to add no more freedom to the system since they do not alter the limits on the old. If ne_j and po_j are minimal and maximal respectively in terms of the limits imposed by (27), then it would be reasonable to conclude that the freedom available in the new system should equal that available in the old.

But from (20) we know this is not the case. Letting $N = M + K$, it is clear that $FG_N < FG_M$, entirely because of the higher exponent in FG_N. A partial correction for this may be achieved by using the M-1th and N-1th roots of FG_M and FG_N respectively. Equality is attained by this measure only when all terms in (20) that include any po_i equal 0, so that the first term is all that is left.

There is a rough analogy between this norming procedure and the computation of the geometric mean, since for systems with no constraints on the pi the resulting measure is almost identical to the geometric mean. This measure may be intuitively thought of as a "mean" or "normed" measure of relative freedom. For convenience, let us denote the M-1th (or Mth, in the case of FG_M as defined by (26) or the unconstrained case) root of FG_M by $S(FG_M)$ and call it the "standardized freed om" measure. Its main applications probably will arise in problems where researchers wish to directly compare freedom measures between systems of unequal dimensionality.

A somewhat related issue concerns the meaningfulness of computing freedom coefficients for systems that have an indefinite (unknown) number of options. The case of interest is when some finite subset of those options is known and the po_i and ne_i bounds on their respective p_i have been established. Call this a "finite subset" (FS) case. For all FS of dimension M-1, a freedom coefficient may be defined by treating the FS as a subsystem and creating dummy po_M and ne_M for the remaining unknown options, but with ne_M and po_M set according to the limits prescribed in (27).

This procedure amounts to assuming that these unknown options are not any more restricted than their complementary (known) counterparts. However, it does permit knowledge engineers and expert systems builders, for instance, to relax the assumption of "logical omniscience" (cf. Horvitz et. al., 1986) which requires that all conceivable options be known before evaluating belief, likelihood, or uncertainty of a proposition. The relaxation of that assumption not only accords with many real-world situations, but also with the spirit of Shackle's (1969) rationale for surprise measures and Zadeh's arguments for possibility theory.

Before leaving these topics, let us consider an example. In a pilot empirical study of possibility (Smithson, 1987, p. 18-22), I had subjects rate the possibility that "several" and "few" could refer to various integers. Table 2 displays some hypothetical possibility ratings for the integers. Is the distribution for "several$_1$" or "several$_2$" looser (freer) than that for "a few"? On what basis could we compare them; should we use FG_M or $S(FG_M)$?

TABLE 2
Possibility Distributions for "Several" and "A Few"

Integer	A Few	Several1	Several2
0	0.00	0.00	0.00
1	0.00	0.00	0.00
2	0.50	0.00	0.00
3	1.00	0.10	0.31
4	1.00	0.40	0.31
5	0.40	0.80	0.50
6	0.00	1.00	1.00
7	0.00	1.00	1.00
8	0.00	1.00	0.40
9	0.00	0.80	0.31
10	0.00	0.20	0.31
11	0.00	0.00	0.00

A conventional approach to comparing these possibility distributions would be to assess their "heights", perhaps by taking the arithmetic mean. In our calculus, the geometric mean would be appropriate as long as the p_i for these distributions did not have any further constraints placed on them. The geometric mean for the possibility distribution of "a few" is 0.669, while for "several$_1$" it is 0.517 and for "several$_2$" it is 0.455. The possibility distribution of "a few" is freer than that for "several$_1$", which in turn is slightly less constrained than "several$_2$".

However, if we view the possibility distributions as upper bounds on subjective probabilities that must sum to 1 and hence impose the strict summation constraint on the pi, a rather different picture emerges. Clearly we may use (20) to compute freedom coefficients for these distributions, and for the distribution in "a few" $FG_4 = 0.660$, while for "several$_1$" $FG_8 = 0.376$ and for "several$_2$" $FG_8 = 0.666$. According to the measure defined in (20), "a few" and "several$_2$" have nearly the same amount of relative freedom, while "several$_1$" is substantially more constrained. If we wish to use the standardized freedom measure, then $S(FG_4) = 0.871$, for "several$_1$" $S(FG_8) = 0.870$, and $S(FG_8) = 0.944$ for "several$_2$". The standardized measure indicates that, when the number of integers is adjusted for, "a few" and "several$_1$" are nearly identical in their amounts of freedom. The obvious lesson here is that the rank order of freedom in systems may be altered depending on the measure adopted, and so the choice of a measure of relative freedom is not trivial in its impact.

A number of formal consequences flow from these definitions of relative freedom. I shall mention only one of them, namely the issue of whether necessity constrains more than impossibility. This issue has already been raised in an earlier example. In unconstrained systems where FG is defined simply by an appropriate version of (14), necessity decrements FG as much as an equivalent amount of impossibility. For sum-constrained systems where FG is defined by (20) or (26), however, this is not the case. Details and proofs of the theorems described below may be found in Smithson (1988).

Theorem 1

Let FG be defined as in (26) but with an appropriate extension to incorporate nei constraints. For any set B contained in $i: i = 1, 2, ..., r$ let po_i and ne_i be assigned in any manner that befits their definitions. Now, let po_j be chosen for all j in B' and $ne_j = 0$, and denote the freedom of this system by FG_p. Likewise, let $ne_j = 1 - po_j$, then define all $po_j = 1$, and denote the freedom of this alternative system by FG_n. Then $FG_p > FG_n$.

This theorem stipulates that necessity constrains freedom more than an equivalent amount of impossibility, in systems modeled by (26). For systems modeled by (20) the picture is not quite so simple. Necessity constrains more than impossibility except under the conditions specified in the following theorem:

Theorem 2

Let FG be defined as in (20), and consider FG_p and FG_n as defined in Theorem 1. Then $FG_p < FG_n$ if and only if:

(1) B' consists of only one j; and

(2) $\sum_{i \in B} ne_i$ $1 - \sum_{i \in B} po_i$.

This case-by-case discussion is intended to give the reader a feeling for the overall approach to a calculus of relative freedom rather than a rigorous introduction. A full elaboration of freedom measures must await another occasion. More important is an introduction to weak prediction and entailment models, which is the final primary component of our alternative framework for quantitative analysis.

Weak Entailment and Prediction

Conventional statistical models are designed to make pointwise predictions or to test for one-to-one associations. The exceptions to this tendency are scattered throughout the literature and rarely articulated for what they are. Although the conceptual and mathematical machinery of statistics is sophisticated and powerful, it is inappropriate for modeling systems in which people have a range of alternatives or choices. This section presents several models of entailment, prediction, and association based on (fuzzy) set inclusion, overlap, and logic which are compatible with the concept of partially restricted choice.

Starting with the 2x2 setup, as various authors have pointed out (e.g., Hildebrand, Laing, & Rosenthal, 1977; in an oddly unfashionable and neglected book), there are two fundamentally different kinds of "pure" relationship between two binary variable's. One is the "perfect" association in which all the observations fall in the diagonal cells, as shown in the first part of Figure 4. This is the 2x2 analog to complete correlation, and in an entailment or prediction scheme constitutes perfect prediction. The other is the so-called "weak perfect" association shown in the second two parts of Figure 4, in which observations may fall in three of the four cells. This is, of course, not a one-to-one association and does not yield perfect predictability. Actors in this scheme are restricted but still have options.

$$
\begin{array}{c|cc|c}
 & \multicolumn{2}{c}{X} & \\
 & 1' & 1 & \\
\hline
2 & 0 & pr_{12} & pr_2 \\
\hline
2' & pr_{1'2'} & 0 & pr_{2'} \\
\hline
 & pr_{1'} & pr_1 &
\end{array}
$$

$$
\begin{array}{c|cc|c}
 & \multicolumn{2}{c}{X} & \\
 & 1' & 1 & \\
\hline
2 & pr_{1'2} & pr_{12} & pr_2 \\
\hline
2' & pr_{1'2'} & 0 & pr_{2'} \\
\hline
 & pr_{1'} & pr_1 &
\end{array}
$$

$$
\begin{array}{c|cc|c}
 & \multicolumn{2}{c}{X} & \\
 & 1' & 1 & \\
\hline
2 & 0 & pr_{12} & pr_2 \\
\hline
2' & pr_{1'2'} & pr_{12'} & pr_{2'} \\
\hline
 & pr_{1'} & pr_1 &
\end{array}
$$

FIGURE 4
Weak and Perfect Association

The traditional armory of measures of association and prediction for 2x2 tables is oriented primarily toward the "perfect association" model. Chi-square based measures (e.g., phi) and proportional reduction of error (PRE) measures such as Goodman and Kruskal's lambda achieve their maximum values only under perfect assocation. The odds ratio is sensitive to imperfect association but it does not provide an interpretive framework that is congruent with theories of action under partial constraint. There are alternatives, some of which lead to useful characterizations of weak prediction and entailment in contingency tables.

There are two direct, simple interpretations for weak perfect associations between two variables. The first one is based on set inclusion. Let the options on X_1 be indexed by 1 and 1', and the options on X_2 by 2 and 2'. If pr_1 and $pr_{1'} = 1-pr_1$ are the marginal portions for X_1 and pr_2 and $pr_{2'} = 1-pr_2$ are the marginals for X_2, then the middle part of Figure 4 corresponds to the inclusion of option 1 by option 2, while the third part shows the inclusion of option 2 by option 1. From the definition of joint possibility we know that $po_{12} = \min(pr_1, pr_2)$, and a measure of set inclusion is

$$I_{1/2} = pr_{12}/pr_1, \text{ and}$$

$$I_{2/1} = pr_{12}/pr_2. \tag{28}$$

The greatest possible association that can occur between two binary variables is a strict inclusion relationship between them. The only circumstance under which "perfect" association (in set theoretic terms, setwise equivalence) can possibly occur is when the marginals of the two variables are identical. Therefore, the corresponding measure of association that we require is the extent to which one variable includes or is included by the other, which shall be called set overlap:

$$OL_{12} = \max(I_{2/1}, I_{1/2}) = pr_{12}/po_{12}. \tag{29}$$

Applications of these inclusion and overlap measures have been detailed in Smithson (1987).

A PRE measure may be developed using $I_{2/1}$, $I_{1/2}$, or OL_{12} to compare the observed degree of inclusion or overlap with that expected when X_1 and X_2 are statistically independent:

$$PRE_{1/2} = 1 - (1-I_{1/2}) / [1-(pr_1pr_2/pr_1)],$$

$$PRE_{2/1} = 1 - (1-I_{2/1}) / [1-(pr_1pr_2/pr_2)], \text{ and}$$

$$PRE_{12} = 1 - (1-OL_{12}) / [1-(pr_1pr_2/po_{12})]. \tag{30}$$

Boolean logic also may be used to characterize weak perfect associations between two binary variables. The middle part of Figure 4 corresponds to the proposition that option 1 implies option 2, while the third part represents the proposition that option 2 implies option 1. Unlike set inclusion, standard logic counts "nonblack noncrows" as evidence for the claim "all crows are black". Hence only one cell contains counterindicative observations for a simple "if-then" hypothesis, as Hildebrand et. al., (1977) explain. They develop a PRE measure, del, of the extent to which the number of expected counterindicative observations under statistical independence between X and Y is reduced by the actual data. In the scheme shown in Figure 4,

$$del_{1/2} = 1 - (pr_{12'}/(pr_1pr_{2'})), \text{ and}$$

$$del_{2/1} = 1 - (pr_{1'2}/(pr_{1'}pr_2)). \tag{31}$$

This measure also may be described in terms of a priori conditional possibilities. If the proposition that 1 implies 2 is true, then the conditional possibilities of the four cells are $po(2/1) = 1$, $po(2/1') = 1$, $po(2'/1') = 1$, and $po(2'/1) = 0$. Thus, $del_{1/2}$ compares the observed with the expected conditional impossibilities, which are 0 for all but $1-po(2'/1) = 1$. An identical argument connects $del_{2/1}$ with appropriate conditional possibility.

A less intuitively accessible but important framework comes from information theory via "proportional reduction of uncertainty" (PRU) coefficients of association proposed by Kullback (1959), Theil (1972), and Kim (1984). Instead of either a set theoretic or logical interpretation, the PRU framework focuses on the amount of "surprise value" that data have, given an hypothesized relationship between the variables. I will not present it here, but the interested reader may find a development of this approach for numerical and fuzzy variables in Gaines and Shaw (1986) and an extension and comparison between the PRU and PRE frameworks in Smithson (1987, ch.7).

The important point to these measures of association and prediction is that they operationalize one-to-many predictions or entailments, which are compatible with a possibilistic approach to data analysis. The simple "if-then" proposition, in a sense, constrains people in one column of the 2x2 table but permits them choice in the other. The generalization of either PRE or del to larger bivariate or multivariate tables is not difficult, and will not be elaborated here. Instead, I shall briefly discuss its generalization to numerical variables.

The scatterplots in Figure 1 are both examples of simple "if-then" entailments involving numerical variables, in the sense that they conform to the statement "if X is high then Y is high." The sets "high X" and "high Y" are fuzzy sets because "high" is a graded concept, and the key to developing a PRE type measure of association to test weak predictions such as these is inducing either a fuzzy set membership scale or possibility scale (in the sense of "possible that this value of the scale could be referred to by the term 'high'") from the raw scales for X and Y. This may be done absolutely or by using z-scores. Having done this, then $I_{1/2}$, $del_{1/2}$, and the rest may be defined using definitions for fuzzy set inclusion and fuzzy logical implication (for details on these see Smithson & Knibb, 1986; and Smithson, 1987, ch.1,3, and 7). The same mathematical tools may be used to operationalize other kinds of weak entailment models (WEMs) whose predictive regions are also regions of conditional possibility for free choice.

Let a bivariate WEM be represented by a simple region in XxY, with X entailing Y. Many such regions may be described by lower and upper bound functions $L(X) \leq Y$ and $Y \leq U(X)$, respectively, as shown in Figure 5. Then the characterization of the conditional possibility distribution po(Y/X) is deducible from the fuzzy logical operationalization of the statement "$L(X) \leq Y$ and $Y \leq (X)$":

$$po(Y_i/X_i) = \min[po(Y_i/L(X_i) \leq Y), po(Y_i/Y \leq (X_i))]. \qquad (32)$$

Given a form and parameters for the functions U and L, numerical methods may be employed to specify them by optimizing on del and precision measures as defined in Smithson (1987). It is not difficult to generalize this scheme to multiple region WEMs or models involving multiple constraining variables which in turn also constrain one another.

The fuzzy set theoretic, fuzzy logical, and information theoretic frameworks are not the only formal models that could be used to characterize one-to-many entailments such as those in Figure 1. It is quite possible to use rather conventional statistical tools to do the same job, namely the "dual equation" approach referred to at the beginning of this paper. The main advantage of the dual-equation approach over the fuzzy set, logical, or information theoretic approaches is its use of conventional and rather generalized concepts, albeit in a slightly deviant guise. The dual-equation models can easily capture a wide variety of relationships, while the other approaches require more computing time in order to perform the same feats. Its primary disadvantage, however, is that it does not connect particularly well with the possibilistic framework, whereas fuzzy set and possibility theory are intimately related. Fuzzy set theory has been used to provide a mathematical measure theoretic basis for possibility theory, and they share a "natural" link. As yet, there is no articulated linkage between possibility theory and constraints on variance.

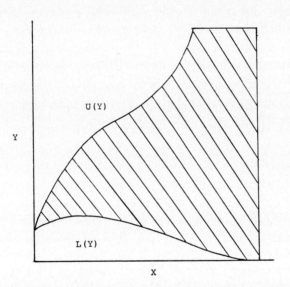

FIGURE 5
Regional Prediction of Y

Weak, one-to-many, or regional prediction models are not new. They are at least embedded in formalisms such as mathematical decision theory with its characterization of "admissible solutions", (non)linear programming, and in social and economic perspectives such as Simon's "satisficing" economic individual. However, the material introduced here refers to the first serious attempts to ground them in a statistical association framework that also can incorporate the concepts of possibility and choice.

The combination of possibilistic and weak predictive models opens a potentially valuable toolbox for data analysis in an agentive mode. The possibilistic calculus enables us to compute the relative loss of freedom imposed by a particular structural, predictive, or entailing restriction on action. The fuzzy logic and inclusion basis for weak prediction permits the operationalization of a wide variety of structural, psychological, or material constraints in possibilistic terms. The two immediately suggest ways of modeling possibilistic and weakly constrained dynamic systems whose states depend on choices or decisions made at various points in time. It is not difficult to imagine pseudo-Markovian models using "transition possibilities" or models in which possibilities and/or constraints are choice-dependent.

Possibilism in Psychological Theory and Research

In several areas of psychology, the elaboration of possibility compels a rereading and, to some extent, a reconstruction of theory. All that can be provided here, however, is an illustrated outline of that reconstruction. I shall focus mainly on social psychological examples. A considerable proportion of psychologists utilize possibility in an implicit, perhaps even unconscious manner, although it is seldom carried into the level of research. One of the first tasks in putting possibility to work is to uncover its role(s) in theoretical perspectives and disentangle it from related concepts.

Many researchers, for instance, confuse possibility with conceivability, which leads to the objection that "any conceivable outcome is possible", or "the possibilities are endless". However, that often is not the case in either the de re or de dicto senses of possibility. Many conceivable things are not physically or socially possible. Furthermore, people are literally forced by the limitations of their minds and the finitistic nature of the social stock of knowledge to make decisions or choices from a finite field of subjective possibilities. In practical social action, the field of possibilities usually is a finite, small subset of the field of conceivable alternatives.

academic with dwindling enrolments in her third year subject. She may attempt to increase the likelihood, pr_i, that students will select hers out of the range of elective subjects, or even try to decrease their preference for other subjects, thereby lowering $pr_{i'}$ (persuasion). On the other hand, she may opt for altering the freedom allowed for third year students. One strategy is to try to make her class required for third year psychology students (increasing ne_i, which increases coercion). A second alternative would be to decrease $ne_{i'}$ (thereby increasing po_i), which amounts to making some other currently required third year subjects nonrequired. Third, she might try to reduce the range of third year options offered in psychology without eliminating her subject, thereby decreasing $po_{i'}$ and hence increasing ne_i indirectly. Academics anywhere will recognize all of these power strategies, and also know the distinctive advantages, costs, and political roles involved in each of them.

Possibility enters into accounts and theories as an ingredient of meaning. Many psychologists have recognized this at least implicitly. Social comparison and judgment, the perception of relative deprivation, reference group processes, reactance, self-enhancement (including several varieties of therapy), and skills acquisition are examples of processes that include a possibilistic component. Taylor's (1983) account of how cancer patients cope by using downward comparison (i.e., "things could be worse —— on is worse off than I am") is one of the most explicit demonstrations that possibility plays a crucial role in the construction of meanings and the assertion of control.

Finally, possibility may constitute an explanatory component in a theory. Some allusions to this have already been made in the remarks concerning decision processes earlier in this section. Clearly either de dicto or de re possibility could be used in an explanatory capacity. Examples of de re possibilistic explanations are plentiful in everyday life: the wealthy and poor approach the same problems of living in dramatically different ways because they also differ widely on what is possible and necessary; transportation strategies adopted by youth alter suddenly because they become legally old enough to get driver's licenses; and so on. Likewise, actions whose explanations depend on an appeal to what could happen, what could be done instead, what someone else might do, or what consequences might follow, often invoke a de dicto possibilistic account. The Prisoner's Dilemma and the arms race share this property, among others.

A good example of the systematic (albeit qualitative) use of de re possibilistic concepts in a middle-range theory of strategic behavior is Mars' account of workplace crime (1982, ch.1). He begins with Douglas' (1978) grid and group constructs as a way of classifying occupations, but his purpose is to explain the rise of characteristic patterns of workplace extralegal activities. Accordingly, he points out that weak grid workplaces exert weak constraints, permit greater freedoms, and thereby allow more scope for extralegal activities on the part of the individual worker. Weak group occupations also involve fewer constraints on the individual from his or her work-group, but strong work-groups also have the potential for exercising teamwork in the service of group-level pilferage. Mars' explanatory typology involves possibility dimensions at both the individual and group levels.

The explanatory linkages among possibility, choice, and constraint may best be appreciated by examining those aspects of the human condition that are making the transition from a deterministic/probabilistic process to a psychological and social process involving possibility and choice. Consider, for instance, the determination of a baby's gender. Until recently the underlying process was best modeled as a simple binomial distribution with a probability of nearly 1/2 for a girl or boy. Sociobiologists have been fond of adding deterministic refinements to this probabilistic model, whereby more boys are produced when food is scarce and more girls when food is plentiful. This process does not involve possibility or choice, but social constraint also is absent. If we adopt the sociobiological model, the possibility, probability, and necessity of having a girl (or boy) are identical; and the constraints are biological.

Enter choice of gender technology, in which claims are made that parents may reliably choose whether or not to have a boy or girl. The question of whether such technology is available now is the center of some controversy, and the actual reliability for current practices is unknown. But suppose a widely available technique were to be created that had a reliability of 0.9. In other words, the expected po(boy) = po(girl) = 0.9, while the expected ne(boy) = ne(girl) = 0.1. Clearly the process for determining the gender of the newborn would then shift from a probabilistic biological to a possibilistic social process, with preference patterns initially influencing the female/male ratio of the newborn.

These preference patterns would suddenly assume the status of influencing factors. Some studies indicate that not only does a majority of parents prefer boy babies to girl babies (this tendency is stronger among men than women), but an overwhelming percentage (91% of the men and 84% of the women, in a study referred to by Fraser, 1986) prefer the firstborn to be a boy. How far away from the nearly even boy-girl split might such preferences take us? Let us conservatively assume that 16% of the women would try for firstborn girls. Taking these preferences on face value, the expected ratio of firstborn boys to girls would be at minimum $[(.84)(.9) + (.16)(.1)] / [(.16)(.9) + (.84)(.1)] = 3.386$. If the 16% who do not insist that a boy be the firstborn include some uncommitted women, then the ratio would be higher. The 84% preference figure also is a least upper bound; to the extent that couples' decisions on the firstborn's gender were made jointly or dominated by men, the ratio could be higher.

What social responses might we expect to this (even prospectively)? Some people might see this as a classic Commons dilemma, in which individual preferences exercised collectively endanger the collectivity as a whole. Spokespeople from several interest groups already have commented unfavorably on the prospect of a widespread pattern of "big brotherism" by birth order (cf. Fraser, 1986). The Luddite option (banning the technology) seems unlikely. More likely are various attempts to socially "regulate" gender choice, or to narrow the range between possibility and necessity. Here, then, is a concrete realization of the oft-repeated refrain in sociological theory that freedom and constraint go hand-in-glove, but with the roles of possibility, choice, and constraint made more explicit and analyzable. Rather than Zinoviev's vision of a replacement of "possible not to" by "not possible to", we should expect that a sudden increase in "possible not to" would be counteracted with a corresponding increase in "necessary to" via social or even legal regulations. Exit a strictly probabilistic biological explanation. Enter moral choice, the imposition of social constraints, and a possibilistic theoretical and quantitative framework. As Hacking (1981, p. 26) so aptly puts it, "The less determinism, the more the possibilities for constraint."

In summary, possibility plays a part in several crucial aspects of psychological theories: (1) As a backgrounding for counterfactuals in hypotheses, prescriptions, and measurement; (2) As an ingredient in meanings; and (3) As an explanatory or analytic variable. A well-formed conceptual and mathematical basis for possibility should enable theorists and researchers alike to amplify each of these capacities. Likewise, the advent of a mathematical framework for possibility improves our ability to address central questions concerning person-situation research, the nature of freedom and control, the operation of power and domination, and how people make decisions or exercise preferences. It does not, of course, solve fundamental philosophical problems associated with the free will versus determinism debate. No simple appeals to language can do that, although at least distinctions such as subjective and objective possibility enable theorists to avoid the pitfalls pointed out by writers like Skinner (e.g., 1978, p. 31) in confusing the feeling of freedom with "real" fredom.

While it is far too early to assess the payoffs from a theory of possibility and weak entailment framework at the theoretical and conceptual level, some immediate claims can be made for benefits at the level of research design and data analysis. No longer need quantitative researchers resort to deterministic statistical mathematics by default, and modern theorists need not inevitably turn knowledgeable, intentional actors into judgmental dopes whenever they quantitatively operationalize any component of their theories. Furthermore, weak entailment models offer a badly needed alternative to "causal analyses" in which a dozen or so ill-defined variables are inappropriately laundered through the GLM to "explain" minute portions of variance in outcome variables. It places the burden of proof, for once, on the determinists.

Although the framework presented here is incomplete and tentative, it clarifies one of the most difficult problems in the human sciences: the conceptual rift between modern psychological theory and quantitative methods. It also offers some foundational concepts and methods which may point a way out of this difficulty, by applying fuzzy logic and possibility theory to models of prediction that allow choice under partial restriction. Finally, those foundations suggest a method for measuring the relative freedom of choice accorded either inviduals or groups which in turn may prove valuable in both normative and explanatory paradigms of decision making and choice behavior.

References

Carroll, K. K. Experimental evidence of dietary factors and hormone dependent cancers. *Cancer Research,* 1974, *35,* 33-77.

Cheeseman, P. *In defense of probability.* Procedings of the 9th International Joint Conference on Artificial Intelligence, Los Angeles, CA., 1985.

Civanlar, M. R. & Trussell, H. J. Constructing membership functions using statistical data. *Fuzzy Sets and Systems,* 1986, *18,* 1-13.

Clarke, D. D. *Language and Action.* Oxford: Pergamon, 1983.

Cohen, L. J. *The Probable and the Provable.* Oxford: Clarendon, 1977.

Cohen, J. & Hickey, T. Two algorithms for determining volumes of convex polyhedra. *Journal of the Association for Computing Machinery,* 1979, *26,* 401-414.

Coleman, J. S. Social theory, social research, and a theory of action. *American Journal of Sociology,* 1986, *91,* 1309-1335.

Collingridge, D. *The Social Control of Technology.* London: Frances Pinter, 1980.

Collingridge, D. *Critical Decision Making.* London: Frances Pinter, 1982.

Collins, R. Statistics versus Words. In R. Collins (Ed.), *Sociological Theory.* San Francisco, Jossey-Bass, 1984.

Douglas, M. *Cultural Bias.* London: Royal Anthropological Institute, 1978.

Elster, J. *Logic and Society: Contradictions and Possible Worlds.* Chichester: Wiley, 1978.

Elster, J. *Sour Grapes: Studies in the Subversion of Rationality.* Cambridge: Cambridge University Press, 1983.

Fraser, L. Oh boy, let's have a baby! *Mother Jones*, 1986, *11*, 16-17.

Gaines, B. R. & Shaw, M. L. G. Induction of inference rules for expert systems. *Fuzzy Sets and Systems,* 1986, 18, 315-328.

Ginsburg, G. P. Introduction and overview. In P. Ginsburg (Ed.), *Emerging Strategies in Social Psychological Research.* New York: Wiley, 1979.

Hacking, I. *The Emergence of Probability.* Cambridge: Cambridge University Press, 1975.

Hildebrand, D. K., Laing, J. D., & Rosenthal, H. *Prediction Analysis of Cross Classifications.* New York: Wiley, 1977.

Horvitz, E. J., Heckerman, D. E., & Langlotz, C. P. *A framework for comparing alternative formalisms for plausible reasoning.* Memo KSL-86-25, Stanford, Ca.: Stanford University School of Medicine, 1986.

Kelly, G. A. *The Psychology of Personal Constructs.* New York: Norton, 1955.

Kelly, G. A. *Clinical Psychology and Personality: The Selected Papers of George Kelly.* (Ed. P. Maher) New York: Wiley, 1969.

Kim, J. PRU measures of association for contingency table analysis. *Sociological Methods and Research,* 1985, *113*, 3-44.

Kullback, S. *Information Theory and Statistics.* New York: Wiley, 1959.

Layder, D. Power, structure, and agency. *Journal for the Theory of Social Behaviour,* 1985, *15*, 131-149.

Lukes, S. *Essays in Social Theory.* London: MacMillan, 1977.

Lukes, S. *Unpublished paper* delivered at the British Sociological Association Conference. Guilford: University of Surrey, 18-19 September 1978.

March, J. G. & Olsen, J. P. *Ambiguity and Choice in Organizations.* Oslo: Universitetsforlaget, 1979.

Mars, G. *Cheats at Work.* London: Allen and Unwin, 1982.

Miller, G. A., Galanter, E., & Pribram, K.L. *Plans and the Structure of Behavior.* New York: Holt, Rinehart, and Winston, 1960.

Muller, E. N. A test of a partial theory of potential for political violence. *American Political Science Review,* 1972, *66,* 928-959.

Schwartz, B. *The Battle for Human Nature.* New York: Norton, 1986.

Shackle, G. L. S. *Decision, Order, and Time in Human Affairs.* New York: Cambridge University Press, 1969.

Shafer, G. *A Mathematical Theory of Evidence.* Princeton: Princeton University Press, 1976.

Skinner, B. F. *Reflections on Behaviorism and Society.* Englewood Cliffs, N.J.: Prentice-Hall, 1978.

Smithson, M. *Accommodation and Transportation for the Elderly in Townsville.* Queensland: James Cook University, 1983.

Smithson, M. *Fuzzy Set Analysis for Behavioral and Social Sciences.* New York: Springer Verlag, 1987.

Smithson, M. *A calculus of freedom.* 1988. (under review)

Smithson, M. & Knibb, K. New measures of association for numerical variables. *Journal of Mathematical Social Sciences,* 1986, *11,* 161-182.

Taylor, S. E. Adjustment to threatening events: A theory of cognitive adaptation. *American Psychologist,* 1983, *38,* 1161-1173.

Theil, H. *Statistical decomposition analysis.* Amsterdam: North-Holland, 1972.

Tversky, A. & Kahneman, D. The framing of decision and rationality of choice. *Science,* 1981, *211,* 453-458.

Wallsten, T. S., Budescu, D. V., Rapoport, A., Zwick, R., & Forsyth, B. Measuring the vague meaning of probability terms. *Journal of experimental Psychology: General,* 1986, *115,* 348-365.

Wilkinson, S. J. Constructs, counterfactuals and fictions: Elaborating the concept of "possibility" in science. In H. Bonarius, R. Holland, & S. Rosenberg (Eds.), *Personal Construct Psychology: Recent Advances in Theory Practice.* Sutton: MacMillan, 1981.

Zadeh, L. A. Fuzzy sets as a basis for a theory of possibility. *Fuzzy Sets and Systems,* 1978, *1,* 3-28.

Zadeh, L. A. Fuzzy sets and probability. *Proceedings of the I.E.E.E.,* 1980, *68,* 421.

Zwick, R., & Wallsten, T. S. *Combining stochastic uncertainty and linguistic inexactness: Theory and experimental evaluation.* Paper presented at NAFIPS workshop, Purdue University, May, 1987.

QUANTIFIERS AS FUZZY CONCEPTS

Stephen E. NEWSTEAD

Department of Psychology, Plymouth Polytechnic
Drake Circus, Plymouth PL4 8AA, England

As their name suggests, quantifiers are used to indicate quantities along certain dimensions, whether of amount (e.g., all, some) or frequency (e.g., never, often). Such terms are among the most commonly used in English and other languages. Thorndike and Lorge (1944) in their list of the 500 most common words in English include more than a dozen such terms, including the four examples given in the previous sentence.

Some writers have assumed that the quantities expressed can be precisely determined. For example, according to logic all means every single member of a set, and some means at least one and possibly all members of a set. Similarly the query languages developed by computer programmers give precise definitions to the quantifiers used in searching through a database. It will be argued in this chapter, however, that outside such artificial languages quantifiers are inherently fuzzy concepts. Furthermore it will also be argued that their meaning is variable depending on the situation in which the words are used.

Previous research on quantifiers has taken place within a variety of different *frameworks. Logicians and reasoning researchers have investigated the interpretation of quantifiers as a possible explanation for errors in syllogistic reasoning; applied psychologists have studied the scale values of quantifiers used in rating scales; and psycholinguists have investigated the representations of quantifiers in both comprehension and memory. Each of these areas of research has its own focus of interest and its own methods of investigation. In this chapter an attempt will be made to review this work and to synthesise the findings from the disparate areas.*

QUANTIFIERS AND SYLLOGISTIC REASONING

According to the Greek philosopher Aristotle, syllogisms are the essence of human rational thought. Syllogisms consist of two premises and a conclusion, each of which contains one of the quantifiers *all, no, some*, and *some ... not*. An example would be:

All men are mortal
<u>No men are women</u>
Therefore some women are mortal

All syllogisms are like this in that they contain a conclusion consisting of two terms (women, mortal) which are linked in the premises by a common middle term (men). Since any one of the quantifiers can be used in each premise and the conclusion, and since the end and middle terms can be ordered in different ways, there are a large number of possible syllogisms. Only a small number of conclusions are valid (the one given above is not) and people are known to make a large number of errors when they attempt to solve syllogisms (see Evans, 1982, for a review).

The errors that people make in syllogistic reasoning could occur at any one of a number of different stages. They could occur at the encoding stage; they could occur when the information from the two premises is combined;

or they could arise at the decision stage. Research to date has failed to provide any conclusive evidence as to which of these stages is the source of errors, and indeed it is possible that all might contribute (Fisher, 1981). However, for the present purposes we shall concentrate on interpretational explanations, i.e. those alleging that the premises are misinterpreted.

Two main errors have been indicated in the literature, *conversion errors* and *incomplete representations*. Conversion errors involve interpreting the quantifiers *all* and *some...not* as if they implied their converses, for example, believing that *All As are Bs* implies *All Bs are As*. That this is not the case can be easily seen by using the real-life example All dogs are animals. Conversion errors could explain a number (though not all) of the errors that are typically made (Dickstein, 1981). Incomplete representations arise when premises are represented in such a way that some logical possibilities are excluded. For example, *Some As are Bs* does not logically preclude all As being Bs, but subjects might represent this premise in such a way that this possibility is precluded.

The fact that these misinterpretations can explain many of the errors made does not, of course, mean that this explanation is correct, since other explanations might fare equally well. Independent evidence that such interpretational errors occur is needed. Such evidence as exists comes from two main experimental methods, Euler diagrams and immediate inferences, and the findings from these two tasks will be considered in turn.

Euler diagrams (frequently confused with Venn diagrams) depict the possible relationships between the two sets mentioned in the proposition. The five possibilities are presented in Figure 1, ranging from identity (diagram 1) to exclusion (diagram 5). In the bottom part of Figure 1 is an indication of which quantified statements are true of which Euler diagrams. In a typical experimental task, subjects would be asked to indicate which statements can be paired with which diagrams, and their performance compared to the logically correct answers given in this figure.

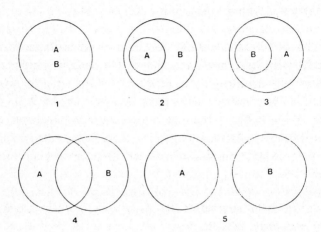

	Diagram number				
	1	2	3	4	5
All As are Bs	v	v	x	x	x
No As are Bs	x	x	x	x	v
Some As are Bs	v	v	v	v	x
Some As are not Bs	x	x	v	v	v

Note: A "v" idicates that a diagram is true with respect to the quantified statement on the left, an "x" that it is false.

FIGURE 1
Euler diagrams

If subjects are making conversion errors with the quantifier *all,* then they should presumably select only diagram 1 (the identity relationship, for which conversion is valid) and not diagram 2 (in which As are a subset of B and conversion is invalid). Unfortunately for conversion theorists, such errors are rare. Griggs and Warner (1982) found only one response out of 32 which came into this category, and unpublished research by the present author confirms the rarity of this choice. If subjects are converting *some...not* then they should choose diagrams 4 and 5, for which both *Some As are not Bs* and *Some Bs are not As* are true, but not diagram 3 which is true for *Some As are not Bs* but not its converse. Once again, however, choice of just these two diagrams is rare (<5% in the unpublished study by the present author). Subjects commonly select just diagram 4, but omit both diagrams 3 and 5. It must be concluded from research on Euler diagrams that conversion is not common.

The evidence for incomplete representations is much stronger, since subjects frequently indicate fewer diagrams than they should. More importantly, there is a pattern to these omissions. Errors with *some* frequently involve omitting diagrams 1 and 2 and errors with *some...not* frequently involve omitting diagram 5 (Neimark & Chapman, 1975). These are precisely those diagrams which are true only because some does not preclude *all,* and *some...not* does not preclude *no*; in other words these diagrams are true for the universal relationships (*all* and *no*). Hence this finding would suggest that subjects interpret some as meaning *some but not all*.

According to conversational implicatures (Grice, 1975), this is precisely what one would expect, since it would be misleading to use the term *some* if one believed all to be the case. The evidence strongly suggests, then, that subjects underinterpret *some* and *some...not* in predictable ways (see Begg & Harris, 1982). Such a finding is a little surprising since subjects are usually specifically instructed to give the quantifiers their logical interpretations.

The second major way of looking at the interpretation of syllogistic quantifiers is the immediate inference task, in which subjects are asked to indicate which statements are logically implied by a given quantified statement. For example, subjects might be given the statement *All As are Bs* and asked if a statement such as *Some As are Bs* logically follows from this (Begg & Harris, 1982; Newstead & Griggs, 1983). Conversion errors would be demonstrated if subjects believed that *all* and *some...not* statements implied the truth of their converses. There is evidence that a significant minority of subjects (30-40%) believe that *all* is convertible, and an even greater number believe *some...not* to be so (Newstead & Griggs, 1983).

There is also significant evidence immediate inference task for incomplete representations. As in the Euler diagram task, many subjects believe that particular statements (i.e., those using *some* and *some...not*) are not consistent with their universal counterparts. One example will serve to illustrate this point; of the subjects given the statement *All As are Bs*, only about 70% believed that *Some As are Bs* was definitely true, despite the fact that the truth of this statement is implied by logic. Further, a significant number of subjects (more than 20% in each of three separate studies) believed that the *some* statement was definitely false (Newstead & Griggs, 1983).

Within the context of syllogistic reasoning, it is also relevant to mention extended syllogisms or set inclusions. These are chained premises of the kind *All As are Bs, All Bs are Cs, All Cs are Ds*. Subjects find such relationships extraordinarily difficult, possibly because they make conversion errors and fail to realise the transivity of the relationship (Griggs, 1976). Newstead and Griggs (1984) have suggested that errors might be caused by subjects interpreting *all* fuzzily, in order words treating it as if it meant almost all. They attempted to demonstrate such fuzzy interpretations by asking subjects to rate the appropriateness of *all* when there were just a few exceptions to a general rule. An example would be if subjects were asked to rate the appropriateness of the quantified sentence *All the people are aged under 100* when in fact there was one person (out of several thousand) aged over 100. The authors obtained moderately high appropriateness rating for such items.

In itself this evidence is far from conclusive. There are obvious demand characteristics in a task such as that used so that subjects might feel obliged to give non-minimal appropriateness ratings when there are just a few exceptions. Fortunately for the authors, there was one aspect of their results which serendipitously supported their hypothesis. In addition to asking subjects to rate *all* they also asked them to rate other quantifiers, including *no*. Presumably the same demand characteristics apply to ratings of *all* and *no*, and yet *all* was rated significantly more appropriate than *no* when there were just one or two exceptions to a general rule. Hence at the very least we can conclude that *all* is interpreted more fuzzily than *no*.

The research on syllogistic reasoning has produced conflicting results, but some general conclusions can still be drawn.

(a) Conversion errors probably do occur, despite the fact that they do not show up in Euler diagram tasks. However, they are manifested only by a minority of subjects.

(b) Incomplete representations occur frequently. An especially prevalent error is that of adopting the everyday interpretation of particular quantifiers rather than that dictated by logic.

(c) There is some preliminary evidence that all may be interpreted fuzzily.

(d) There are large individual differences in the interpretation of quantifiers.

QUANTIFIERS AND RATING SCALES

If syllogistic reasoning is historically the most important context in which quantifiers have been studied, the most important in terms of weight of recent research is the use of quantifiers in rating scales. In rating scales, people are asked to assess somebody or something along a labelled dimension. In many cases this dimension involves frequency or amount and the labels used are quantifiers. As an example, people might be asked in a questionnaire whether they enjoy parties *always, often, occasionally, seldom* or *never*.

It is normal practice to convert responses on such scales into numbers, for instance assigning 5 to a response of *always*, 4 to *often*, and so on. In this way the results from different questions can be pooled and compared.

However, this conversion assumes that the responses are intervally scaled, such that the distance between any two adjacent responses is the same; thus in the above example the distance between *always* and *often* should be the same as that between *seldom* and *never*. Unless these responses achieve interval scaling it is inadmissible to convert them into numbers (which are, of course, intervally scaled). Hence one of the principal aims of research in this area has been to find quantifiers which possess interval scaling properties. One of the best known studies here is that of Bass, Cascio and O'Connor (1974). They used the scaling method of magnitude estimation (Stevens, 1971) to assess the scale values of 44 quantifiers of amount and 39 of frequency. On the basis of their findings they were able to list optimum sets of quantifiers, depending upon the number required for a particular scale. They claimed that these scale values were relatively robust since they were unaffected by the importance of the topic described (the War in Vietnam vs rainfall in Nepal). Some researchers have suggested minor alterations to the proposed optimum lists (e.g., Schriesheim & Schriesheim, 1978; Pohl, 1981) but in general they have stood the test of time.

Despite the relative stability of these scale values, there is a growing body of evidence that they can be substantially affected by a number of factors. Perhaps the best documented of these is the *expected frequency* of the event described. Pepper and Prytulak (1974) found that quantifiers were interpreted as indicating higher frequencies in contexts which were themselves of high expected frequency. To illustrate, *frequently* was taken to mean approximately 70% of the time when used to describe the frequency with which Miss Sweden was found attractive (high expected frequency) but only about 20% of the time when describing the frequency of air crashes (low expected frequency). These results have been replicated on a number of occasions (eg. Newstead & Collis, 1987; Wallsten, Fillenbaum & Cox, 1986). Each of these studies also found that the effects of expected frequency were much more marked with high than with low frequency quantifiers. Indeed quantifiers denoting less than half seemed unaffected by base rate.

There is also some evidence that *attitudes and experience* can influence interpretation of quantifiers. Goocher (1965) asked subjects to assign quantifiers to various activities (such as going dancing) which occurred with specified frequencies. Their experience and liking of such activities was also assessed. Subjects who disliked or had little experience of an activity tended to assign higher frequency quantifiers than subjects with greater liking or experience. Thus a subject who disliked dancing might think going dancing three times a month to be "often" whereas a keen dancer might call this "seldom". It is possible that this finding is a special case of the expected frequency effect, since those who like an activity or indulge in it regularly might have a higher expected frequency for such an event.

There is also evidence that *type of activity or object* being described can influence the interpretation of quantifiers. As has already been mentioned, Bass et al., found no differences in scale values dependent on the importance of the activity described. Similarly, Newstead and Collis (1987) found that scale values showed little variation over a variety of different activities, ranging from emotional feelings to overseas visits. However subjects seemed less inclined to use the extreme ends of the scales when emotions were being described; for example *always* feeling happy is interpreted as implying being happy little more than 90% of the time. But perhaps the most impressive evidence comes from an interesting series of experiments reported by Hörmann (1983). He found that each of the following factors could influence the amount indicated by a quantifier:

(a) Object described: a few crumbs means more than 8, whereas a few shirts means approximately 4.

(b) Size of object: a few *large* cars suggests a smaller number than a few cars.

(c) Spatial location: a few people standing in front of a hut is not as many as a few people standing in front of a building.

(d) Size of field of vision: a few people seen through a peephole indicates a smaller number than a few people seen through a window.

It is possible once again that these findings are related to expected frequency since one might expect to see fewer large objects than small ones, or to see more objects through a window than a peephole.

Another factor which has been claimed by some to affect interpretation is the *set size* that is being described. The first study to investigate this was carried out by Borges and Sawyers (1974), who asked subjects to choose the appropriate number of marbles suggested by a quantifier. They varied the size of the set of marbles and were thus able to compare for example the proportions indicated by *some* with set sizes of 12 and 108. They reported no effects of set size, though this may be because they used an insensitive statistical technique. In two replications of their study, Newstead, Pollard and Riezebos (1987) were able to demonstrate significant effects, such that low magnitude quantifiers (e.g., *few*) indicated larger proportions with small set sizes than they did with large set sizes. However Newstead and Collis (1987) found much smaller (and nonsignificant) effects when quantifiers of frequency rather than amount were used. It would appear that set size is an important but somewhat unpredictable determinant of interpretation.

The final factor which seems to affect scale values is the *range of other qualifiers available*. Chase (1969) found that *ocassionally* was assumed to indicate a higher frequency when presented in the context of low frequency quantifiers (*seldom, rarely*) than when embedded in high frequency quantifiers (*usually, often*). Similarly Newstead and Griggs (1984) found that *some* was rated as more appropriate with proportions between 50%-100% when it was the only quantifier available than when the quantifier *most* was also available. However there is evidence that the number of other quantifiers has little effect providing that the quantifiers there are give a reasonable spread across the dimension (Newstead & Collis, 1987).

It should be clear from the work on rating scales that the interpretation of quantifiers is very much affected by *context*; the expected frequency of the event, the type of activity, the set size described and the range of alternative quantifiers available can all affect scale values. What this implies, of course, is that we must not look for *the* meaning of a quantifier; rather we must acknowledge that there is a range of meanings and the problem thus becomes one of characterizing this variable meaning. Some possible ways of doing this are indicated in the final section.

A further point to note is the considerable variability in the interpretation of quantifiers. This can be readily detected in those experiments in which subjects have been asked to indicate ranges for quantifiers (e.g., Wallsten, Fillenbaum & Cox, 1986; Newstead & Collis, 1987). Interestingly, Wallsten et al., found that quantifiers in the middle of the range were most variable, followed by high frequency quantifiers and then low frequency terms.

Another aspect of variability can be seen in the large individual differences that emerge in many of these studies. That these are not simply attributable to experimental error has been shown by Budescu and Wallsten (1985) who found that individual subjects were consistent over time but significantly different to other subjects.

Psycholinguistic studies of quantifiers

The comprehension and representation of quantifiers have challenged psychologists for many years, but despite extensive research and theorizing there is still little consensus in the area. A number of different approaches have figured in the literature, and this chapter will briefly summarise four of the more influential of these: the featural theory of Just (1974): the analogue theory of Holyoak and Glass (1978): the propositional theory of Anderson (1981); and fuzzy set theory.

Featural theories are based on the assumption that words can be analysed into unique combinations of features. A commonly-cited textbook example is the word *bachelor*, which can be defined in terms of possessing the feature "male" on the male/female dimension and the feature "unmarried" on the married/unmarried dimension. With respect to quantifiers, dimensions such as universal/particular, large/small and negative/positive can be used to categorise quantifiers. In a series of experiments involving sentence-picture verification, Just (1974) was able to provide evidence in favour of such a featural analysis. It is worth noting, however, that these results were obtained with highly-practised subjects who showed a consistency and accuracy that would probably not be found with naive subjects. In addition it is now generally accepted that feature theories are oversimplistic; while certain core aspects of meaning can be represented in this way for some words, some of the subtleties and variations of meaning cannot be captured.

According to theorists such as Holyoak and Glass (1978), quantifiers can be characterised as points along an analogue scale. This can be visualised as a line with *all* at one end and *no* at the other, with each quantifier occupying a location between these two extremes. The experiment carried out by Holyoak and Glass (1978) indicated that quantifiers located close together on this scale were more likely to be confused with each other in memory than were more distant terms. This is strong evidence to support the analogue theory but, as the authors themselves point out, would also be predicted by certain featural theories and, indeed, other theoretical apporaches.

The propositional theory of Anderson (1981) claims that quantifiers are given abstract representations in the form of propositions. In the particular version of the theory espoused by Anderson, quantifiers have an effect on whole sentences, altering the type and number of propositions into which they can be analysed. Quantifiers which produce similar propositional analyses (e.g. *all* and *some*) should be prone to confusions in memory, and this is precisely the result obtained by Anderson. However, it must be noted once again that such a result could also be predicted by feature theories (*all* and *some* are both positive) and by certain versions of the analogue theory.

Fuzzy set theory has been investigated only recently in psycholinguistic research on quantifiers. According to this theory, quantified terms can be represented as membership functions over the range from 0 to 1. For any quantifier there will be some values that are definitely outside the membership function (to which a value of 0 is assigned), others that are definitely within it (which receive a value of 1), and others which have intermediate values. Consider for example the quantifier *some;* the value of 0% would certainly be outside the membership function for this term; 30% would definitely be in it; but values such as 90% might be regarded as having intermediate values, not being definitely within or definitely outside the meaning. Some preliminary work has suggested that this approach may be fruitful (e.g., Wallsten, Budescu, Rapaport, Zwick & Forsyth, 1986; Farkas & Englander, 1987; Ekberg & Lopes, 1980), but as yet little work has been done to test the predictions of this theory against those of other approaches.

Perhaps the only certain conclusion that can be drawn from this psycholinguistic research is that no certain conclusion can be drawn. The empirical evidence that has been accumulated can be adduced to support any of the theories but to disprove none of them. The predictions of several of the theories are a little obscure and it is difficult to be optimistic of an early resolution of the conflicting views.

Overview and Synthesis

To a certain extent, the division of research on quantifiers into three areas - syllogistic reasoning, rating scales and psycholinguistics - is arbitrary. To give just one example, Hörmann's (1983) study was included in the section on rating scales because it related so closely to work on the effects of context on interpretation; in reality, the research was motivated more by psycholinguistic considerations. Despite such cross-references, in general researchers in each area have pursued their enquiries independently of each other. Nevertheless there are important and illuminating links that can be made, and in this section an attempt will be made to draw some general conclusions about the interpretation of quantifiers.

(1) Quantifiers are Fuzzy

The conclusion that quantifiers are best regarded as fuzzy concepts seems inescapable. Attempts to determine *the* meaning of a quantifier have been noticeably unsuccessful in all areas of research. Clearly quantifiers have a range of meanings and there are no clear-cut boundaries between one term and another; it is difficult to see how this could be captured other than by notions such as those derived from fuzzy set theory. One aim of future research should clearly be to determine membership functions for each quantifier. Some studies of rating scales have provided useful information here since subjects have been asked to give maximum and minimum values for quantifiers. However much the most sophisticated work in this area so far has been carried out by Wallsten et al., (1986) who have empirically obtained membership functions for frequency and probability terms. The experimental test they used involved choosing which of two terms best described the proportions indicated on a pie-chart. It would be interesting to know if similar results can be obtained using different methods; although they were used in rather different paradigms, the techniques of allocating 100 points between different quantifiers (Begg & Harris, 1982), or giving appropriateness ratings to quantifiers (Newstead & Griggs, 1984) might both be adaptable to investigating this question.

It is perhaps not too surprising that quantifiers such as *some* have meanings covering a fairly wide range. However it has also been claimed that universal quantifiers might be fuzzy in nature. Clearly the range of end terms such as *all* and *always* is much more restricted than for intermediate terms; Budescu and Wallsten (1985) found interquartile ranges close to zero for end terms but in excess of 20% for other terms. However, Newstead and Griggs (1984) provided evidence that *all* is sometimes seen as appropriate even when there are one or two exceptions. Furthermore, an incidental finding of Newstead and Colis (1987) was that *always,* when applied to statements concerning emotional states, may mean little more than 90% of the time. Hence there is evidence that *all* and *always* can be interpreted fuzzily, though interestingly there is little evidence for fuzziness in the interpretation of *no* and *never*.

(2) Meaning is context dependent

Once again, the conclusion that the meaning of quantifiers varies depending upon the context in which they occur is totally inescapable. It emerges most clearly in work on rating scales but is evident also in other research areas. The most widely researched type of context is base rate or expected frequency. Pepper and Prytulak (1974) have shown quite clearly that terms such as *often* suggest a higher frequency when applied to expected or common events rather than unexpected ones. What is more, a number of other context effects can be explained using this concept. Effects of liking and familiarity (Goocher, 1965) can be explained in terms of familiar or enjoyable events having a higher expected frequency. Similarly, effects of size of object or size of field of vision (Hörmann, 1983) may be due to an expectancy of seeing fewer large objects or fewer objects in a restricted field of vision.

Another aspect of context that can influence interpretation is the size of the set described. Although the effect is somewhat variable, it does apear that at the very least the proportion signified by low frequency quantifiers of amount decreases as set size is increased.

One finding to emerge in both studies of expected frequency and of set size is that low frequency quantifiers are more fixed and less variable in their meaning than high frequency quantifiers. The effects of expected frequency have repeatedly been found to be more marked with high frequency terms, and it is actually disputable whether any effect exists with low frequency terms (Wallsten, Fillenbaum & Cox, 1986). At first sight the effects of set size contradict this, since it is the proportion signified by *low* frequency terms that varies with set size. However, this contradiction is more apparent than real. Low frequency terms signify a greater proportion with small set sizes than with large set sizes, and hence the actual amount suggested by the quantifier varies less than with high magnitude quantifiers. Thus *few* might suggest 4 out of 10 and 10 out of 100, a relatively small difference, while *many* might signify 7 out of 10 and 70 out of l00, a proportionally greater difference. It is interesting to recall that terms at the bottom extreme of the scale (*no, never*) also seem to be less variable (fuzzy) than their counterparts at the top end of the scale (*all, always*).

Another aspect of context which seems to affect interpretation is the number of other quantifiers available. Chase (1969) has shown that *occassionally* means rather more when embedded in low magnitude quantifiers than when embedded in high magnitude ones. Similarly, in an experiment partly reported by Newstead and Griggs (1984), it was found that *some* was regarded as highly appropriate for any proportion between 10% and 90% when the only other quantifiers available were *all* and *none*; but when most was added to the list, the appropriateness of *some* between 50% - 100% declined markedly. It would appear that quantifiers can expand or contract to fill the available semantic space; they will be interpreted narrowly if many other quantifiers are available, but much more broadly if there are few other quantifiers.

It was concluded in the previous section that quantifiers are best characteristed as fuzzy concepts. In view of the marked effects of context, this conclusion needs to be reviewed. Since the scale value of quantifiers can be influenced by context, t hen presumably each quantifier can be characterised by a whole range of fuzzy sets, the appropriate one being determined by context. However, this is a rather unsatisfactory state of affairs since it implies an essentially infinite number of fuzzy sets a ssociated with each quantifier. The issue is complicated even further by the finding that meaning depends on the other quantifiers available. Not only does the scale value around which the set is centred vary, but also the size of the set itself; the spread will be large when only a few other quantifiers are available, but smaller when there are more available.

Clearly a straightforward fuzzy set approach is too simplistic. The complexity of quantifiers can perhaps be captured by assuming that they can be represented as fuzzy sets which can vary along an analogue scale. To illustrate this, in the absence of context, *often* might have a membership function corresponding to a normal curve with a mean of 65% and a standard deviation of ±10%. The effect of context would be to alter the mean value, but the membership function itself might be relatively unchanged.

This is an improvement, but there are still problems. As quantifiers get pushed towards the extremes of the scale, the set must be compressed to prevent it going beyond the end-points. In addition, some quantifiers (the low magnitude ones) are more fixed than others, and hence their corresponding sets cannot be moved up and down the scale in the same way. And finally, as we have seen, the range of quantifiers is determined by what other quantifiers are available; hence the sets indicated by quantifiers must be able to expand and contract according to the context in which they occur. Hence the analogue scale must make provision for some quantifiers being more fixed than others in their position on the scale, and for the sets themselves being variable in range depending on their position in the scale and on which other quantifiers are represented on the scale.

Thus far the picture that has been painted might appear to be one of almost unlimited variability. Fortunately, there are a number of aspects of meaning that do not change; in particular the *rank order* of quantifiers seems to remain virtually constant (Pepper, 1981). There are some terms that are synonymous (e.g. *seldom* and *rarely*), and the rank order of these can vary from study to study; but if these special cases are ignored, rank order correlations approach unity. Thus the meaning of a quantifier such as *few* can vary so that it means as much as 50%; but there seems to be no way in which it can mean more than *some* does in the same context, even though the meaning of this latter, out of context, is only about 30%.

Hence there is here an important regularity that can be built in to a model of quantifiers. We are left with a complex but coherent conceptualisation: quantifiers can be characterised as fuzzy sets located in fixed order along an analogue scale. The size of the sets is variable, and some quantifiers are more mobile than others in their position along this scale.

It would be pretentious to elevate these ideas into a "theory" of quantifiers; rather they provide a framework for summarising existing research findings. Nevertheless the conclusions, drawn from existing data, are all open to empirical disproof. More importantly, the ideas might indicate where future researchers should look for lawful relationships. It is likely that the meaning of quantifiers varies regularly with expected frequency and set size; and the range of quantifiers probably varies in a predictable way with the number of other quantifiers available and proximity to the end-points of the scale. There are hints of such relationships in the literature, but only a richer pool of experimental data could confirm or disprove their existence.

(3) There are large individual differences in interpretation

Research on syllogistic reasoning and rating scales has revealed marked individual differences in interpretation. Some subjects (approximately 40%) regard *all* as convertible, the remainder do not; some subjects adopt the interpretation dictated by logic, others are swayed by conversational implicatures; and there are substantial individual differences in the scale values given to frequency terms. These individual differences are not due simply to experimental errors since both Newstead and Griggs (1984) and Budescu and Wallsten (1985) have, in very different paradigms, shown consistency in subjects' interpretations.

There are presumably a number of different sources for individual differences, and some possibilities are raised by the literature reviewed in this chapter. It has been shown that quantifiers can expand or contract to fill the available semantic space. It follows from this that someone with a large vocabulary of quantifiers will assign rather more specific meanings to these terms than someone with a less rich vocabulary. In addition, there are almost certainly differences in subjects' willingness to adopt fuzzy interpretations of universal quantifiers, and it has been found that fuzzy interpreters are more likely to make conversion errors (Newstead & Griggs, 1984). Perhaps some subjects are simply less precise in their use of terms, and allow for a certain amount of flexibility in meaning. It would be interesting to know if these subjects are the same ones as those who respond in experiments as predicted by conversational implicatures, refusing to be tied down to the highly specific meanings prescribed by logic.

SUMMARY AND CONCLUSIONS

This chapter has reviewed research on quantifiers from three quite distinct areas: syllogistic reasoning, rating scales and psycholinguistics. It is quite clear that such terms are imprecise and ill-defined, and mean rather different things to different people. It is also true that their meaning varies markedly from one context to another, depending on such factors as expected frequency, size of set described and range of other quantifiers available. However, the order of quantifiers seems to be fixed. These properties are captured if quantifiers are visualised as fuzzy sets of variable size which are in fixed order relative to each other but which can be moved up and down a continuous scale.

REFERENCES

Anderson, J. R. Memory for logical quantifiers. *Journal of Verbal Learning and Verbal Behavior*, 1981, *20*, 306-321.

Bass, B. M., Cascio, W. F., & O'Connor, E. J. Magnitude estimations of expressions of frequency and amount. *Journal of Applied Psychology*, 1974, *59*, 313-320.

Begg, I., & Harris, G. On the interpretation of syllogisms. *Journal of Verbal Learning and Verbal Behavior*, 1982, *21*, 595-620.

Borges, M. A., & Sawyers, B. K. Common verbal quantifiers: Usage and interpretation. *Journal of Experimental Psychology*, 1974, *102*, 335-338.

Budescu, D. V., & Wallsten, T. S. Consistency in interpretation of probabilistic phrases. *Organisational Behaviour and Human Decision Proceses*, 1985, *36*, 391-405.

Chase, C. I., Often is where you find it. *American Psychologist*, 1969, *24*, 1043.

Dickstein, L. S. The meaning of conversion in syllogistic reasoning. *Bulletin of the Psychonomic Society*, 1981, *18*, 135-138.

Ekberg, P. H. S., & Lopes, L. L. Fuzzy quantifiers in syllogistic reasoning. *Goteborg Psychological Reports*, 1981, *10*, 1-16.

Evans, J. St. B. T. *The Psychology of deductive reasoning.* London: Routledge and Kegan Paul, 1982.

Farkas, A., & Englander, T. *Experimental investigation of fuzzy quantifiers.* Paper delivered to the Hungarian Psychological Congress, 1987.

Fisher, D. L. A three-factor model of syllogistic reasoning: A study of isolable stages. *Memory and Cognition*, 1981, *9*, 496-514.

Goocher, B. E. Effects of attitude and experience on the selection of frequency adverbs. *Journal of Verbal Learning and Verbal Behavior*, 1965, *4*, 193-195.

Grice, H. P. Logic and conversation. In P. Cole and J. L. Morgan (Eds.), *Syntax and Semantics*, Vol.3: Speech Acts. New York: Seminar Press, 1975.

Griggs, R. A. Logical processing of set inclusion relations in meaningful text. *Memory and Cognition*, 1976, *4*, 730-740.

Griggs, R. A., & Warner, S. A. Processing artificial set inclusion relations: Educing the appropriate schema. *Journal of Experimental Psychology: Learning Memory and Cognition*, 1982, *7*, 51-65.

Holyoak, K. J., & Glass, A. L. Recognition confusions among quantifiers. *Journal of Verbal Learning and Verbal Behavior*, 1978, *17*, 249-264.

Hörmann, H. The calculating listener, or how many are einige, mehrere and ein paar (some, several and a few). In R. Bauerle, C. Schwarze and A. von Stechow (Eds.), *Meaning, use and interpretation of language.* Berlin: De Gruyter, 1983.

Just, M. Comprehending quantified sentences: The relation between sentence-picture and semantic memory verification. *Cognitive Psychology*, 1974, *6*, 216-236.

Neimark, E. D., & Chapman, R. H. Development of the comprehension of logical quantifiers. In R. J. Falmagne (Ed.), *Reasoning: Representation and process*. Hillsdale, N.J: Erlbaum, 1975.

Newstead, S. E. & Collis, J. Context and the interpretation of quantifiers of frequency. *Ergonomics*. (in press)

Newstead, S. E., & Griggs, R. A. Drawing inferences from quantified statements: A study of the Square of Opposition. *Journal of Verbal Learning and Verbal Behavior*, 1983, *22*, 535-546.

Newstead, S. E., & Griggs, R. A. Fuzzy quantifiers as an explanation of set inclusion performance. *Psychological Reseasarch*, 1984, *46*, 377-388.

Newstead, S. E., Pollard, P., & Riezebos, D. The effect of set size on the interpretation of quantifiers used in rating scales. *Applied Ergonomics*, 1987, *18*, 178-182.

Pepper, S. Problems in the quantification of frequency expressions. In D. Fiske (Ed.), *New directions for the methodology of social and behavioural science: Problems with language imprecision*. San Francisco: Jossey-Bass, 1981.

Pepper, S., & Prytulak, L. S. Sometimes frequently means seldom: Context effects in the interpretation of quantitative expressions. *Journal of Research in Personality*, 1974, *8*, 95-101.

Pohl, N. F. Scale considerations using vague quantifiers. *Journal of Experimental Education*, 1981, *49*, 235-240.

Schriesheim, C., & Schriesheim, J. The invariance of anchor points obtained by magnitude estimation and pair-comparison treatment of complete ranks scaling procedures: An empirical comparison and implication for validity of measurement. *Educational and Psychological Measurement*, 1978, *38*, 977-983.

Stevens, S. S. Issues in psychophysical measurement. *Psychological Review*, 1971, *78*, 426-450.

Thorndike, E. L., & Lorge, I. *The teachers word book of 30,000 words.* New York: Teacher's College, 1944.

Wallsten, T. S., Budescu, D. V., Rapaport, A., Zwick, R., & Forsyth, B. Measuring the range meanings of probability terms. *Journal of Experimental Psychology: General,* 1986, *115,* 348-365.

Wallsten, T. S., Fillenbaum, S., & Cox, J. A. Base rate effects on the interpretation of probability and frequency expressions. *Journal of Memory and Language,* 1986, *25,* 571-587.

A COMMON FRAMEWORK FOR COLLOQUIAL QUANTIFIERS AND PROBABILITY TERMS

A. C. ZIMMER

University of Regensburg, FRG

The importance of modal qualifiers for argumentative reasoning is investigated and it is shown that colloquial quantifiers and uncertainty expressions can be interpreted as fuzzy numbers in the interval [0,1]. Empirical procedures are suggested for the determination of these fuzzy numbers.

The empirical results reveal that for propositions on a defined level of abstraction colloquial quantifiers and probability terms can not only be expressed as fuzzy numbers but furthermore can be used in according to the rules for fuzzy combination numbers.

For specific areas of content well defined scope functions can be empirically determined. These influence the meaning of colloquial quantifiers systematically, that is, they catch the contextual meaning. A somewhat related effect is observed in probability terms for conditional propositions.

1. INTRODUCTION: Schemes for Reasoning and Argumentation

In the history of Western thought, starting with Aristotles Organon, mechanical procedures for reasoning have been devised serving as normative theories for human reasoning and at the same time as tools for the processing of evidence. The rise of psychological investigations of thought processes has debunked the notion of logic as an - albeit normative - theory of human reasoning. The question, however, if and how formal approaches of reasoning and human reasoning can be brought together, remains open. The approach proposed here is intended to close the gap somewhat by proposing what could be termed an approach to a formal theory of informal reasoning (Zimmer 1984a).

In order to make more specific what such an approach is intended to achieve, it seems appropriate to compare classical formal approaches, that is, predicate calculus and probability theory, with what is known about everyday reasoning. In Table 1 the positions of predicate calculus, probability theory, and everyday reasoning regarding central problems of reasoning are compared.

The inspection of Table 1 highlights the fact that, in general, predicate calculus and probability theory take very similar approaches towards problems and modes of reasoning despite their different structure. Exceptions, however, are the evaluation of partial or circumstantial evidence and in the weighing of evidence by probabilities or by divers colloquial qualifiers; here probability theory and everyday reasoning take similar positions. What distinguishes these points from the rest of Table 1? They apply to situations where only approximate solutions are possible, where the reasoning is invalid or does not lead either to the alternative "true/false", but where only degrees of plausibility or veridicality can be reached.

As Toulmin (1964) observes, standard logic has been developed with an eye on mathematics where such ambiguous situations are to be avoided at nearly any cost (but see, for instance, Kline, 1980). In order to liberate logic from this Procrustean bed, Toulmin suggests the reconstruction of logic

TABLE 1
Modes of reasoning

	Predicate calculus	Probability Theory	Everyday Reasoning
law of the excluded middle	<u>valid</u> without specification	<u>valid</u> in the definition of the event space	<u>usually not valid</u> exept for easily enumerable emsembles
Modes of reasoning: - deduction	<u>valid</u> without exception	<u>valid</u> without exception	<u>valid</u> but the result looses plausibility with the length of the deductive chain
- induction	<u>not valid</u> (the method of proof by complete induction is a deductive method)	<u>not valid</u> in classical approaches (v. Mises, Popper) <u>valid with restriction</u> in non-standard approaches	<u>valid</u> if many convincing analogies can be brought forward
evaluation of partial or circumstantial evidence	<u>not possible</u> except for non-standard approaches (non-monotonic reasoning, default reasoning	<u>possible</u> by means of Bayesian or information-theoretic schemes	<u>possible but biased</u> because of heuristic and/or characteristic of memory (primacy and recency
use of qualifiers	only standard quantifier (all, some, not all, none)	not defined except for weighing by probability	applicable without restrictions

according to the model of legal argumentation. He suggests a scheme of syllogistic reasoning with the following components:

(i) *claims* propositions that are supposed to be true or to be at least plausible to a certain degree
(ii) *grounds* or reasons for believing the claims to be valid (the usual form is that of explicitly or implicitly quantified statements)
(iii) *warrants*, statements about the relations between grounds and claims (e.g. causality, necessity, sufficiency, contingency)
(iv) *backing* commonly shared knowledge (Smith, 1982) which provides the rules for a combination of grounds and warrants in order to justify the claims (e.g., rules of syllogistic reasoning or statistical inference)
(v) *modal qualifications* general quantifiers and uncertainty expressions, such as possibly, usually, necessarily. They apply to to the propositions and to the inferential process.
(vi) *rebuttals* alternative claims which can also be inferred from the grounds, warrants, and the backing because of the modal quantification of propositions and inferential rules. Rebuttals can be overcome by either showing that they imply a smaller set of consistent propositions than the claim or by comparing the overall modal qualification of the rebuttals with the evaluation of the claims.

These components are combined as shown in Figure 1 (Toulmin, 1964, p. 104; the figure has been slightly changed in order to avoid inconsistencies).

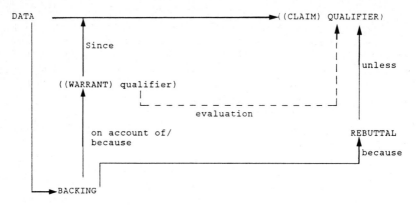

FIGURE 1
A modified version of Toulmin's (1964) model for syllogisms in argumentation. Toulmin's original model is indicated by upper-case letters and bold lines.

A Framework for Colloquial Quantifiers and Probabilistic Terms

An example for this kind of syllogistic reasoning using an analysis of the chains of arguments in the determination of the probable price of a used book (see Figure 2).

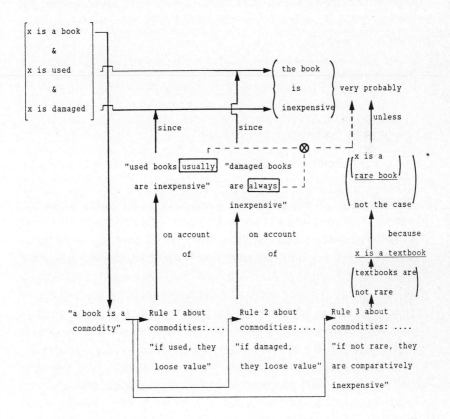

FIGURE 2
The application of the modified Toulmin model.

Figure 2 especially reveals the importance of implicit (...) or explicit (usually, always etc.) quantifiers, for the evaluation (qualification) of the claim. In order to develop a formalized version of Toulmins approach to plausible reasoning, it is necessary to develop a common framework for the interpretation of explicit and implicit quantifiers and furthermore an algorithm for their concatenation. Zadeh (1983) has suggested the interpretation of quantifiers as fuzzy numbers in the [0,1] interval and the use of the operations of fuzzy numbers (Dubois & Prade, 1980) as the algorithm for their concatenation.

2. FUZZY NUMBERS AND QUALIFIERS

The meaning of quantities like "about 50 %" or "slightly below 0.3" (Smithson, 1987) and adding "about 50 %" and "a bit more than 10 %" with the result "probably somewhat more than 60 %" seem to make sense immediately. However, it is not clear how this intuitive meaning is reflected in the formal definitions of fuzzy numbers and their rules of concatenation (Dubois & Prade, 1980). The formal definitions allow for the proving of abstract theorems in fuzzy number theory and for checks of consistency but these formal definitions do not provide any guidelines for the mapping of imprecise observable quantities into the different types of fuzzy numbers (Π, s, z, s/z, or z\s numbers) and for the setting of parameters. On the other hand, Smithsons (1987) and others purely empirical approach characterizing a fuzzy number by a listing of relative frequencies is not sufficient either, because he does not propose empirically testable rules for the concatenation of these numbers. Such rules however can be derived from results on approximate calculation in the areas of foreign exchange (Zimmer, 1984b) and of the stock market (Zimmer, in preparation). For merely illustrative purposes, let us start with fuzzy numbers of the form "standard number + qualification" (e.g. "approximately 0.7").

Figure 3(a and b) represents this fuzzy number where the core (0.7), and the fuzzy upper and lower boundaries, (the fuzziness due to the qualification) can be discriminated. The fuzzy number can now be represented by the following triple: lower boundary relative to the core, core, upper boundary relative to the core (0.1/0.7, 0.7, 0.15/0.7).

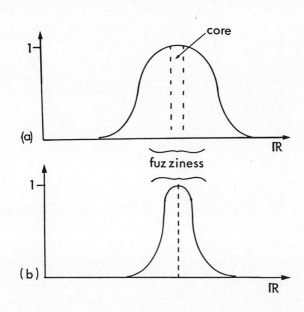

FIGURE 3
Fuzzy numbers consisting of a core (or prototypical) meaning of 0.7 and fuzzy upper and lower boundaries. (a) is a fuzzy number with core interval, (b) is a fuzzy number with a point-wise core.

Any two fuzzy numbers can be concatenated following these steps: (i) calculating the resulting core by means of standard arithmetics, by (ii) averaging the respective upper and lower boundaries, and by (iii) determining the resulting boundaries from the averaged boundaries in relation to the resulting core. The operations with fuzzy numbers corresponding to the standard operations in arithmetics are illustrated in Figure 4(a-d).

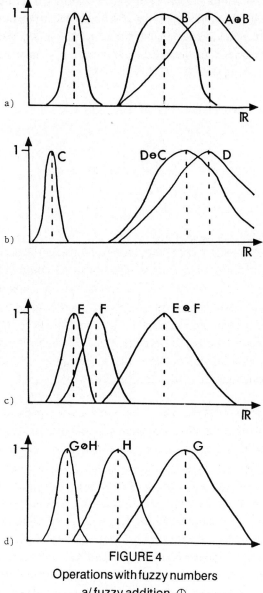

FIGURE 4
Operations with fuzzy numbers
a/ fuzzy addition ⊕
b/ fuzzy multiplication ⊗
c/ fuzzy substraction ⊖
d/ fuzzy division ⊘

The described approach of handling the core and the fuzziness seperately has been established empirically. Specifically, the think-aloud protocols of the subjects reflect this procedure. For fuzzy numbers like several, most or likely the same procedure can be applied provided the core has been determined empirically.

3. FUZZY ARITHMETIC AS A MODEL FOR REASONING

Starting with the experimental studies by Zimmer (1982, 1984a, 1984b) empirical evidence has been amassed for Yagers (1980) and Zadehs' (1983ab,(1984) claim that fuzzy numbers can be used for the representation of generalized quantifiers (Barwise & Cooper, 1981; Peterson, 1979) and furthermore that human reasoning with these quantifiers can be modelled according to the operations with fuzzy numbers. As noted above, further experimental studies have led to modifications in the definition of fuzzy numbers as well as of operations with them. These modifications, however, are not crucial for the general claim.

From a formal point of view, quantifiers expressed as fuzzy numbers in the interval [0,1] and uncertainty expressions represented as fuzzy probabilities are comperable. Furthermore, in chains of argumentation (see Figure 2) both kinds of qualification can be found and should therefore be represented in a common framework for the use in intelligent or expert systems (Zadeh, 1983c). Empirical analyses of uncertainty expressions (Zimmer, 1983, 1986a; Wallsten, Budescu, Rapoport, Zwick, & Forsyth 1986; Zwick & Wallsten, 1987) have consistently shown that verbal expressions like "probable", "likely", or "toss-up" can be expressed as fuzzy numbers. Different experimental techniques (e.g. pair comparison vs. staircase estimation), different forms of display s (e.g. circle segments vs. random dots), and different samples of uncertainty expressions (all expressions of a language community vs. only those expressions that a subject has in his/her personal active vocabulary) have led to seemingly conflicting results about the consistency of estimates and therefore the applicability of verbal uncertainty expressions in decision support or expert systems. There are two solutions for this problem: One consists in the Wallsten et al., (1986) approach of determining the fuzzy numbers for a complete lexicon of uncertainty expressions. This leads to *averaged meanings* that can be assumed to be valid for an entire language community. By means of iterative methods (Zimmer, 1986b), ambiguous meanings (fuzzy numbers with more than one peak) can be resolved. The problem with this approach is that the individuals lexicon of uncertainty expressions might differ from that of the language community. The important advantage of this approach, however, is its generality. The other solution for the problem consists in concentrating on the *individual's* lexicon of uncertainty expressions.

Calibrating individual vocabularies of uncertainty expressions by means of staircase methods with random-dot displays Zimmer and Körndle (1987) has resulted in fuzzy numbers that can be represented by (i) single-peaked membership functions, (ii) of comparable shape and (iii) with the tendency towards a proportional relation between the value of the core and the fuzziness. To be more precise: in contrast to fuzzy numbers without interval bounds, the fuzziness in the closed interval [0,1] is relative to the smaller distance of each core from the upper or lower limit. These qualitative aspects of the fuzzy numbers are consistent with the model described in Part 2 above. It should be kept in mind that (iii) contradicts one of the theoretical assumptions of Zimmer (1982, 1983), namely the assumption of equal informativeness on the entire scale of judgment, and therefore equal fuzziness for all uncertainty expressions in an individual active vocabulary. However, the consequence of the unequal informativeness (low in the central part and high in the extremes) is in accordance to the results reported by Wallsten and his group (Wallsten et al., 1986).

The major disadvantage of this individualistic approach, specifically its lack of generalizability, can be overcome by the procedure described in Zimmer (1986b). Starting with the individuals expressions but then mapping them into the general lexicon. If a mapping does not result in a single-peaked fuzzy number or if the fuzziness is excessive, it is iteratively searched for the non-degenerate expression which captures best the initially intended meaning.

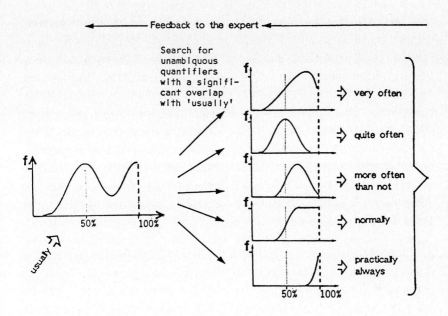

FIGURE 5
Interactive determnination of unambiguous uncertainty expressions
(Zimmer 1986b).

During and by the interaction with this computer-controlled procedure, subjects learn to use only those expressions which have a meaning meaning in accordance with that of the language community. However, this method restricts the expressive power of the indivial lexicon.

The common framework for quantifiers and uncertainty expressions as established by the assignation of fuzzy numbers in [0,1] has to be complemented by a comparison of the algorithms of inference. The standard algorithms, that is, syllogistic resolution and Bayesian weighing of evidence are seemingly incomparable. However, since Zadeh (1983a) has shown that any form of syllogistic resolution can be modelled by fuzzy quantifiers and fuzzy operators (addition, multiplication, and conjunction), it is possible to use the same operators for Bayesian inference. There is only one additional operator necessary, division, needed for working with conditional probabilities. On the first glance this operator does not fit into the reasoning with quantifiers. However, as Hörmann (1983) and Zimmer (1986a) have shown, in the colloquial usage of quantifiers these are quite often conditioned on the background knowledge about the situation. For instance, the utterance "many of the convertibles" can only be properly modelled if the general meaning of many is taken into account as well as the fact that convertibles form a very small subset of all cars. The implicit reasoning runs as follows: if the cars in question are convertibles, then even a small proportion of all cars fulfills the condition of applying "many". This construction of a conditional quantifier is completely compatible with the notion of conditional probabilities. Using this result, it is now possible not only to assign fuzzy numbers to the qualifiers in Toulmin's model (Figure 1 and 2) but also to interp and their combination as fuzzy evaluations of operators (e.g. $\otimes$ means fuzzy multiplication). Furthermore, the relations between the backing and the warrants becomes straightforward fuzzy arithmetic.

4. CONTEXT SPECIFICITY OF QUALIFIERS

Yagers (1980) as well as Zadehs (1983a) models for the interpretation of quantifiers in natural language as fuzzy numbers assume implicitly that there is a one-to-one relation between quantifiers and fuzzy numbers (for redundant sets of quantifiers the relation might be many to one). One major experimental result of Zimmer (1982, 1984b) was that for sufficiently rich contexts (natural sciences vs. social sciences and everyday events) this simplifying assumption does not hold. In these contexts the standard meaning of quantifiers (see Figure 6) is modified by the subjects knowledge about the normal scope of discourse in these contexts, specifically, how often events are mentioned with an occurrence rate of x %. It has been shown that the scope functions for contexts can be determined independently from the quantifiers (Figure 7a) and that the context-denedant quantifiers (Figure 7b) result from the fuzzy conjunction of the standard quantifiers and of the respective scope functions (the MIN-operator).

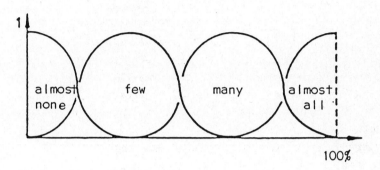

FIGURE 6
Standard meaning of colloquial quantifiers (Zimmer, 1984b).

Slightly different is the situation for conditional quantifiers (see above) where the background knowledge is taken care of by dividing the fuzzy number for the standard meaning of the quantifier question by the fuzzy munber for occurence rate for the eve nt in question. As mentioned above, the notion of conditional quantifiers bridges the gap between the apparently disjoint types of modal qualifiers, namely, quantifiers and uncertainty expressions or qualitative fuzzy probabilities. The context dependability of explicit or implicit conditioning. The procedures of Wallsten et al., (1986) as well as of Zimmer (1983) and Zimmer and Körndle (1987) assume implicitly that uncertainty and frequency expressions can be estimated independently from contextual inf luences. This is apparently true for impoed contexts like circle segments and random dots but remains - at least - questionable for more realistic situations.

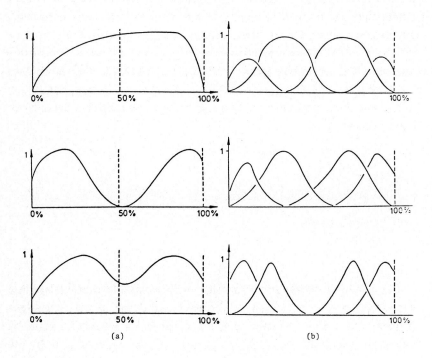

FIGURE 7
Context dependability of quantifiers: (a) scope functions for contexts /from above: everyday events, natural sciences, social sciences/ (b) resulting context-dependent quantifiers /contexts as above/

From Hörmann s (1983) results one might conjecture that in situations where the individuals have background knowledge, the meaning of their observable lexicon is the result of conditioning the observable frequency on the possible frequency. Despite the fact that the operators are different (fuzzy conjunction vs. fuzzy division) the result for quantifiers and uncertainty expressions is comparable: The observable usage of modal qualifiers can be deduced from a fuzzy operation on the standard meaning and the occurrance rate of the background knowledge. In Figure 1 this context dependability is indicated by the arrow from DATA to BACKING which results in data-dependent constraints on the background knowledge (e.g. the rules for commodities).

5. CONCLUSIONS

The common framework for the two major kinds of modal qualification allows for a modelling of intricate nets of arguments (see Figure 2) by means of fuzzy arithmetics. Furthermore, the variability of meaning caused by individual differences and different context can be taken care of by the generalization of this framework. This consists in a decomposition of the observable meaning into the standard meaning and into the contextual or individual constraints.

REFERENCES

Barwise, J., & Cooper, R. Generalized quantifiers and natural language. *Lingustics and Philosophy,* 1981, *4,* 159-219.

Hörmann, H. *Was tun die wörter miteinander im satz? Oder wieviele sind einige, mehrere und ein paar?* Göttingen: Verlag für Psychologie, 1983.

Kline, M. Mathematics - *The loss of certainty*. Oxford: Oxford University Press, 1980.

Peterson, P. On the logic of few, many and most. *Notre Dame Journal of Formal Logic,* 1979, *20,* 155-179.

Smith, N. V. *Mutual Knowledge.* London: Academic Press, 1982.

Smithson, M. *Fuzzy set analysis for behavioral and social sciences.* Heidelberg: Springer-Verlag, 1987.

Toulmin, S. E. *The uses of argument.* Cambridge: The University Press, 1964.

Wallsten, T.S., Budescu, D.V., Rapoport, A., Zwick, R., & Forsyth, B. Measuring the vague meanings of probability terms. *Journal of Experimental Psychology: General,* 1986, *115,* 348-365.

Yager, R. R. (1980) Quantified propositions in a linguistic logic. In E. P. Klement (Ed.) *Proceeedings of the 2nd International Seminar on Fuzzy Set Theory.* Linz: Johannes Kepler Universitat, 1980.

Zadeh, L. A. A computational approach to fuzzy quantifiers in natural languages. *Computers and Mathematics,* 1983, *9,* 149-184. (a)

Zadeh, L. A. Linguistic variables, approximate reasoning and dispositions. *Medical Informatics.* 1983, *8,* 173-186. (b)

Zadeh, L.A. Fuzzy logic as a basis for the management of uncertainty in expert systems. *Fuzzy Sets and Systems.* 1983, *11,* 199-227.

Zadeh, L.A. A theory of commonsense knowledge. In H. J. Skala, S. Tremini, and Trillas (Eds.) *Aspects of Vagueness.* Dodrecht: Reidel, 1984.

Zimmer, A. Some experiments on the meaning quantifiers in normal language. In L. Troncale (Ed.) *General Systems Science.* Louisville, Ky:. Society for General Systems Research, 1982.

Zimmer, A. Verbal vs. numerical processing of subjective probabilities. In R. W. Scholz (Ed.) *Decision Making Under Uncertainty.* Amsterdam: North-Holland, 1983.

Zimmer, A. A formal theory of informal reasoning. paper read at psychometric Socety Meeting Santa Barbara, CA., 1984 (a) Zimmer, A. A model for the interpretation of verbal predictions. *International Journal of Man-Machine Studies,* 1984, 20, 121-134. (b)

Zimmer, A. What uncertainty judgments can tell about the underlying subjective probabilities. In L. N. Kanal and J.F. Lemmer (Eds.) *Uncertainty in Artificial Intelligence.* Amsterdam: Elsevier Science, 1986. (a)

Zimmer, A. A fuzzy model for the accumulation of judgments by human experts. In W. Karwowski and A. Mital (Eds.) *Applications of Fuzzy Set Theory in Human Factors.* Amsterdam: Elsevier, 1986. (b)

Zimmer, A. *Fuzzy arithmetics as a model for approximate calculations in foreign exchange and stock market.* (in preparation)

Zimmer, A. & Körndle, H. Schematheoriesche begründungen für die ordung unsicheren wissen. In J. Krems and G Heyer (Hsrg.) *Wissensarten und ihere darstellung. Ergebnisse eines Workshop des Arbeitskreises Kognition im Fachausschuss 1.2 der Gl.* Heidelberg: Springer, 1987.

Zwick, R., & Wallsten, T.S. *Combining stochastic uncertainty and linguistic inexactness: Theory, and experimental evaluation.* Workshop of North America Fuzzy Information Processing Society (NAFIPS), Purdue University, 1987.

AN EMPIRICAL STUDY OF THE INTEGRATION OF LINGUISTIC PROBABILITIES

Rami ZWICK
Carnegie Mellon University Pittsburgh, PA. USA

David V. BUDESCU
University of Haifa, Israel

Thomas S. WALLSTEN
University of North Carolina at Chapel Hill, USA

In many situations we use the probabilistic judgments and forecasts of external agents in forming our own opinions or reaching decisions. For example, we depend on forecasts of financial experts regarding interest rates and market fluctuations when selecting among investments and on physicians' forecasts of prognosis when choosing treatments for a given disease. Thus our own decisions and choices are affected by the ability of others to properly express and communicate their beliefs, and of ourselves to understand and use them. A well-known phenomenon in the forecasting literature (e.g. Beyth-Marom, 1982) is that many experts, like laypeople, prefer to communicate their opinions by means of verbal probabilistic phrases rather than numbers. Typically, this preference for words over numbers is linked to the vagueness these studies are that most probabilistic phrases cover large ranges of probabilities and overlap to a considerable *degree, and that there exist consistent individual differences in the us Beyth-Marom, 1982; Nakao & Axelrod, 1983). Recently, Wallsten, Budescu, Rapoport, Zwick and Forsyth (1986) and Rapoport, Wallsten and Cox (1987) have shown that the vague meanings of probability phrases can be captured and described reliably by means of membership functions over the 0,1 interval. In the present investigation we propose to employ the methodology used by Rapoport, et al. (1987) and the theoretical framework of fuzzy set theory with regard to the various aggregation connectives (e.g., Dubois and Prade, 1985) to understand how people combine and aggregate a number of vague probability judgments. For example, what does the patient think when one physician says that it is "likely" that the pain will disappear, and another considers this eventuality "highly probable"? More specifically, we are interested in how subjects combine the opinions of two observers with regard to the possibility of an event. These opinions are expressed linguistically and are independent (in the sense that there is no communication between the observers prior to their expressing their opinions). Further, there is no reason for the receiver to believe that one observer is more reliable than another. This situation was study extensively from a prescriptive point of view (e.g. Winkler & Makridakis, 1983; Cholewa, 1985) but little is known about the descriptive power of the aggregation models and even less about the case when the opinions themselves are expressed in an inexact linguistic manner. In the next section we describe the various theoretical models which have been proposed in the fuzzy set literature for the aggregation operations, and review the empirical work in this area. Finally, we describe an experiment in which subjects were required to consider simultaneously two vague probabilistic judgments and we compare their responses with predictions generated by fuzzy set theory.*

Fuzzy Set Aggregation Connectives

Given a generalized concept of membership, operations on sets are no longer restricted to the boolean binary algebra, with the result that a much larger and richer class of operations can be defined. This richness has caused some confusion about which are the "correct" operators for fuzzy sets. A large variety of definitions have been suggested in the literature (see Czogala & Zimmermann, 1984; Smithson, 1984, 1987; and Dubois and Prade, 1985) on the basis of various normative, empirical, or purely mathematical considerations for the logical operators "and" and "or". A large number of functions can be considered proper, in that they all yield the correct results for classical crisp sets as a special case. However, it is logically and mathematically impossible to determine the appropriate "general" rule on the basis of a special case, central as it may be. Indeed, since fuzzy set theory was offered as a more flexible model that is better suited for "humanistic systems" (e.g. Zadeh, 1974), which are affected by human judgment, perception and emotion, it is a mistake to search for a single best set of definitions of the "and" and "or" operations. In fact, Zadeh's (e.g., 1976) position has always been that the choice of these definitions should be left to the users so that they best reflect the unique characteristics of the particular situation on hand. However, this position should not be interpreted as a call to cease research on the different operators. It is important for future applications, and theoretical developments to identify individual and situational factors that are associated with the choice of a given class of operators. Obviously, experimental psychologists and psychometricians can, and should, play a major role in this research by developing measurement methods and appropriate experimental procedures.

As mentioned by Dubois and Prade (1985), a consensus has formed in the literature that the concept of triangular norms and conorms proposed by Menger (1942) is appropriate for representing the logical operators "and" and "or" (respectively) applied to fuzzy sets (Weber, 1983).

Definition 1. A t-norm is a binary operator

$T:[0,1] \times [0,1] \rightarrow [0,1]$

such that

(1) $T(0,0) = 0$ and $T(a,1) = a$ (boundary conditions);
(2) $T(a,b) \leq T(c,d)$ if $a \leq c$ and $b \leq d$ (monotonicity),
(3) $T(a,b) = T(b,a)$ (symmetry),
(4) $T(a,T(b,c)) = T(T(a,b),c)$ (associativity).

A strict t-norm is a continuous t-norm which is increasing in both arguments.

Definition 2. A co-t-norm S is a binary operator

$S:[0,1] \times [0,1] \rightarrow [0,1]$

such that

(1) $S(1,1) = 1$ and $S(a,0) = a$ (boundary condition),
(2) $S(a,b) \leq S(c,d)$ if $a \leq c$ and $b \leq d$ (monotonicity),
(3) $S(a,b) = S(b,a)$ (symmetry)
(4) $S(a,S(b,c)) = S(S(a,b),c)$ (associativity).

A strict co-t-norm is a continuous co-t-norm which is increasing in both arguments. Associated with every t-norm T there exists a co-t-norm S, called its dual, such that $S(a,b) = 1-T(1-a, 1-b)$. Examples of t-norms and their duals that have been proposed in the literature to represent the pointwise fuzzy set-theoretic intersection or union are:

$T^{(1)} = \min(a,b)$ $S^{(1)} = \max(a,b)$ (Zadeh, 1965) *(1)*

$T^{(2)} = ab$ $S^{(2)} = a + b - ab$ (Bellman and Zadeh, 1970) *(2)*

$T^{(3)} = \max(0, a+b-1)$, $S^{(3)} = \min(1, a+b)$ (Bellman and Zadeh, 1977) *(3)*

Smithson (1984, 1985) pointed out that these three classes can be thought of representing various degrees of extremity since for all $(a,b) \in [0,1] \times [0,1]$

$$T^{(1)} \geq T^{(2)} \geq T^{(3)}$$

and

$$S^{(1)} \leq S^{(2)} \leq S^{(3)}.$$

It was therefore suggested that all three classes can be incorporated into one general family of connectives with one or more free parameters. For an excellent review of the class of fuzzy set aggregation connectives based on triangular norms see Dubois and Prade (1982, 1985) and for specific examples of general rules see Czogala and Zimmerman (1984), Smithson (1984, 1985), and Yager (1980).

From a psychological perspective, little, if any understanding of the parameters can be claimed. It is not clear whether they depend on individual and/or situational factors, how stable and reliable they are, or what other factors may affect them. In short, it is not clear what usage could be made once the parameters are estimated. In fact, estimation procedures for most of the parameters have not been developed. Smithson (1984, 1985) presented a least-squares approach to estimate a linear transform of the free parameter that can sometimes yield values outside the admissible range (i.e., smaller than 0 or larger than 1). For example, when this technique was applied to our data 20% of the estimates were inadmissible. However the experimental significance of this approach is that one can test simultaneously the appropriateness of a large class of possible dual connective operators by testing the requirements that must hold if the dual rules are the t-norm and the corresponding co-t-norm (see definitions 1 and 2).

Another class of connectiveness is based on the notion that in everyday life people rarely use "and" and "or" in their respective strict noncompensatory and fully compensatory senses. Rather, people's judgments are based on a partially compensatory interpretation, represented by a free parameter, q, which can be called "grade of compensation." The first to introduce this idea were Zimmerman and Zysno (1980) who suggested a weighted geometric mean:

$$\mu_{A\&B}(x) = [\mu_{A \cap B}(x)]^{1-\gamma} [\mu_{A \cup B}(x)]^{\gamma}. \tag{4}$$

Here (A&B) denotes the generalized compensating connective, which varies between the regular "or" when $\gamma = 1$ and the regular "and" when $\gamma = 0$. Note, that this approach does not eliminate the need to decide on a definition of the strict "or" and "and", for which any of the operations described earlier can be used. In a similar fashion Luchandjula (1982), and Smithson (1984) proposed different forms of generalized connectives that include the pure union and intersection formulas as special cases. Finally, in a recent development Dyckhoff and Pedrycz (1984) suggest use of generalized means to model the compensatory connective:

$$\mu_{A\&B}(x) = [\omega_1 \mu^r_A(x) + \omega_2 \mu^r_B(x)]^{1/r} \tag{5}$$

where ω_1 and ω_2 are weights which add to 1, and r can be any real non-zero number. A variety of special cases can be considered. Taking yields $r \to \pm \infty$ the minimum and maximum, respectively, $r = 1$ gives a weighted mean of the membership functions, $r \to 0$ produces their weighted geometric mean, and $r = -1$ their harmonic mean.

The interesting novelty of this approach is that it offers a definition of a compensatory connective without explicitly adopting a set of definitions for the strict "and" and "or." Rather the (A&B) operator is defined as a direct function of the individual memberships $\mu A(x)$ and $\mu B(x)$. Although this approach simplifies the situation conceptually and facilitates calculations and estimation, it is questionable whether the Dyckhoff and Pedrycz model is comparable to the others. It is possible to apply monotonic transformations to the exponent, r, to obtain a bounded compensation parameter (e.g. $\gamma = [1 + |r|/(1 +)]/2$) but this obviously does not represent a compensation between "and" and "or." It appears that this model should be considered a general model of combination and aggregation of levels of membership.

It is also possible to consider other forms of averaging of the two individual membership functions as models of aggregation. In particular, we would like to emphasize one possibility that turns out to be consistent with some of our data - the mean of two fuzzy numbers - "the fuzzy mean." Dubois and Prade (1980) define a real fuzzy number as a fuzzy subset with a continuous membership function satisfying some mild regularity conditions. The mean of two fuzzy numbers (A and B) is defined as:

$$\mu_{Mean(A,B)}(X) = \max_{x=(y+z)/2} [\min(\mu_A(y), \mu_B(z))] \tag{6}$$

Previous Empirical Results

In reviewing this literature, it is important to focus on two issues: measurement of the membership functions, and experimental designs used to test the various models for operators on fuzzy sets. In the past, two methods have been used to quantify degrees of membership. Hersh and Caramazza (1976) and Hersh, Caramazza and Brownell (1979) used what Hillsdal (1985) labelled the "yes-no" paradigm, in which x, an element of E, is presented to the subject and he/she has to decide whether the element is a member of A, the supposedly fuzzy subset. The fraction of positive responses across replications (within or across subjects) is considered a measure of $\mu A(x)$. Rubin (1979) has pointed out the main problems of this method, that it confounds fuzziness with response variability and that it can be interpreted as indicative that words have various, but nevertheless precise, meanings to different people and/or at different times. The other standard paradigm has been direct estimation. Oden (1977a) used direct scaling along a scale with verbally anchored end points and Thöhle, Zimmerman and Zysno (1979) and Zimmerman and Zysno (1980) employed a similar procedure, but with numerically anchored ends. However, these procedures lack solid foundations, and their results are difficult to interpret unless they are embedded within a firm theoretical structure.

Such a framework was outlined by Norwich and Turksen (1982, 1984), by Wallsten, et al. (1986) and by Zwick (1987), and by Wallsten, et al. (1986), who pointed out the relationships between the axiomatic formulation of the algebraic difference (ratio) structure (e.g. Krantz, Luce, Suppes, & Tversky, 1971) and the measurement of memberships functions. This approach was recently refined and successfully tested by Wallsten, et al. (1986) who developed a graded pair-comparisons procedure (similar to the one used by Oden 1977b), which allows simultaneous testing of the necessary axioms, scaling of the responses in order to obtain memberships, and tests of goodness of fit of the scale values. In a subsequent study Rapoport, Wallsten and Cox (1987) demonstrated a high level of similarity between membership values determined through graded pair comparison and direct magnitude estimation. Thus, a sound theoretical justification was established for the quantification of the vague meanings of inexact linguistic terms by means of direct scaling.

We turn now to tests of the combination rules. Hersch and Caramazza (1976) tested the max rule (Equation 1) for the union operator. Subjects had to judge whether squares of various sizes qualified as "either small or large." Both group and individual analyses indicated that the shape of the "or" membership function resembled the max prediction, but was consistently lower. Oden (1977a, 1979) contrasted the min-max and sum-product operators for the two connective. His subjects were presented with pairs of statements such as "A chair is a furniture" and "A robin is a bird" and asked to perform several judgments. For the conjunction task they were asked to judge the "degree to which both statements were true" and for the disjunction task they judged "the degree to which one statement or the other was true." Note that these instructions could be easily interpreted as referring to an "exclusive or," i.e. one of the statements is true, but not both. It is also interesting to note that half of the comparisons involved pairs of statements pertaining to two sets (e.g. birds and furniture), so the union/intersection instructions called for judgment of joint/disjoint membership in the fuzzy set of "true" statements, and not in the fuzzy sets of "birds" or "furnitures." Group and individual analyses based on mean deviations from the predicted rules, and a functional measurement analysis (Anderson, 1981, 1982) indicated that both classes of operators fit the data quite well, but that the sum-product connectives clearly outperformed the min-max rules.

Thöle et al. (1979) studied the intersection operators. Sixty subjects judged the degree to which various objects were "containers", "metallic objects", and "metallic containers." The assumption was that this latter judgment would necessarily reflect an intersection of the two others, i.e. the degree to which an object is a metallic container is solely a function of the degree to which it is metallic and the degree to which it is a container. The analyses were performed at the group level and slightly favored the min rule over the product operator. It is important to note that the subjects did not actually see the objects, but were presented with cards containing their names. It is possible that such a procedure induces an additional source of (unintended) fuzziness, i.e. the vagueness associated with the definition of such classes of objects as bag, car, fridge, etc.

Zimmermann and Zysno's study (1980) has the distinction of being the most realistic. Sixty subjects were presented with 24 exemplars of fire resistant tiles and were instructed that their quality is judged according to two criteria; solidity (reflected in their grayness) and dovetailing (reflected in the correspondence between their shape and a standard shape). Using direct scaling methods group membership functions were obtained for "solidity", "dovetailing" and also for "ideal tile." The authors showed that in such a situation the subjects' judgments of "ideal" are best represented as a combination of the two membership functions that does not fit any of the rules proposed for conjunctions and disjunction. They ended up advocating the generalized γ-operator described earlier (Equation 4).

To summarize, the evidence regarding the and/or operations is inconclusive. Oden (1977a, 1979) advocates the sum-product rule, Thöle et al. (1979) favor the min rule for "and", and Hersh and Caramazza report (at least weak) support for the max rule for "or". Furthermore, given the various methods used to quantify the degrees of membership and the variety of experimental procedures used to elicit the two types of judgments, it is impossible to determine whether the different results are really contradictory, or just reflect the use of different methodologies and different situations. Finally, Zimmermann and Zysno's (1980) work clearly illustrates that in real life situations information is sometimes combined according to an averaging rule that allows for compensation between the extreme alternatives. This result is consistent with a large class of empirical studies conducted by Anderson and his associates (e.g. 1981, 1982) which singles out the averaging combination rule as the preferred mode of information integration.

Zimmermann and Zysno's results are important since they illustrate that although the duality between the union and intersection operations, and the corresponding "or" and "and," may reflect a nice mathematical normative structure (imposed by DeMorgan's theorem), they do not necessarily reflect a psychological reality. It is one major purpose of this work to further investigate this possibility. This was achieved by considering a series of testable requirements which must hold if the dual rules for intersection and union are based on t-norm and the corresponding co-t-norm respectively. Under any of the dual rules

(i) $T(a,b) \leq \min(a,b) \leq \max(a,b) \leq S(a,b)$,

(ii) The boundary conditions for both T and S must hold (see definitions 1 and 2).

(iii) The monotonicity conditions of both T and S must hold. Under any strict dual rules.

(iv) T and S are strictly increasing in both arguments.

And finally if T belongs to the Frank family[1] (Frank, 1979) and S is its dual then:

(v) $T(a,b) + S(a,b) = a + b$.

The last condition is interesting since many of the "classical" definitions of intersection such as the min the product and the bounded sum are included in this family.

The second stage in our analysis, as in most previous studies, was an attempt to identify the best fitting model for the data, restricting ourselves, quite arbitrarily, to tests of simple models which do not require estimation of free parameters. This was done in order to avoid problems related to parameter estimation and interpretation, and comparison of goodness-of-fit of models involving different numbers of parameters estimated by various methods. Thus, only the three basic dual models (min-max, sum-product and bounded sums) and simple parameter-free compensatory rules (pointwise mean, pointwise geometric mean, and fuzzy mean) were examined.

The methodology used is derived from our previous work on probability phrases and fuzzy sets (Wallsten et al., 1986; Rapoport et al., 1987). First, individualized membership functions for the various phrases were obtained by a direct magnitude procedure. The novelty of the present design is the simultaneous presentation of two phrases (supposedly generated by two sources) and a given probabilistic display, so that the subject could judge the degree to which *both* phrases, or *at least one* phrase describe the probabilistic display.

---

[1] The Frank family (Frank, 1979)

$$T_S(a,b) = Log_S \left[1 + \frac{(s^a-1)(s^b-1)}{s-1} \right] \quad s > 0$$

T_S is a strict t-norm for ∞ s > 0. T_0 = min; T_1 = product, and $F_\infty +$ = bounded -sum.

METHOD

Subjects: Sixteen graduate students in social sciences and business were recruited by placing notices advertising the experiment in the departmental mailboxes. They received $20 for participation in all three sessions of the experiment.

General Procedure: All subjects participated in one practice session and two actual data collection sessions, and performed four judgmental tasks: direct estimation of membership functions of individual probability phrases, judgment of pairs of phrases connected by "and," judgment of intersection of pairs of phrases connected by "or," and judgment of similarity of phrases. The experiment was controlled by an IBM/PC computer.

Membership Estimation: The instructions for this task read in part:

> You are to imagine that you are to predict whether a spinner you cannot see will land on white on the next random spin. A friend of yours can see the spinner, although not too well, because it is rotating at a moderate rate. Your friend will use a non-numerical probability phrase to give you his or her best opinion about the chances of the spinner landing on white. This gives you some basis for judging the probability of that event.
> Thus a phrase (from your friend), a spinner, and a response line will come on the screen for a single trial. You are to indicate how close the displayed probability of landing on white is to the judgment you would form upon hearing the particular phrase. If the displayed probability is not at all close to your judgment, move the arrow all the way to the left. If it is a close as possible to your judgment, move the arrow all the way to the left. If it is a close as possible to your judgment, move the arrow all the way to the right. If the displayed probability matches your judgment to some degree, then place the arrow accordingly.

The top panel of Figure 1 presents one of the displays used in this task. This example shows a spinner with Pr(white) = 0.15, presented in conjunction with the word doubtful and is used to quantify doubtful(0.15). Subjects responded by moving the arrow by means of a joystick. The precision of the response was determined by the resolution of the monitor, and allowed for almost 200 locations anchored by two verbal labels: "absolutely" and "not at all".

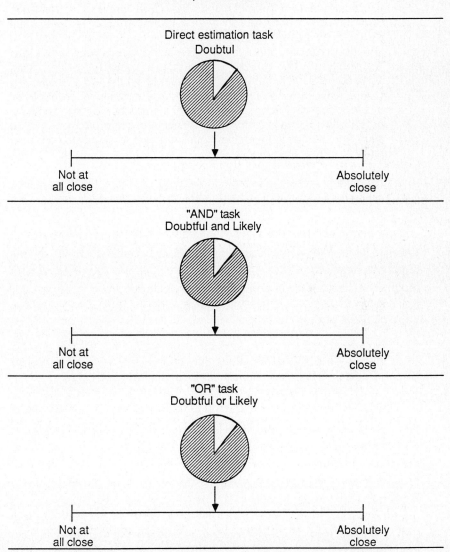

FIGURE 1
Three experimental tasks

"And" Judgment: The instructions for this task read in part:

> "Two friends have viewed the rotating spinner. They have not spoken to each other, so neither knows what the other thinks, but each uses a non-numerical phrase to tell you his or her opinion about the chances of the spinner landing on white. They may use the same or different phrases. Now, we are interested in the judgment you form about the probability of the spinner landing on white on the basis of the reports you receive from *both* of your friends. Thus, two phrases (one from each friend), a spinner, and a response line will come on the screen for a single trial. On the basis of the two phrases, you can form some judgment about the probability of landing on white for the spinner your friends saw. The question is, how close does the displayed probability of landing on white come to the judgment you form based on the two phrases? As before, move the arrow on the line all the way to the left for probabilities that are not at all close to your judgment, and all the way to the right for probabilities that are as close as possible to your judgment. If the displayed probability matches your judgment to some degree, then place the arrow accordingly."

An example is presented in the middle panel of Figure 1 with the words *doubtful* and *likely* connected by "and".

"Or" Judgment: The display and instructions for this task were identical to the ones for the intersection task, except for the key sentence describing the judgment required. In this case it read: "Now we are interested in your judgment of how well *at least* one of the two people described the probability of the spinner landing on white." The bottom panel of Figure 1 presents an example of this display.

"Similarity" Judgment: These data were analyzed and reported elsewhere (Zwick, Carlstein & Budescu, 1987) and will not be described here.

Stimuli: Six probability phrases and 11 probabilities were used in this study. All combinations are presented in Table 1. Half the phrases represented probabilities generally judged to be Higher than 0.5 (H), and the remaining described probabilities usually judged to be Lower than 0.5 (L). The probabilities were equally spaced across the 0.02 - 0.98 range. All (11 x 6 =) 66 combinations were judged in the estimation task. However, only 5 probabilities (see Table 1) were used in the other tasks. They were presented with each of the (6x6 =) 36 pairs of words for a total of 180 judgments in each of the two tasks.

TABLE 1
Expressions and Probability Values Used in the Design

Probability of the Spinner

	Phrase	.02	.12	.21	.31	.40	.50	.60	.69	.79	.88	.98
	Doubtful	--	x	--	x	--	x	--	x	--	x	--
L	Slight Chance	--	x	--	x	--	x	--	x	--	x	--
	Improbable	--	x	--	x	--	x	--	x	--	x	--
	Likely	--	x	--	x	--	x	--	x	--	x	--
H	Good Chance	--	x	--	x	--	x	--	x	--	x	--
	Fairly Certain	--	x	--	x	--	x	--	x	--	x	--

Note: "--" Combination used only in the estimation task.
"x" Combination used in all three tasks.

Procedure: The practice session consisted of a full set of 66 estimation judgments followed by short versions of the union and intersection tasks. Each of the 36 pairs of words was presented only once, in conjunction with a display randomly selected from the relevant 5 probabilities. Each of the two data collection sessions consisted of a full set of 66 estimation judgments followed by a full se of 180 union or intersection judgments. Half the subjects judged unions on the first data session and intersections on the second session. The other half performed the judgments in reversed order. The order of stimuli presentation within each task and session was randomly determined by the computer program. Also, in all union/intersection judgments the left/right location of the two target words was randomly counterbalanced for each subject. All subjects completed the three sessions and were paid at the conclusions of the last one. Because of a computer malfunction all 180 union judgments of one subject, and 30 intersection judgments of a different subject were lost.

Results

Direct Estimation: This task was replicated by each subject in each of the data collection sessions. Prior to finding the membership functions the two sets of responses were correlated to yield measures of reliability. The results are displayed in the first column of Table 2 and reflect a high level of stability -- 13 of the 16 subjects have reliabilities above .85. No consistent differences between the reliabilities of the responses to the six words were found.

TABLE 2
Reliabilities of Judgments in the Three Tasks

Subject	Estimation	Task Union	Intersection
1.	981	.914	.937
2	.925	.884	.606
3	.934	.847	.787
4	.894	.918	.802
5	.960	.960	.902
6	.903	.863	.730
7	.936	.931	.791
8	.972	.973	.965
9	.979	----	.981
10	.783	.942	.851
11	.930	.956	.894
12	.912	.923	.683
13	.908	.952	.971
14	.837	.608	.581
15	.801	.314	.519
16	.880	.878	.800
Mean	.928	.903	.857

Note: "---" Subject did not perform this task

The actual membership functions were obtained by rescaling to the 0-1 range the average response across the two sessions. Values of 0 and 1 represent cases in which the arrow was moved to the extremes of the response line on *both* sessions. Most subjects display reasonable membership functions that within a subject are similar for all three H words and for all three L words. The most dominant feature is the distinction between the membership values of the H and L words which display generally increasing and decreasing patterns, respectively. This result is best illustrated in Table 3 which shows very high correlations between the membership values of the 3 H words and the 3 L words, and high negative correlations in the LH pairs for all subjects. (These correlations were computed pointwise along the 11 common spinner probabilities in the estimation task.)

TABLE 3

Mean Correlations Between Membership Functions in Pairs of Words

Subject	HH	Pair LL	HL
1	.929	.967	-.954
2	.976	.967	-.965
3	.884	.939	-.932
4	.923	.956	-.857
5	.740	.934	-.828
6	.562	.953	-.793
7	.924	.979	-.929
8	.990	.972	-.920
9	.981	.989	-.932
10	.899	.880	-.523
11	.905	.929	-.915
12	.793	.930	-.865
13	.959	.873	-.833
14	.735	.881	-.680
15	.922	.860	-.904
16	.444	.801	-.547
Mean	.905	.942	-.874

Union and Intersection Judgments: Since the order of presentation of the two words was reversed in the two replications, the mean difference in the replicated responses for each subject across all pairs of words and probabilities was computed and tested against the assumption that the means do not equal zero. No significant violations of the prediction were recorded, i.e. the judgments appear to be reliable and unaffected by the order of presentation of the stimuli.

The last two columns of Table 2 present the correlations between the replicated judgments of these tasks. Note, that unlike the results for the direct estimation task, these values are based on within-session replications. In general these reliabilities are satisfactory, and are higher for the union judgments (only 3 of the 15 subjects have higher reliabilities for the intersection judgments). A closer examination of the reliabilities indicates that in practically all cases judgments involving two H or two L words were less reliable than those involving one H and one L word. (Only one subject had a reliability under 0.8 for the union task and only four subjects had reliabilities lower than this value for the intersection judgments when all H-L judgments were excluded from the calculations.)

Tests of the Duality Predictions: We turn now to tests of the five conditions for any dual union/intersection rule based on the t-norm and the corresponding t-co-norm respectively. The first condition requires that the two individual membership functions be at least as large as their intersection, but not larger than their union. Table 4 presents mean differences between the relevant membership functions for each subject across all pairs of words and probabilities. A negative entry in the table identifies a violation of the prediction, and the number in parentheses indicates the proportion of violations (of a total of 105 judgments) for each subject. In order to control the hypothesis error rate at $\alpha = 0.05$, the results of the individual t-tests were declared significant at $\alpha^* = 0.05/15 = 0.003$. The results are clear and consistent -- all subjects judge the intersection to be larger than the smaller of the two memberships functions (see the first column in Table 4). All subjects judge unions to be at least as large as the intersection (column three), and, with two exceptions, the union does not appear to be significantly different from the larger of the two membership functions (column two).

TABLE 4
Mean Differences Between Membership Functions (and Proportion of Violation of Expected Order)

Inequality

$Min(\mu_A(x), \mu_B(x)) - \mu_{A \cap B}(x)$

$\mu_{A \cap B}(x) - Max(\mu_A(x), \mu_B(x))$

$\mu_{A \cup B}(x) - \mu_{A \cap B}(x)$

	$Min(\mu_A(x),\mu_B(x))-\mu_{A \cap B}(x)$	$\mu_{A \cap B}(x)-Max(\mu_A(x),\mu_B(x))$	$\mu_{A \cup B}(x)-\mu_{A \cap B}(x)$
1	$-.201^x$ (.78)	$-.037^x$ (.54)	.091 (.37)
2	$-.077^x$ (.71)	$-.012$ (.56)	$.149^x$ (.30)
3	$-.184^x$ (.71)	.012 (.44)	$.121^x$ (.30)
4	$-.099^x$ (.66)	.007 (.60)	$.229^x$ (.31)
5	$-.148^x$ (.67)	$-.018$ (.60)	$.144^x$ (.29)
6	$-.208^x$ (.75)	$-.018$ (.57)	.031 (.43)
7	$-.163^x$ (.70)	$-.055^x$ (.58)	.080 (.45)
8	$-.150^x$ (.74)	$-.005$ (.44)	$.154^x$ (.30)
9	$-.143^x$ (.60)	---- (---)	---- (---)
10	$-.135^x$ (.84)	$-.007$ (.57)	.115 (.36)
11	$-.100^x$ (.68)	$-.013$ (.55)	$.167^x$ (.22)
12	$-.263^x$ (.84)	$-.027$ (.46)	.014 (.43)
13	$-.101^x$ (.66)	$-.003$ (.50)	$.153^x$ (.30)
14	$-.137^x$ (.84)	$-.023$ (.55)	.061 (.43)
15	$-.181^x$ (.79)	$-.017$ (.58)	.037 (.49)
16	$-.147^x$ (.85)	.026 (.42)	$.121^x$ (.39)
Mean	$-.152x$ (.73)	$-.013$ (.53)	.110 (.36)

Not: "-" Subject did not perform union task
"x" Significant at a = 0.05/15 ' 0.003 (two tailed tests)

The same data can be used to evaluate the fifth condition, namely that the sum of the union and intersection judgments equal the sum of the individual membership values. The mean difference for each subject can be obtained by subtracting the first column in Table 4 from the second one. Tests of these differences show that for all subjects the sum of the conjunction and disjunction judgments is significantly larger ($p < 0.001$) than the sum of the individual memberships. This result is due largely to the high responses for the intersections.

Next we tested the monotonicity requirement of the union/intersection membership functions in each of the individual functions by examining whether the union/intersection memberships are monotonical non-decreasing functions of the six individual membership functions for each of the five fixed probabilities. In Table 5 the percentage of violations of weak order is presented for each subject and task. For each task, an average of 27% of the cases violate the requirement. However, this number is misleading since it is based on the extremely stringent requirement that the order of judgments in the "or" and "and" tasks duplicate exactly the order of the individual term judgment, with precision implied from a response scale of almost 200 values. A weaker, and more reasonable test of the requirement is obtained by relaxing the definition of a violation. Thus, any pair of judgments was considered to be in the required order when the violations of the order were smaller than a predetermined "unit of measurement error." This unit was calculated for each subject by calculating the mean absolute deviation between the two replications of each of the three tasks. Across subjects, this unit of error varies between 0.062 (subject 9) to 0.209 (Subject 15) with a mean of 0.108. When violations of order are recalculated, under this "approximate" definition of equality, their overall level drops to 6% for the "or" and 11% for the "and" tasks. Thus, the requirement is satisfied at an acceptable level, particularly for the "or" judgments. It is also interesting to note that when judgments are averaged across subjects a consistent pattern emerges. The highest levels of violations are recorded around probabilities of 0.5, the degree of violation decreases for more extreme probabilities, and the degree of violation is consistently higher for "and" than for "or" judgments.

TABLE 5
Percent of Weak Order Violations Under Strict and Approximate Equality

	Union		Intersection	
Subject	Strict Equality	Approximate Equality	Strict Equality	Approximate Equality
1	18	6	23	11
2	32	5	27	8
3	27	6	28	13
4	32	7	40	23
5	24	5	27	15
6	25	6	24	5
7	26	5	23	10
8	22	6	21	12
9	--	-	32	8
10	33	2	30	11
11	24	5	22	13
12	22	4	26	14
13	24	6	21	9
14	32	7	33	12
15	38	11	27	6
16	24	5	24	9
Mean	27	6	27	11

Note: "-" Subject did not perform task

The monotonicity condition for strict t and co-t-norms requires the judged union and intersection memberships of a given word with itself (e.g. likely and/or likely) to be a strictly increasing function of its individual membership function. Table 6 presents the mean (across words) rank correlations (Kendall's Tb) between the individual membership functions and the appropriate union/intersection judgments.

TABLE 6
Mean Rank Order Correlations Between Individual, Union and Intersection Membership Functions for a Given Word and Itself (Proportion of Violations of Strong Monotonicity)

Subject	Intersection	Union	Intersection x Union
1	.925 (.12)	.967 (.04)	.958
2	.858 (.10)	.967 (.02)	.891
3	.867 (.14)	.867 (.08)	.933
4	.606 (.29)	.779 (.26)	.708
5	.891 (.12)	.900 (.06)	.858
6	.933 (.12)	.967 (.02)	.900
7	.940 (.16)	.904 (.10)	.892
8	.974 (.18)	.991 (.02)	.983
9	.862 (.13)	---- (---)	----
10	.600 (.24)	.833 (.10)	.567
11	.967 (.02)	.967 (.02)	.933
12	.512 (.38)	.850 (.12)	.558
13	.867 (.10)	.867 (.08)	1.000
14	.767 (.14)	.656 (.22)	.690
15	.733 (.16)	.700 (.18)	.633
16	.833 (.14)	.739 (.20)	.706
Mean	.821 (.16)	.872 (.10)	.826

Note: "---" Subject did not perform union task

Also presented in the same table is the proportion of cases (out of 50) in which the strict monotonicity assumption was violated. With a few notable exceptions the rank correlations are high (11 subjects have rank correlations above .8 for each task), and the proportion of order violations is low. In most cases the violations can be linked either to ties, or to a single unusual response.

Again note that the data from the union task are more consistent with the requirement. Finally, since the union and intersections of a term with itself are monotonically increasing under any strict dual rules then the union and intersection judgments should also be monotonically related. The third column of Table 6 presents the rank correlations between the judgments of the corresponding union and intersection (i.e., "likely and likely" with "likely or likely"). This requirement is strongly satisfied in all cases.

Next we focus on the boundary conditions, i.e., cases where at least one of the individual membership functions is 0 or 1. This condition cannot be tested for all subjects since not everyone recorded such extreme responses for the probabilities involved in union/intersection judgments. Table 7 presents the mean deviation from the expected response for each of the four relevant conditions only for those subjects whose individual memberships allowed test.

TABLE 7

Mean Deviation from Expected Judgments in Boundary Conditions

Subject	Zero Membership			Unit Membership		
	n	Intersection	Union	n	Intersection	Union
1	17	.060*	-.047*	12.	089.	000
4	0	-----	-----	12	.183	-.111
6	17	.317*	-.016	0	----	-----
7	6	.105	.084	6	.171	-.089
8	6	.117	.006	11	.112	.004
11	0	-----	-----	6	.023	.025
12	15	.216	-.123	0	----	-----
15	0	-----	-----	6	.136	.000
Mean	61	.180	.039	52	.121	-.035

Note: *Significantly different from zero at a = 0.05/22 : 0.0025

These mean differences were tested for departure from 0 by means of t-tests, with a hypothesis error rate of $\alpha = .05$, ($\alpha^* = 0.05/22 : 0.0025$). In 19 of the 22 cases the null hypothesis could not be rejected. Note that in general the departure from the expected results is much larger for the intersection judgments.

TABLE 8
Goodness of Fit of Best Model for the Intersection Task

Subject	Model	Criterion of fit	
		Root Mean Square	Mean Absolute Deviation
1	Fuzzy M	.130	.103
2	Geometric M	.135	.117
3	Mean	.242	.183
4	Geometric M	.163	.131
5	Fuzzy M	.194	.144
6	Mean	.192	.167
7	Fuzzy M	.188	.154
8	Fuzzy M	.263	.191
9	Mean	.359	.331
10	Fuzzy M	.314	.239*
11	Fuzzy M	.160	.117
12	Fuzzy M	.310*	.238
13	Fuzzy M	.149	.113
14	Geometric M	.149	.117
15	Mean	.191	.153
16	Fuzzy M	.159	.132

Note: *Not the best fitting model by this criterion

Metric Comparisons of the Various Model: In comparing the fits of the various theoretical models to the data we focus on the union and intersection judgments involving one L and one H phrase. As shown in Table 3, membership functions within each class of phrases are very similar to each other. Therefore, the predictions of the various theoretical models for union or intersection of two H or two L phrases are highly correlated, so it would be impossible to reach any general conclusions about the relative merit of the

TABLE 9
Goodness of Fit of Best Model for the Union Task

Subject	Model	Criterion of fit	
		Root Mean Square	Mean Absolute Deviation
1	Max	.141	.083
2	Max	.091	.072
3	Sum Product	.176	.118
4	Max	.208*	.141
5	Max	.109	.080
6	Max	.118	.097
7	Max	.195	.144
8	Max	.087	.061
9	---	----	----
10	Max	.159	.109
11	Max	.059	.074
12	Max	.193	.140
13	Max	.098	.065
14	Max	.158	.122
15	Max	.216	.178
16	Max	.122	.090

Note: Subject 9 did not perform this task. *Not the best fitting model by this criterion

various models on the basis of these data. Two criteria were used to determine the goodness-of-fit of the various models, the root mean square deviation (RMS), and the mean absolute deviation (MAD). Tables 8 and 9 identify the best fitting model for any given task for each of the subjects. In most cases the two criteria identified the same "best" model (they correlate 0.97) and in the three cases where slightly different results were obtained, the determination of the "best" model was based on the magnitude of the badness of fit of the various models according to the two criteria.

The results are clear and consistent -- the max rule best describes the union judgments for all subjects, but one, and some "averaging" model yields always the best description of the intersection judgments. The judgments of a majority of subjects (9 of 16) are best captured by the fuzzy mean model and the remaining are better described by simple pointwise arithmetic or geometric means models. The variety of shapes of union/intersection memberships is illustrated in figures 2, 3 and 4, which present selected empirical results and best fitting models from 3 different subjects.

Subject 1 (Figure 2), who has single peaked individual membership functions displays single peakedness for the "and" judgments and multiple peaks for the "or" judgments. Note the close similarity between the "or" judgments and the maximal value of either phrase, and the obvious departure of the "and" judgments from the two functions. Obviously, two different types of judgments are involved in the two tasks. Subject 2 (Figure 3) has strictly monotonic functions for both words. As a result his "or" judgments are single dipped, in close correspondence with the max rule. His "and" judgments, best captured by the pointwise geometric mean, represent a clear compromise between the two individual functions.

Subject 6 is slightly less predictable than 1 and 2. The judgments for the two pairs of words displayed in Figure 4 are of interest for two reasons. First, note the similarity in the shape of his union and intersection data, which, suggests that he was using similar decision processes in both tasks. The judgments for the pair "improbable/fairly certain" in both tasks are essentially monotonically decreasing and appear to be mainly influenced by the characteristics of "improbable." This subject's data could be better fitted by a weighted average model in which "improbable" is overweighted.

The Integration of Linguistic Probabilities

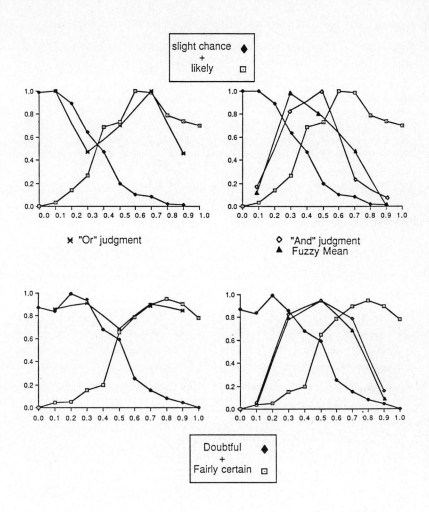

FIGURE 2
Subject 1

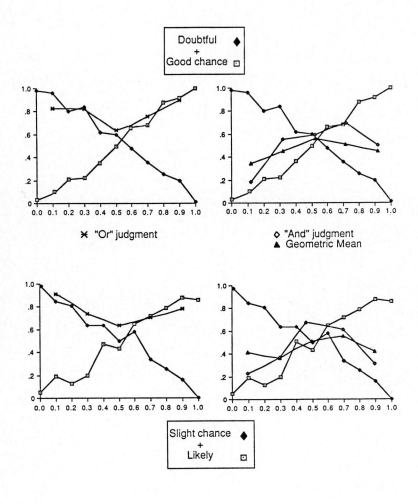

FIGURE 3
Subject 2

The Integration of Linguistic Probabilities 119

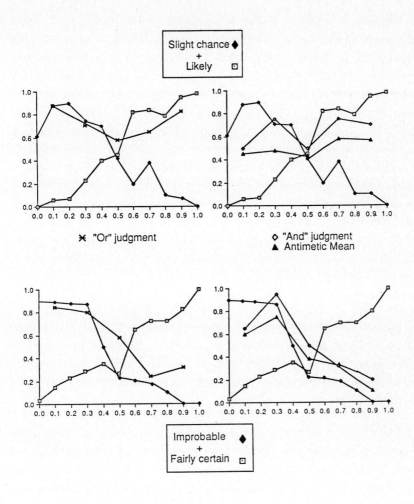

FIGURE 4
Subject 6

DISCUSSION

In previous research (Hersh & Caramazza, 1976; Hersh et al., 1979; Zimmermann and Zysno, 1980; Thöle et al., 1979; Oden, 1977a,b, 1979; and Kulka & Novak, 1984) it was found that the logical (fully noncompensatory) "and" operator does not describe the linguistic "and" as applied to two linguistic categories referring to different scales (e.g., a *high* brewing *speed* and *low capacity* coffeemaker). In these cases "not only are the idempotent intersection and union not a good model, but one may drift out of the domain of triangular norms and conorms, and get operations such as the arithmetic or the geometric means as proper models" (Dubois & Prade, 1985 p.97). We also found this phenomenon, although this time the two linguistic categories refer to the same scale (i.e., probability). We have shown, by testing several requirements that any such dual should fulfill, that no dual t- and co-t-norm can simultaneously fit our data. Most of these requirements were severely violated in our data. Furthermore, analysis of the two tasks ("and" and "or") independently reveals that the violations are due mainly to the compensatory nature of the linguistic "and" operator.

Summarizing the overall results, judgment in the "or" task is well described by the max rule. Judgment in the more common "and" task is best described by some sort of a compensatory averaging model. The fuzzy mean model is best for nine subjects, the arithmetic mean for four, and the geometric mean for three.

The implications of the present data are very interesting. First, when subjects judge whether a probability value is well described by *at least one* expert (the "or" task), they effectively decide which person had come closest and then ignore the input from the other person. Although the task was somewhat contrived, this kind of thinking may occasionally occur when an individual is being especially risk adverse and wishes to focus only on the maximum possibility of extreme probabilities describing an event.

The more common situation occurs when an individual integrates the verbal probabilistic inputs of two judges to form a single overall judgment. The present results suggest that this occurs according to an averaging process, by which the resulting judgment may be very different from either of the two inputs. This result is clearly reminiscent of the many other situations in which an averaging rule has been demonstrated (e.g., Anderson, 1981, 1982), but an important difference must be noted. Namely, previous investigations have assumed precise underlying values, whereas vague judgments are at the core of the present research.

The "and" responses of a majority of the present subjects are well described by an operation that takes the mean of two fuzzy numbers. It may be that these subjects actually engaged in information processing analogous to this operation, or it may be that they engaged in rather different processing that nevertheless yielded similar results. These distinctions can only be explored after alternative models are developed, and then only if the alternative and the fuzzy mean models differ in some testable ways.

We are currently running an additional experiment involving only the "and" task, in which a greater number of replications is being obtained in order to achieve more reliable estimates. We hope to confirm in this study that the "and" operation is indeed described by an averaging operation, and if so, to determine which operation is taking place.

This research was supported in part by Grant BNS 8608692 from the U.S. National Science Foundation and in part by contract MDA 903-83-K-0347 from the U.S. Army Research Institute. The views, opinions, and findings contained in this paper are those of the authors and should not be construed as official Department of the army position, policy, or decision.

The authors contributed equally to the project.

REFERENCES

Anderson, N. H. *Foundations of information integration theory*. New York: Academic, 1981.

Anderson, N. H. *Methods of information integration theory.* New York: Academic, 1982.

Bellman, R. E., & Zadeh, L. A. (1977). Local and fuzzy logics. In J. M. Dunn and E. Epstein, (Eds.), *Modern Uses of Multiple Valued Logic.* Dodrecht: Reidel, 1977.

Bellman, R. E., & Zadeh, L. A. Decision-making in fuzzy environment. *Management Science,* 1970, *17,* 141-164.

Beyth-Marom, R. How probable is probable? Numerical translation of verbal probability expressions. *Journal of Forecasting,* 1982, *1*, 267-269.

Budescu, D. V., & Wallsten, T. S. (1985). Consistency in interpretation of probabilistic phrases. *Organizational Behavior and Human Decision Processes,* 1985, *36,* 391-405.

Budescu, D. V., & Wallsten, T. S. Subjective estimation of precise and vague uncertainties. In G. Wright and P. Ayton (Eds.), *Judgmental Forecasting.* New York: Wiley, 1987.

Cholewa, W. Aggregation of fuzzy opinions - an axiomatic approach. *Fuzzy Sets and Systems,* 1985, *17,* 249-258.

Czogala, E., & Zimmermann, H.-J. The aggregation operations for decision making in probabilistic fuzzy environment. *Fuzzy Sets and Systems,* 1984, *13,* 223-239.

Dubois, D., & Prade, H. *Fuzzy sets and systems: Theory and applications.* New York: Academic, 1980.

Dubois, D., & Prade, H. A class of fuzzy measures based on triangular norms. *International Journal of General Systems,* 1982, *8,* 43-61.

Dubois, D., & Prade, H. A review of fuzzy set aggregation connectives. *Information Sciences,* 1985, *36,* 85-121.

Dyckhoff, H., & Pedrycz, W. Generalized means as model of compensative connectives. *Fuzzy Sets and Systems,* 1984, *14,* 143-154.

Frank, M. J. On the simultaneous associativity of F(x,y) and x + y- F(x,y). *Aequationes Mathematicae,* 1979, *19,* 194-226.

Hersh, H. M., & Caramazza, A. A fuzzy set approach to modifiers and vagueness in natural language. *Journal of Experimental Psychology: General,* 1976, *105,* 254-276.

Hersh, H. M., Caramazza, A., & Brownell, H.H. Effects of context on fuzzy membership functions. In M. M. Gupta, R. K. Ragde, and R. R. Yager (Eds.), *Advances in fuzzy set theory and applications.* Amsterdam: North-Holland, 1979.

Hillsdal, E. *Are grades of membership probabilities?* Paper presented at the 1st IFSA Congress, Palma de Mallorca, 1985.

Krantz, D. H., Luce, R. D., Suppes, P., & Tversky, A. *Foundations of measurement* (Vol. 1). New York: Academic, 1971.

Kulka, J., & Novak, V. Have fuzzy operations a psychological correspondence? *Studia Psychologica,* 1984, *26,* 131-140.

Luchandjula, M. K. Compensatory operators in fuzzy linear programming with multiple objective. *Fuzzy Sets and Systems,* 1982, *8,* 245-252.

Menger, K. (1942). Statistical metrics. *Proceedings of the National Academy of Sciences U.S.A.,* 1942, *28,* 535-537.

Nakao, M. A., & Axelrod, S. Numbers are better than words. Verbal specifications of frequency have no place in medicine. *The American Journal of Medicine,* 1983, *74,* 1061-1065.

Norwich, A. M., & Turksen, I. B. The fundamental measurement of fuzziness. In R. R. Yager (Ed.), *Fuzzy set and possibility theory.* New York: Pergamon, 1982.

Norwich, A. M., & Turksen, I. B. A model for the measurement of membership and the consequences of its empirical implementation. *Fuzzy Sets and Systems,* 1984, *12,* 1-25.

Oden, G. C. Fuzziness in semantic memory: Choosing exemplars of subjective categories. *Memory and Cognition*, 1977, 5, 198-204. (a)

Oden, G.C. Integration of fuzzy logical information. *Journal of Experimental Psychology: Human Perception and Performance*, 1977, 3, 565-575. (b)

Oden, G.C. Fuzzy propositional approach to psycholinguistic problems: an application of fuzzy set theory in cognitive science. In M. M. Gupta, R. K. Ragde, and R. R. Yager (Eds.), *Advances in fuzzy set theory and applications*, Amsterdam: North Holland, 1979.

Pepper, S. Problems in the quantification of frequency expressions. In D. Fiske (Ed.), *New Directions for Methodology of Behavioral Science Problems with Language Imprecision*, No 9. 1981.

Rapoport, A., Wallsten, T. S., & Cox, J.A. Direct and indirect scaling of membership functions of probability phrases. *Mathematical Modelling*, 1987, 9, 397-411.

Rubin, D. C. On measuring fuzziness: A comment on "A fuzzy set approach to modifiers and vagueness in natural language". *Journal of Experimental Psychology: General*, 1979, 108, 486-489.

Smithson, M. Multivariate analysis using "And" and "Or". *Mathematical Social Sciences*, 1984, 7, 231-251.

Smithson, M. *Fitting "And" and "Or" models to data.* Paper presented at 1st IFSA Congress, Palma de Mallorca, 1985.

Smithson, M. *Fuzzy set analysis for behavioral and social sciences*. New York: Springer, 1987.

Thöhle, U., Zimmermann, H.-J., & Zysno, P. On the suitability of minimum and product operators for the intersection of fuzzy sets. *Fuzzy Sets and Systems*, 1979, 2, 167-180.

Wallsten, T. S., Budescu, D. V., Rapoport, A., Zwick, R., & Forsyth, B. Measuring the vague meaning of probability terms. *Journal of Experimental Psychology: General*, 1986, 115, 348-365.

Weber, S. A general concept of fuzzy connectives, negations and implication based on t-norms and t-conorms. *Fuzzy Sets and Systems,* 1983, *11,* 115-134.

Winkler, R. L., & Makridakis, S. The combination of forecasts. *Journal of the Royal Statistical Society,* 1983, *A-146,* 150-157.

Yager, R. R. On a general class of fuzzy connectives. *Fuzzy Sets and Systems,* 1980, *4,* 235-242.

Zadeh, L. A. Fuzzy sets. *Information and Control,* 1965, *8,* 338-353.

Zadeh, L. A. The concept of a linguistic variable and its application to approximate reasoning. In K. S. Fu and J. T. Tou (Eds.), *Learning systems and intelligent robots.* New York: Plenium, 1974.

Zadeh, L. A. A fuzzy algorithmic approach to the definition of complex or imprecise concepts. *International Journal of Man-Machine Studies,* 1976, *8,* 249-291.

Zimmerman, H.-J., & Zysno, P. Latent connectives in human decision making. *Fuzzy Sets and systems,* 1980, *4,* 37-51.

Zwick, R., Carlstein, E., & Budescu, D. V. Measures of similarity among fuzzy concepts: A comparative analysis. *International Journal of Approximate Reasoning,* 1987, *1,* 221-242.

Zwick, R. A note on random sets and Thurstonian scaling methods. *Fuzzy Sets and Systems,* 1987, *21,* 351-356.

A FUZZY PROPOSITIONAL ACCOUNT OF CONTEXTUAL EFFECTS ON WORD RECOGNITION

Jay RUECKL

Harvard University, Department of Psychology,
Cambridge, Mass 02138 USA

A longstanding issue in cognitive psychology concerns the nature of contextual effects in word identification paradigms. While some theorists have seen these effects as reflecting the influence of contextual information on the identification process itself, other theorists have argued that word identification occurs independently of contextual information, and that the contextual effects usually observed are brought about by processes that occur after the word identification process has been completed. The conception of the word identification process offered by the fuzzy propositional framework suggests a new experimental paradigm for studying contextual effects. The results of experiments within this paradigm strongly suggest that contextual information does influence the word identification process. These results are consistent with the predictions of the fuzzy propositional model of word identification, which explicitly describes how visual and contextual information can be integrated in the process of identifying a word.

It is commonly agreed that understanding a written sentence or passage involves processes that apply orthographic, syntactic, semantic and pragmatic knowledge to the analysis of the visual input. One can think of this analysis as occuring in a series of stages, with each successive stage interpreting the input at a "higher" level of analysis. For example, at an early stage of processing orthographic knowledge would be used to identify each individual word in the input. The resulting word identity information would then be passed on to a process that constructs as interpretation of the syntactic structure of the input, and in turn the output of the syntactic process would serve as input for a process concerned with understanding the meaning of the input.

Although there is general agreement that language comprehension involves processes at a number of levels of analysis, there is sharp disagreement about the structure of the processing system that supports this analysis. On the other hand, a number of theorists, including Forster (1979, 1981) and others (e.g., Seidenberg, Tanenhaus, Leitman, & Bienkowski, 1982; Stanovich & Wets, 1983), have argued that the language comprehension system consists of a hierarchy of processing stages, with information flowing only upwards through the system. Thus, in this view language comprehension is seen as a series of discrete decisions. The decision made at a given level depends only on the results of processes at lower levels of the hierarchy, and cannot be influenced by processes that occur farther up in the system.

On the other hand, other theorists, including Morton (1979), Rumelhart (1977), and Marslen-Wilson and Welsh (1978), have argued in favor of a different conception of the language processing system. In the *interactive* processing framework proposed by these theorists, processing is again assumed to occur at a number of levels of analysis. In contrast to the above *autonomous* framework, however, in the interactive framework it is assumed that there is both a bottom-up and a top-down flow of information. The top-down flow of information allows initial results at higher levels of processing to influence decisions made at lower levels of analysis. It is argued that a top-down flow of information allows for more informed interpretations of what can often be ambiguous bottom-up information (Rumelhart, 1977; McClelland & Rumelhart, 1981).

Thus, the interactive and autonomous views make different predictions about situations in which the role of top-down information is explored. The autonomous view predicts that top-down information will have little impact on the outcome of language comprehension processes, while the interactive view suggests that the outcome of these processes will be influenced by top-down information.

CONTEXTUAL EFFECTS ON WORD IDENTIFICATION

Although the interactive and autonomous models make conflicting predictions about top-down effects across a variety of processes, these predictions have been put to the test most often in experiments studying the word identification process. In the most common paradigm for studying top-down effects on word identification, subjects read a context sentence and then either name a target word or make a lexical decision about it. The influence of higher-level context on the word identification process is studied by manipulating the semantic relationship between the target and its context sentence. For example, the target word "desk" might appear in the context of a sentence with which it is congruent, such as "Mary's books were piled up on her ......", or in the context of a sentence with which it is incongruent, such as "last night Mary read a good ...... ". The results of studies employing this paradigm are consistent and robust. Words are identified more accurately (Morton, 1964; Tulving & Gold, 1963) and responses are made more quickly (Fischler & Bloom, 1979; Forster, 1981; Stanovich & Wets, 1981) when the target is congruent with the context than when it is not.

At the first glance, this pattern of results seems to support the intercative view at the expense of the autonomous framework. The finding that semantic context can influence performance on word identification tasks is precisely what one should expect if there is a top-down flow of information within the language processing system. On the other hand, if it is assumed that information has a purely bottom-up flow, then contextual information should not influence the duration or outcome of the word identification process, as the results cited seem to suggest.

However, it is often the case that the obvious interpretation is not necessarily the correct one. In this particular case the reported contextual effects can be taken to be consistent with the autonomous framework if it is assumed that the locus of the contextual effects is not the word identification process itself, but instead some other process that occurs after the word identification process has been completed. For example, one might suppose that the detection of semantic incongruity triggers an error-checking mechanism. If this were the case, response times might be slower in the incongruent condition even though the duration of the word identification process itself was the same across conditions. Thus, both the interactive and autonomous frameworks can account for the finding that responses occur more quickly for congruent targets than for incongruent targets.

In order to tease apart these competing accounts, response times for congruent and incongruent targets can be compared to response times for words that occur in neutral contexts. In this case, the alternative accounts make clear and contradictory predictions. In the interactive framework it is assumed that a top-down flow of information facilitates the word identification process. Thus, according to this view word identification should occur more quickly for words that appear in a congruent semantic context than for words that appear in a neutral context. On the other hand, according to the autonomous framework contextual information has no influence on the word identification process. Under this account, response times in the congruent condition should be no faster than response times for neutral targets. The only contextual effect consistent with the autonomous view is a slowing down of responses for incongruent targets brought about by post-identification error-checking mechanisms.

Unfortunately, although the logic underlying this method for testing between the interactive and autonomous models is straightforward, the implementation of this method has proven to be anything but straightforward. The problem is that the term "neutral context" can be construed in several ways. For example, Stanovich and West (1981,1983) used the word strings "They said it was the ...... " and "the the the ...... " as their neutral contexts, and found that responses were made more quickly for words that appeared in a congruent context than for words that appeared in these neutral contexts. However, Forster (1981, Experiment 1), using a random word list as the neutral context, found only an inhibitory effect on response latencies for words appearing in incongruent contexts. To resolve the conflict between the results of these studies, Forster (1981, Experiment 5) directly compared the "It was the ...... " and random word list contexts. Forster found that the congruent targets were named significantly faster than words appearing in the "It was the ...... " neutral context, but that when a random word list was used as neutral context, the contextual effect was inhibitory.

Clearly, then, the interpretation of these depends on which neutral condition is accepted as truly "neutral". On the one hand, it can be argued that if neither condition provides information about the identity of the target, the faster one (the random word list) should be chosen. On the other hand, if sentence-level processing is both obligatory and resource demanding, then the random word list, not being as resource demanding as sentence contexts, would be artificially fast, and the "It was the ......" condition would be the more appropriate choice. Thus, although there is at least some reason to suppose that either sort of neutral context would serve as the appropriate baseline, neither choice is obviously a better choice than its alternative. Given this problem in identifying the appropriate baseline, the results of the reaction time studies discussed above remain ambiguous. Although these results can be taken as supporting the hypothesis that contextual information facilitates word identification, they can also be taken as supporting the claim that the locus of the contextual effect is an error-checking mechanism that operates after the word identification process has been completed.

AN ALTERNATIVE PARADIGM

There are two aspects of the reaction time paradigm discussed above that make the results of experiments using that paradigm difficult to interpret. The problem brought out in the previous section is that different baseline contexts allow for different conclusions to be drawn, and there are reasonable arguments for assuming that each of these different baselines is the appropriate one. A second problem is that, in the face of the lack of an appropriate neutral condition, both the autonomous and interactive frameworks can account for the finding that responses are faster in a congruent context than in an incongruent context. Thus, in order to test between the alternative hypotheses about the role of contextual information in word identification, one must either resolve the baseline issue or manipulate the context in such a way that the autonomous and interactive frameworks make different predictions.

Gregg Oden and I (Rueckl & Oden, 1986) performed a sequence of experiments in which we took the latter route. The fundamental assumption underlying these experiments is the claim that the congruity of a word with its context is not an all-or-none thing, but is instead a matter of degree. For example, consider the sentence frame "The ...... had a pair in his hand". This sentence frame can be sensibly completed by a number of words, including "cardplayer" and "arthritic". Note, however, that although either word sensibly completes the sentence, it seem intuitively obvious that the sentence is more sensible when completed by the word "cardplayer". That is, it is more likely that a cardplayer would have a pair in his hand than an arthritic would have a pair in his, even though it is perfectly possible for an arthritic to be holding a pair.

By making the assumption that a word can be more or less sensible with a given context, it is possible to create an experimental situation in which the autonomous and interactive frameworks make different predictions. In particular, if the word identification process is influenced by the degree to which a target word is congruent with its context, as might be expected within the interactive framework, then performance in an identification task should differ with differences in the degree to which a target word is a sensible completion of its sentence context. On the other hand, if the only effect of target/context sensibleness is to trigger an error-checking mechanism when the target word is incongruent with its context, as is posited by the autonomous framework, then differences in target/context sensibleness should have no effect on word identification, provided that the target is a sensible completion of the context sentence.

In the experiments Oden and I conducted, contextual information was manipulated by creating a set of sentence contexts that differed in the degree to which they were sensibly completed by either of two orthographically similar target words. For example, in the sentence frame "The ...... had a (pair/pain) in his hand", the blank was replaced by "cardplayer", "shoemaker", "piano player", and "arthritic". Note that the relative sensibleness of the alternative target words ("pair" and "pain") differs across these context. In the "cardplayer" context, "pair" is the more sensible target, while in the "arthritic" context, "pain" is the more sensible target. The other contexts, "shoemaker" and "piano player", are less strongly biased in either direction.

In addition to the contextual manipulation, the physical characteristics of the target words were also manipulated. In particular, pairs of target words were chosen so that these words differed by a single letter feature. For example, the words "pair" and "pain" differ physically solely in the length of the line that distinguishes "r" from "n". The lenght of this line was systematically varied, resulting in a series of stimuli that differ systematically in the degree to which they resemble the alternative target words (see Figure 1). Previous research has found that letter and word identification is a consistent and systematic function of this type of manipulation (Massaro, 1979; Oden, 1979, 1984).

The interactive and autonomous frameworks make clear and distinct predictions about the effects of these manipulations on the outcome of the word identification process. Both views agree that manipulating the physical characteristics of the stimulus will influence the outcome of this process. However, only the interactive view predicts that the contextual manipulation will also influence the outcome of the word identification process. Because the contextual manipulation described above does not produce any sentences that would trigger a post-identification error-checking mechanism, the autonomous view must predict that word identification will not be influenced by that contextual manipulation.

pair	bears	car
pair	bears	car
pair	bears	ear
pain	beans	ear
pain	beans	ear

FIGURE 1

The feature manipulations used in Experiment 1.

EXPERIMENT 1

Two experiments were conducted, each involving the featural and contextual manipulations described above. In the first experiment, the subject was presented with a series of sentences, and was asked to choose which of two test words was in each sentence. Each sentence was briefly presented on a video monitor, and was replaced by a display containing the two response alternatives. The subject was then asked to push a button corresponding to the test word that had appeared in the sentence.

The sentences used in this experiment were drawn from three stimulus matrices that were constructed by combining five levels of a feature manipulation with four different sentence contexts. One matrix was the pair/pain matrix described above. For this matrix, the response alternatives offered the subject were the words "pair" and "pain". The second stimulus matrix was the bear/bean matrix. The featural manipulation for this matrix involved the same r/n continuum used in the pair/pain martix. In this case, the r/n characters were embedded in the letter string "bea_s", resulting in word stimuli that varied in the degree to which they resembled the words "bears" and "beans". The contextual manipulation for this matrix was produced by replacing the blank in the sentence frame "The .... raised (bears/beans) to supplement his income" with "lion tamer", "zookeper", "botanist", and "dairy farmer". Again, these words differ in the degree to which they support the alternative interpretations of the target word. The third matrix was the car/ear matrix. For this matrix, the featural manipulation involved changes in the length of the horizontal bar that distinguishes "c" from the "e". For the contextual manipulation, the words "rusty","dented","bleeding ", and "deformed" were embedded in the sentence frame "The detective carefully watched the man with the ...... (car/ear)".

There were three blocks of trials. Within each block, all 60 test sentences (20 feature X context combinations from each of the three matrices) were presented, as were 15 filler sentences. Some of these filler sentences were closely related to the test sentences. In particular, the filler sentences were based on the sentence frames used to generate the stimulus matrices. However, in the filler sentences the blanks in the sentence frames were filled by context words other than used in the test sentences, and the target word was also changed. (For example, one filler senetence "The old ballplayer had a pain in (his/her) hand".) The purpose of these filler sentences was to discourage the subject from adopting a strategy of only looking at the target , rather than reading the entire sentence.

Results of Experiment 1

The results of the first experiment indicated that both semantic and featural information influenced the outcome of the word identification process. For all three matrices the probability of a given response was a systematic function of both the semantic featural support for that interpretation.

The results for the pair/pain matrix are presented in Figure 2. On the vertical axis is the proportion of trials on which the target word was identified as "pain". The levels of the featural manipulation run along the horizontal axis. The individual curves in the figure correspond to the different semantic contexts. As can be seen in the figure, the feature manipulation produced a large and systematic effect, with the probability of the target word being identified as "pain" increasing as a direct function of the length of the manipulated feature, $F(4,92) = 133.79, p < .001$.

The effect of the contextual manipultion was relatively small, but it was also reliable and systematic, $F(3,69) = 3.22, p < .05$. At each level of the featural dimension, the target word was more likely to be identified as "pain" in the arthritic context than in the card player context, while responses in the other contexts, which seem intuitively to be less biased in either direction, fell in between.

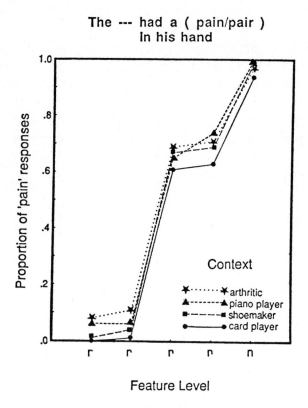

FIGURE 2

The portion of "pain" responses for the pair / pain matrix as a function of sentence context and featural information.

Thus, the results from the pair/pain matrix indicate that featural and semantic information jointly determined word identification. The influence of these factors was systematic, in that the probability that the target was identified as "pain" increased with increases in either the featural or semantic support for that interpretation. Finally, although the featural and contextual manipulations had significant effects, the interaction of these factors was not significant.

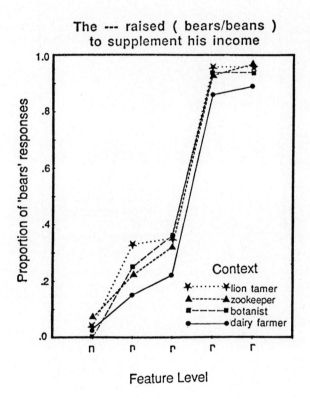

FIGURE 3

The portion of "bears" responses for the bears / beans matrix as a function of sentence context and featural information.

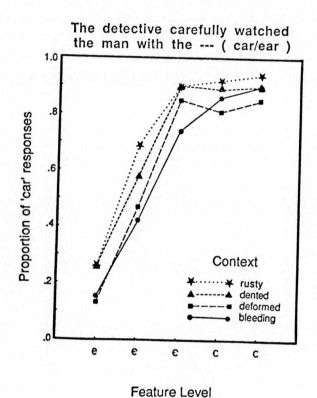

FIGURE 4
The portion of "car" responses for the car / ear matrix as a function of sentence context and featural information.

The results for the other matrices are presented in Figure 3 and 4. The results for these matrices follow the same pattern as was found for the pair/pain matrix. For each matrix, there was a large and systematic effect of the featural manipulation ($F(4,92) = 210.76, 103.93, p < .001$. for the bears/beans and car/ear matrices, respectively). In addition, for each matrix the effect of the contextual manipulation was both significant ($F(3,69) = 4.58, 4.92, p < .05$ for the bears/beans and car/ear matrices, respectively) and sensible. For the bears/beans matrix, responses in the lion tamer context were biased towards the "bears" response, responses in the dairy farmer context were biased towards the "beans" response, and responses in the other context fell in between. For the car/ear matrix, the "rusty" and "dented" contexts were biased towards the "car" response, while the "deformed" and "bleeding" contexts were biased towards "ear". For neither matrix was the interaction of the effects of the context and feature manipulations significant.

EXPERIMENT 2

The results of the first experiment suggest that contextual information influences the word identification process. The results of the second experiment support this conclusion. Although Experiment 2 involved the same semantic and featural manipulations used in Experiment 1, both the design and the procedure of the second experiment differed somewhat from that of Experiment 1. Sixteen stimulus matrices were constructed for Experiment 2, thus allowing us to study featural and contextual effects word identification over a wider variety of materials. In constrast to the increase in the number of matrices, the number of cells within each matrix was reduced: there were three leves of the feature manipulation and two leves of context per matrix. The design of Experiments 1 and 2 thus have a complementary nature. The design of Experiment 1 allowed for a fine-grained examination of the influence of contextual and featural information on word identification in a task involving a small number of stimuli. In contrast, the examination of the role of featural and semantic information was less fine-grained within each matrix of Experiment 2, but the effect of these factors was studied across a greater variety of materials.

The stimulus matrices for this experiment were constructed so that the three levels of the feature manipulation included two intact letters that differed along a single dimension (e.g. "r" and "n") as well as an intermediate stimulus that partially resembled each of the intatct letters (see Figure 5). The two context sentences were chosen so that each interpretation of the target word was the more sensible completion of one of the two contexts. For example, for the tan/fan matrix, the two conte sentences were "They drove to Florida to get a (fan/tan)" and "On these hot days it helps to have a (tan/fan)". Although both "tan" and "fan" sensibly complete either sentence, "tan" is the more sensible completion of the former sentence, while "fan" is the more sensible completion of the latter.

In addition to testing the generality of the results of Experiment 1, the inclusion of more stimulus matrices allows for a better assessment of the actual size of the context effect. In the first experiment the subjects were repeatedly exposed to each of the different contexts for each of the stimulus matrices. It is not unreasonable to suppose that the repeated occurence of each context in Experiment 1 may have reduced effects of diferences in the relative sensibleness of each interpretation of the target word in a given context by making each interpretation highly familiar. Thus, in Experiment 2 each subject was presented with only one of the test sentences from each of the 16 matrices. (In addition, each subject was presented with 32 unrelated filler sentences. Of the filler sentences, 16 were sensible, and 16 were semantically anomalous.)

The task used in Experiment 2 was also different from that used in Experiment 1. Rather than performing what amounted to a forced-choice recognition task, subjects in the second task were asked to read each sentence aloud, and then decide whether or not the sentence made sense. There were three reasons for altering the task in this way. First, this task emphasizes the semantic properties of the sentence. If semantic context does influence word identification, as suggested by the results of Experiment 1, then increasing the degree to which the subject focuses on the semantic properties of the sentence should increase the size of the context effect.

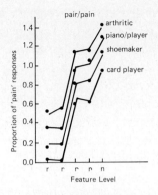
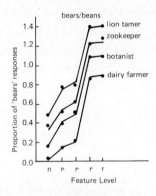
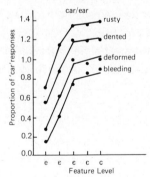

FIGURE 5

The expected and observed values (curves and points, respectively) of the responses for each stimulus matrix. The values for the "shoemaker", "botanist", and "deformed" contexts have been displaced upwards by .15 units. The values for the other contexts were also displaced upwards, by equal increments. (Adapted from Rueckl & Oden, 1986.)

Secondly, having the subject rate the sensibleness of each sentence gives us an independent measure of the congruity between each context and each interpretation of the target word. Although the pattern of results in Experiment 1 followed our intuitions about the relative sensibleness of each target interpretation in a given context, having the subjects make sensibleness judgments about target/context pairs gives us an objective measure of this relationship. If one interpretation of a target really is more sensible with a given context, then that sentence should have been more likely to be judged sensible when the target word was given that interpretation.

Finally, the use of this new task allows us to test an account of the results of Experiment 1 that is consistent with the autonomous framework. Recall that autonomous models could account for contextual effects in reaction time studies by positing that responses to semantically incongruent target words are slowed by an error-checking mechanism triggered by semantic incongruity. The results of Experiment 1 can not be accounted for in this way, however, because in that experiment both interpretations of the target word were congruent with each context. Thus, the error-checking mechanism should not have been triggered, and contextual information should not have influenced the outcome of the word identification process.

However, there is another account of the results of the first experiment that is consistent with the autonomous framework. It might be argued that, because each sentence was presented for a brief amount of time, there may have been a number of trials on which the subject was not able to identify the target word before it was removed from the screen. On these trials the subject's response would be nothing more that a guess about the target word's identity. If the subject had some information about the rest of the sentence, and if this contextual information influenced the subject's guessing strategy on these trials, then it would have been possible to find a contextual effect on the actual responses made, even though contextual information does not influence the word identification process itself. If context influences word identification in Experiment 2, this account can also be ruled out. By simply asking the subject to read a sentence aloud, the possibility that the subject is guessing about the identity of the target word has been removed.

To summarize, in Experiment 2 subjects were presented with one sentence from each of 16 2X3 stimulus matrices, along with 32 filler sentences. The presentation of the test sentences was counterbalanced in such a way that each subject saw approximately the same number of sentences from each combination of feature level and context level (collapsed across matrices). Furthermore, each sentence from each matrix was presented an equal number of times across subjects. The subject's task was to read each sentence aloud as it was presented, and then to push one of the two buttons, indicating whether or not the sentence made sense. The experimental session was tape-recorded, so that the subject's identifications of the target words could be determined.

Results of Experiment 2

In Experiment 1, the data for each matrix were analyzed separately. Because there was far less data per matrix in Experiment 2, the data were collapsed across stimulus matrices. In order to do this, the alternative interpretations of the target word for each matrix were arbitrarily labelled "Word 1" and "Word 2" responses. Each context can then be denoted with reference to the response alternative that was most sensible given that context. For example, in the fan/tan matrix "tan" was assigned label "Word 1", and "fan" was labelled "Word 2". "Tan" is the most sensible completion of the sentence "During spring break they drove to Florida to get a .... ". Thus, this context was labelled "Context 1". Conversely, the context supporting "fan" ("On these hot days it helps to have a ...... ") was labelled "Context 2". Finally, the levels of the feature manipulation were also labelled in a similar manner. The level of the feature manipulations corresponding to an intact "Word 2" stimulus was labelled level "F1", while the feature level corresponding to an intact "Word 2" stimulus was labelled "F2". The intermediate level along the feature dimension was labelled "FA" ("A" for "ambiguous").

Labelling the levels of the feature and context manipulations in this way identifies the cells that correspond to each other across the stimulus matrices. Thus, the number of Word 1 responses for a given cell in a given matrix can be added to the number of Word 1 responses for that same cell in all the other

```
r r n          c c e
t f f          n h h
p b b          l l t
```

FIGURE 6
The feature manipulations used in Experiment 2.

matrices. The result of this process is shown in Figure 6. In this figure, the vertical axis represents the proportion of Word 1 responses. The levels of the feature manipulation run along the horizontal axis, and the separate curves correspond to the different contexts. As can be seen in the figure, the feature manipulation again had a large effect on the outcome of the identification process. The contextual effect was quite small when the feature levels corresponded to intact letters (levels F1 and F2), but was quite small when the physical stimulus was an ambigous letter form (level FA). An analysis of variance found significant effects for both the feature manipulation, $F(2,118) = 877.93$, $p < .001$, and the contextual manipulation, $F(1,59) = 80.22$, $p < .001$. In addition, the interaction of the contextual and featural manipulations was also significant, $F(2,118) = 42.31$, $p < .001$, indicating that the contextual effect was largest when the featural information was least informative.

The results of the sensibleness decision were also analyzed. According to the experimental logic, each interpretation of the target word was relatively more sensible with one of the sentence contexts. If this was the case, then Context 1 sentences should be more likely to be judged sensible when the target word was interpreted as Word 1 than when it was interpreted as Word 2, and vice versa for Context 2 sentences. This predicted pattern of results did obtain. In 28 of the 32 sentence contexts, the sentence was judged as being sensible more often when the target was identified as the interpretation hypothesized to be more sensible in that context, $p < .001$, by sign test.

Futhermore, in only one of the 32 contexts was the biased against interpretation not judged as sensible by any of the subjects. Thus, the contextual manipulation did not result in one sensible sentence and one anomalous sentence. Instead, the contextual manipulation resulted in two sensible sentences that differed in their degree of sensibleness.

GENERAL DISCUSSION

The results of these experiments are consistent with the interactive view of the language processing system, which assumes that decisions made at a given level of analysis can be influenced by information derived by processes at a higher level of analysis. In the present case, the identification of a target word was systematically influenced by the context in which that word appeared. It is difficult to see how this effect could have been produced by the factors that exert their influence after the word identification process has been completed. Identification differs across sentence contexts that were sensibly completed by either interpretation of the target word, thus precluding the possibility that the context effect was brought about by a post-identification error-checking mechanism. Futhermore, because the task used in Experiment 2 involved simply reading each sentence aloud, it is highly unlikely that the context effect found in that experiment could be the result of the context-sensitive guessing strategy. Thus, it is difficult to escape the conclusion that the locus of the context effect is the word identification process itself.

Given this conclusion, we can now turn to an account of the result reported here. What is needed is an account of the process that takes as input visual and contextual information and yields as output a representation of the identity of the word being processed. In fact, a variety of accounts of this process have been suggested (Morton, 1979; Marslen-Wilson & Welsh, 1978; Rumelhart, 1977). These alternative accounts share the property that they allow contextual information to influence the processing of visual information. For example, in Morton's model, contextual information lowers the threshold of word detectors that are congruent with that context, causing a word detector to reach threshold more quickly when the context supports that interpretation of the input that when it does not.

However, although these models account for the general finding that word identification is influenced by contextual information, they cannot explicitly account for the kind of contextual and featural manipulations used in these experiments. That is, these accounts lack the conceptual apparatus needed to understand how the featural and contextual support for a given interpretation of a word can vary sytematically and continuously, as was the case in the present experiments.

One account that does provide the needed conceptual apparatus is the fuzzy propositional framework (Oden, 1979, 1984; Massaro, 1979; Massaro & Oden, 1980). This framework has been successfully applied to a variety of psycholinguistic domains, including letter (Oden, 1979), word (Massaro, 1979; Oden, 1984), and speech sound identification (Massaro & Oden, 1980; Oden & Massaro, 1978), and syntactic ambiguity resolution (Oden, 1978, 1983). In the fuzzy propositional framework, concepts are represented by propositions that can take on continuous truth values. Complex propositions are construed by conjoining simpler propositions with fuzzy connectives (e.g. AND, OR). The truth value of a complex proposition is a systematic function of the truth values of its component propositions (Oden, 1977).

In the fuzzy propositional model of word identification, it is assumed that each word is represented by a fuzzy proposition composed of more elementary propositions. These component propositions represent the letters that make up the word. For example, the word "pain" would be represented by the proposition "p_1 & a_2 & i_3 & n_4", where "p_1" stands for the proposition that the first letter is a "p", and "&" stands for the fuzzy logical "AND" operator. Letter propositions are also complex propositions composed of more elementary propositions. For example, the letter "p" might be represented by the proposition "f_1 & f_2 & f_3 ... ", where each "f" represents a specific letter feature, such as a line or curve segment or a relation between segments.

It is assumed that the truth values of these feature propositions can be determined by low-level perceptual processes. That is, upon the representation of a visual stimulus, low-level perceptual processes determine the degree to which each feature is present in the stimulus. This featural information can then be used to compare the stimulus to lexical representations stored in long-term memory. In order to perform this comparison process, information about the presence of each of the elementary features specified in a word's proposition is integrated according to the rules of fuzzy logic (Oden, 1977). The result of this comparison process is a fuzzy truth value corresponding to the degree to which the stimulus matches the lexical representation. This value will be a systematic function of the degree to which each of the elementary features was present in the stimulus.

According to the model, the stimulus is compared to the proposition representing each word in the reader's vocabulary in parallel. The result of this process is a set of fuzzy truth values corresponding to the degree to which the stimulus matches each word pattern. If the stimulus is an isolated word (that is, if it is not encountered in the context of other words), the probability of identifying the stimulus as a given word is taken to be a direct function of the associated truth value. Thus, the stimulus is most likely to be identified as the word with the highest truth value, and hence the best match with the stimulus.

On the other hand, if the stimulus appears in a context that constraints the likelihood of a given word's occurence, the model will take this source of information into account. Of particular relevance here as a source of information is the sentence context in which a word appears. Because different interpretations of a stimulus word will be more or less sensible given a particular sentence context, this context can serve as a value source of information in choosing the best interpretation of the stimulus. In order to take this information account, it is assumed that in addiditon to computing the degree to which the stimulus matches each lexical representation, the reader concurrently computes the degree to which each meaning of a word makes sense given the context. (See Oden, 1983, for an explanation of how these sensibleness values are determined.)

Thus, in identifying a word in the context of a sentence, it is assumed that the reader has available information about both the degree to which the stimulus matches each alternative interpretation and the degree to which each interpretation makes sense given the context. The overall support for each interpretation is given by the fuzzy conjunction of these two values. The probability of selecting a particular interpretation of the stimulus is a function of the degree of support for that interpretation relative to the degree of support for other interpretations. In particular, it is assumed that response selection follows the equation

$$R1 = (F1\&S1) / (F1\&S1) + (F2\&S2) + \ldots\ldots (Fn\&Sn)$$

where R1 is the relative degree of support for Word 1, F1 is the featural support for Word 1, S1 is the sensibleness of Word 1 given the context, and so on. In this equation, both featural and semantic information influence the word identification decision. Thus, the identification process is able to take advantage of both sources of information, and need not be completely dependent on either source alone.

An account of the experimental results. The fuzzy propositional model gives a clear and natural account of the experimental results reported above. Consider, for example, the results for the pair/pain matrix in Experiment 1 (Figure 2). Note first that, disregarding the contextual manipulation, the proportion of "pain" responses increased steadily as the length of the feature distinguishing "r" and "n" in the last letter increased. According to the model, increasing the length of this feature will increase the featural support for "pain" and decrease the featural support for "pair". Increases in the featural support for "pain" will in turn lead to increases in the size of the numerator (relative to the denominator) in the equation for the relative degree of support for "pain". Thus, the probability of a "pain" response will increase as the featural support for "pain" increases.

On the other hand, disregarding the featural manipulation, the value for the sensibleness of "pain" should be greater in the "arthritic" context than in the "card player" context, and this difference leads to a greater relative degree of support for "pain" in the arthritic context. This difference is again reflected in the data, in that the proportion of "pain" responses is higher in the "arthritic" context at each level of the feature manipulation. A similar account can be given for the results for each of the other matrices in Experiment 1 and 2.

In addition to giving a qualitative account of the effects of both the featural and contextual manipulations, the fuzzy propositional model can be used to derive a quantitative account of the joint influence of these manipulations. In order to provide this account, the iterative model-fitting routine BESTFIT (Wilkinson, 1982) was used to fit the model to the data from Experiment 1 using a least squares criterion of fit. Of the nine parameters of the model (five for the featural dimension and four for the different contexts) eight were free to vary and one was set arbitrarily to establish the scale.

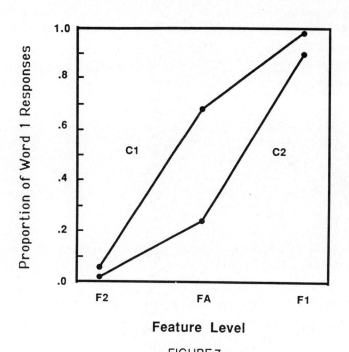

FIGURE 7

The proportion of Word 1 responses in Experiment 2 as a function of semantic context and featural information.

(Adapted from Rueckl & Oden, 1986.)

The resulting parameters were then used to estimate the expected values for each of the 20 data points for each matrix. In general, these expected values were reasonably close to the actual data - the root mean squared derivations between the expected and observed values of the means of the responses were 0.019, 0.020, and 0.027 for the pair/pain, bears/beans, and car/ear matrices, respectively. These values are comparable to results obtained in previous work of this sort.

Figure 7 presents the expected and observed values (curves and points, respectively) for each of the matrices in Experiment 1. In order to clearly depict the relationships between the observed and expected values for each context, successive curves have been displaced upwards by a fixed amount. (For example, in the pair/pain matrix, the values for the "shoemaker", "piano player", and "card player" contexts have been displaced upward by 0.15, 0.30 and 0.45 units, respectively.) This figure illustrates the goodness of the fit between the values produced by the model and the data actually obtained. The deviations are quite small, and do not follow any systematic pattern. Thus, the fuzzy propositional model gives a good quantitative account of the results reported above.

CONCLUSION

The results reported here demonstrate that both featural and contextual information influence the word identification process. In both experiments, the probability that a target word was assigned a given interpretation was a systematic function of both the featural and semantic support for that interpretation. These results are consistent with the assumption that the language processing system is designed to operate in an environment that often provides the system with noisy and somewhat ambiguous input. In such an environment, a system will be best off if takes into account a variety of different, partially redundant sources of information. A corollary of this view is the principle that the influence of one source of information should be greatest when the information provided by the other sources is most ambiguous. This principle is exemplified by the results of Experiment 2, where the effect of the contextual manipulation was greatest when the featural information was most ambigous.

The conclusion that contextual information plays a role in the word identification process is consistent with the interactive processing framework, and in particular, with the fuzzy propositional model of word identification. This model assumes that both semantic and featural information are continuous in nature, and that information of both sorts are integrated during the word identification process. It was argued above that the fuzzy propositional model can account for the present results both qualitatively and quantitatively.

Finally, it should be noted that the experimental paradigm used in the present study differed in several ways from the sort of paradigms more typically used to study contextual effects on word identification. Rather than studying the influence of contextual information on the time needed to make a decision about the identity of a word, we examined the effect of context on the outcome of that process. Futhermore, the experimental logic depended on the assumption that semantic congruity is a matter of degree, and not an all-or-none thing. This fundamental insight of the fuzzy propositional approach played a key role in the development of a contextual manipulation capable of teasing apart the predictions of the intercative and autonomous frameworks.

ACKNOWLEDGMENTS

The experiments presented here were also reported in Rueckl and Oden (1986). Special thanks to Gregg Oden for guidance and support. Thanks also to Joan Schleuder and G. G. Rueckl.

REFERENCES

Fischler, I., & Bloom, P. Automatic and attentional processes in the effects of sentence contexts on word recognition. *Journal of Verbal Learning and Verbal Behavior,* 1979, *18,* 1-20.

Forster, K. Levels of processing and the structure of the language processor. In W. Cooper & E. Walker (Eds.) *Sentence processing: Psycholinguistic studies presented to Merrill Garrett.* Hillsdale, N.J.: Erlbaum, 1981.

Forster, K. Priming and the effects of sentence and lexical contexts on naming time: Evidence for autonomous lexical processing. *Quarterly Journal of Experimental Psychology,* 1981, *33A,* 465-495.

Marslen-Wilson, W., & Welsh, A. Processing interactions and lexical access during word recognition in continuous speech. *Cognitive Psychology,* 1978, *10,* 29-63.

Massaro, D. W. Letter information and orthographic context in word perception. *Journal of Experimental Psychology: Human Perception and Performance,* 1979, *5,* 595-609.

Massaro, D.W., & Oden, G. C. Speech perception: A framework for research and theory. In N. J. Lass (Ed.) *Speech and language: Advances in basic research and practice,* Vol. 3, New York: Academic Press, 1980.

McClelland, J., & Rumelhart, D. An interactive activation model of context effects in letter perception: Part 1. An account of basic findings. *Psychological Review,* 1981, *88,* 375-407.

Morton, J. The effects of context on the visual duration threshold for words. *British Journal of Psychology,* 1964, *55,* 165-180.

Morton, J. Word recognition. In J. Morton & J. C. Marshall (Eds.) *Psycholinguistics,* Series II, London: Elek, 1979.

Oden, G. C. Integration of fuzzy logical information. *Journal of Experimental Psychology: Human Perception and Performance*, 1977, *4*, 565-575.

Oden, G. C. Semantic constraints and judged preference for interpretations of ambiguous sentences. *Memory and Cognition*, 1978, *6*, 26-37.

Oden, G. C. A fuzzy logical model of letter identification. *Journal of Experimental Psychology: Human Perception and Performance*, 1979, *5*, 336-352.

Oden. G. C. On the use of semantic constraints in guiding syntanctic analysis. *International Journal of Man-Machine Studies*, 1983, *19*, 335-357.

Oden, G. C. Dependence, independence, and emergence of word features. *Journal of Experimental Psychology: Human Perception and Performance*, 1984, *10*, 394-405.

Oden, G. C., & Massaro, D. W. Integration of featural information in speech perception. *Psychological Review*, 1978, *85*, 172-191.

Rueckl, J., & Oden, G. C. The integration of contextual and featural information during word identification. *Journal of Memory and Language*, 1986, *25*, 445-460.

Rumelhart, D. Toward an interactive model of reading. In S. Dornic (Ed.) *Attention and Performance* VI. Hillsdale, N.J.: Erlbaum, 1977.

Seidenberg, M. S., Tannenhaus, M. K., Leiman, J. M., & Bienkowski, M. Automatic access to the meaning of ambiguous words in context: Some limitations of knowledge-based processing. *Cognitive Psychology*, 1982, *14*, 489-537.

Stanovich, K. E., & West, R. F. The effct of sentence processing on ongoing word recognition: Tests of a two-process theory. *Journal of Experimental Psychology: Human Perception and Performance*, 1981, *7*, 658-672.

Stanovich, K. E., & West, R. F. On priming by a sentence context. *Journal of Experimental Psychology: General*, 1983, *112*, 1-36.

Tulving, E., & Gold, C. Stimulus information and contextual information as determinants of tachistoscopic recognition of words. *Journal of Experimental Psychology,* 1963, *66*, 319-327.

Wilkinson, A. C. BESTFIT. (unpublished computer program). Madison, WI: University of Wisconsin.

M-FUZZINESS IN BRAIN/MIND MODELLING

György FUHRMANN*

*IGAC Computer Science Laboratory
Technical University of Budapest, Hungary*

Fuzzy sets are argued to be useful though not sufficient for models of concepts: Continuous membership seems essential though fuzzy sets are defined on sets, i.e., on an inherently contradiction-free domain (by the axioms of set theory). Thus any account for the contradictions experimentally proved on conceptual categories inevitably requires a generalization of fuzzy sets. Therefore modified fuzzy sets (m-fuzzy sets, for short) defined on classes (in the sense of set theory) are proposed. The hierarchy of elements in a category is modelled by a structure of classes, m-fuzzy sets, sets, and finite subsets, such that each item represents the preceding one, and the last consists of the prototypes. The hierarchy of the categories is modeled by m-fuzzy grammars which are pure and regulated (in the sense of formal language theory). A related neural network is also explained. In this, the collective induced states of the functional units, called modules, represent concepts (like in Hebbs cell-assemblies). Both the encoding and the drawing of inference are distributed and parallel. Formally, they constitute residue number systems and a kind of syntactic pattern recongition, respectively.

INTRODUCTION

Concepts are fuzzy. As Miller and Johnson-Laird (1976, p.697) put it: "Concepts are invisible, impalpable, ill-defined abstractions that have a nasty way of being whatever a theorist needs them to be at the moment".

In spite of this simple and well-know fact, fuzzy-set theory has not found much fruitful applications in modelling concepts. Although fuzzy sets were originally defined by Zadeh (1965) as a tool to formalize concepts, their possible use in the cognitive sciences is still a vexing question (Cohen & Murphy, 1984). Moreover, there has been a highly influential point of view, proposed by Osherson and Smith (1981), claiming that fuzzy set theory is inherently inadequate for modelling concepts. All this, however, by no means indicates that fuzzy-set theory has been unsuccesful. Indeed, dozens of books and thousands of articles have been published in this field, in other than the full length journal Fuzzy Sets and Systems devoted entirely to it. In this enormous literature, however, a relatively small portion is directed towards the original aim, i.e., to formalize the cognitive sciences. Fuzzy set theory, practically as a whole, developed in areas that had already been formalized, i.e., in mathematics, system theory, computer science, etc.

The present paper departs from these apparent contradictions. First some causes, both virtual and real, that have prevented a fruitful application of fuzzy-set theory in the cognitive science are briefly discussed. Then, by a generalization of the underlying domain, modified fuzzy sets (m-fuzzy sets, for short) are defined.

For modelling the hierarchy of elements in a conceptual category, a formal structure of concepts in their generality (classes, in the sense of axiomatic set theory), m-fuzzy sets, concept cores (sets, in the above sense), and prototypes (finite subsets) is developed such that each item represents the preceding one. In this structure the possible occurence of contradictions proved inevitable in the logic of concepts (e.g., the famous Russel's antinomy cited in section 1.2) is localized to the first two items. Not only fuzzy sets, but also prototypes, are defined in a modified way; as the elements maximizing the mutual representativeness between the element and the category.

For modelling the hierarchy in representativeness of the categories comprising a given element, m-fuzzy grammars are defined. These are pure, i.e., no distinction is made between nonterminal and terminal symbols, and regulated, i.e., different levels in the rewriting involve different collections of the rules. According to this formal linguistic approach, category identification appears as a kind of syntactic pattern recognition.

Since the rewriting rules operate on confined information only, they can be implemented by units of local arborization, e.g., by formalized neurons. On this basis a related network of (a new type of) formalized neurons is proposed. The functional units are called modules referring to the experimentally found neuron-modules (Szentágothai, 1975, 1978; Mountcastle, 1978). Collective induced states of the (formalized) modules are assumed to represent concepts, like Hebb's (1949) cell-assemblies. Since these modular states directly refer to the representation including the learning and recognition of concepts, the proposed approach sharply contrasts with Fodor's (1983) theory on the modular organization in the mind/brain.

Finally some basic properties of m-fuzzy relations, also referring to operations, are briefly discussed. The possible genaralizations of classical set theoretical and logical relations as well as some specific m-fuzzy relations are considered.

1. THE CONCEPTION OF M FUZZINESS

Fuzzy sets (Zadeh, 1965) manifest a substantial step from (classical) sets towards concepts by abolishing the constraints imposed by the law of the excluded middle on category membership. This has generally been recognized in the cognitive sciences so long as membership itself is considered only. Observations on more intricate phenomena, e.g. conceptual combination (Osherson & Smith, 1981), or extension and intension (Cohen & Murphy, 1984), have resulted, however, in serious objections against the acceptance of fuzzy sets as a model of concepts. Although these considerations are criticized below, their conclusion will be shared as the result of basically different observations.

Having explained that the inherent logic of fuzzy sets is not that of concepts, a modified approach to "fuzziness", in the intuitive sense, is proposed for a mathematical model of the hierarchies of concepts.

1.1. Intuitive and mathematical senses of "fuzziness".

The contradiction between the inherent fuzziness of concepts and the tenuous utilization of fuzzy-set theory in the cognitive sciences is apparent only since "fuzziness" is meant in different senses in the two areas. Concepts are certainly fuzzy in the everyday sense of the word as a synonim of "vague", "ambiguous", or the adjectives in the quotation at the beginning of the present paper. On the other hand, fuzziness in fuzzy-set theory is a rigorously defined mathematical property, viz. the continuous variability of the characteristic function over the real interval [0,1]. Such incongruities often occur between the everyday and the mathematical senses of the same terms, e.g. game theory does not apply to (most) games, measure theory does not apply to (most) measures, probability theory does not apply to (most) probabilities, etc. Or in the other way round: most games, measures, probabilities, etc. meant in the everyday sense do not satisfy the respective definitions in the corresponding mathematical theories, i.e., they are not games, measures, probabilities, etc. in the sense of these theories.

This situation is well-known in the fields where mathematical techniques have traditionally been used. Accordingly, in these fields fuzzy-set theory has been applied to a large extent (Kandel, 1986). In the cognitive sciences, however, where mathematical methods are not common, the above two senses of "fuzziness " are not clearly distinguished. What is more, they are often confused not only with each other but also with further distinct ideas.

The most curious as well as most influential notion on fuzziness in the cognitive sciences is the one proposed by Osherson and Smith (1981). To start with, they identify the fuzzy characteristic function with category-representativeness (or typicality) as used in prototype theory (Mervis & Rosch, 1981). Certainly, both fuzzy-set theory and prototype theory address the nonequivalent status of category members, however, their respective approaches are diametrically opposite. One of their main differences is directly related to the above two senses of fuzziness: prototype theory has been basically informal, thus even if it used "fuzziness", that idea could not be a priori identified with the "fuzziness" in fuzzy-set theory. Furthermore, prototype theory does not use "fuzziness" in any sense. (Technically: Osherson and Smith (1981) failed to refer to any item in the literature which mentions both "prototype" and "fuzziness".) Another inherent inconsistency of the integrated "fuzzy typicality" originates from the fact that the characteristic function (whether fuzzy or not) is essentially an element-to-category relation while representativeness is concieved in prototype theory as an increasing, otherwise not specified function of the similarity between the given element and the prototypes of the given category, thus it is basically reduced to an element-to-element relation. (Incidentally, the Osherson and Smith (1981) notions allows only for one prototype in each category, which directly contradicts prototype theory itself.) Furthermore, not only element-to-element and element-to-category relations are confused in the "fuzzy typicality" but also category-to-category relations, since Osherson and Smith (1981) probe their notion on "conceptual combination". Here again they mix up the mathematical and intuitive approaches: they confront some formal expresssions taken from Zadeh's (1965) paper with their own "strong intuitions". What is the most curious in all this notion is that the actual "strong intuitions" rerpesent nothing but Aristotelian logic (viz. the classical AND, OR, and NOT operations, and INCLUSION relation). Thus the criteria for the adequacy of an amalgam of fuzzy-set theory and prototype theory is eventually the law of the excluded middle whose own inadequacy is the very starting point of both "constituent" theories.

In spite of the above shortcomings, Osherson and Smith's (1981) notion has exerted a heavy influence on subsequent approaches to fuzziness in the cognitive sciences (Roth & Mervis, 1983; Armstrong, Gleitman, & Gleitman, 1983; Johnson-Laird, 1983; Cohen & Murphy, 1984; Medin & Smith, 1984). This fact justifies a detailed analysis which is attempted in Fuhrmann (forthcoming d).

Cohen and Murphy (1984) seem to realize that Osherson and Smith's objections are due to their confusion of the intuitive and the mathematical approaches, however, they try to clarify the situation by getting rid of sets altogether whether fuzzy or not. Nevertheless, categories and thus also categorization can hardly be studied without considering sets since "set" in the classical sense used in the cognitive sciences, including Cohen and Murhpy's (1984) approach, is a synonim of "category" (this will be discussed in Section 1.2).

Further confusion about "fuzziness" in the cognitive sciences originates from its comparison to probabilistic modelling. In the intuitive sense the fuzzy and the probabilistic approaches can be identified since both deny the law of the excluded middle.

If, however, probability is mentioned without any specification, it should certainly be assumed in the sense of probability theory and in this mathematical sense, the two approaches are not identical at all: probability theory obeys the strict axioms of measure theory (Halmos, 1950) while fuzzy-set theory has not been developed to an axiomatic theory (in spite of axiomatizing attempts, e.g. Gougen, 1974), thus it can be used without any prescribed mathematical constraint. The two senses are often confused by both proponents and opponents of fuzziness in the cognitive sciences. For example, McClosky and Glucksberg (1979) propose "fuzziness" arguing that human categorization is probabilistic. On the other hand, for example, Johnson-Laird (1983) opposes "fuzziness" in favor of a probabilistic description while, however, he rejects probability theory as the means of calculating joint "probabilities". These and some other curious notions on fuzziness are discussed in more details in Fuhrmann (forthcoming a).

One of the fundamental differences between the intuitive and the mathematical senses of "fuzziness" lies in their respective novelty. That of the latter, i.e. of the generalization of set membership from dichotomous to continuous, is enormous. Indeed, since "set" is an essential idea in mathematics, the abolition of a stringent constraint (i.e. the law of the excluded middle) on set membership obviously provides for various generalizations in practically all fields of pure and applied mathematics. (This is by no means to say, however, that fuzzy sets mark a new era of mathematics. "Fuzzification" may be quite unimportant and uninteresting in many areas, nevetheless it is possible, which means a lot of work whether significant or not.) On the other hand, "set" in the intuitive sense is only a synonym of "category" and studies on both the inherent logic and the practical applications of categories have long been revealed as having graded membership (Rescher, 1969; Simpson, 1961, respectively).

Accordingly, to find the appropriate approach to sets is certainly a crucial point in an understanding of fuzzy sets and all the more for their possible utilization in modelling concepts (or categories). Cohen and Murphy (1984) apparently realized this fact, however, their proposition to omit sets altogether, both classical and fuzzy, cannot be accepted: "Set" has two aspects, one of which is inherently related to "category" (or "concept"). Since the two aspects are also inherently related (to each other), an analysis of their various relationships appears useful. This is attempted in the next section.

1.2 Concepts, sets, fuzzy sets

The question whether sets can be used as a model of concepts, as raised by Cohen and Murphy (1984), should not be seen as a starting point to discuss the relationship between concepts (or categories) and sets. In fact, sets were introduced as a model of concepts by Frege's Axiom of Comprehension (or of Abstraction):

$$\exists y \forall x [x \in y \leftrightarrow \phi(x)] \tag{1}$$

That is, given a predicate $\phi(x)$, there is a set y to which any given entity x belongs (as an element) if and only if $\phi(x)$ holds true. Evidently, this classical idea of sets differs from that of category or concept in denotation only. It is also evident that axiom (1) does not specify set membership apart from its reduction to the truth of predicates: If the latter is dichotomous, then so is the former, if the latter is graded or continuous, then so is the former. It is only a consequence of the general assumption of Aristotelian logic (particularly the law of the excluded middle) on predicates that set membership became also generally assumed to be dichotomous. The introduction of the classical, bivalued characteristic function was based on this mechanically adopted assumption:

$$\mu_y(x) = \begin{cases} 1, & \text{if } x \in y \\ 0, & \text{if } x \notin y \end{cases} \qquad (2)$$

Making use of the original meaning of set membership, expressed by equation (1), we may write:

$$\mu_v(u): U \otimes V \longrightarrow \{0,1\} \qquad (3)$$

where U and V denote the set of all (real or imaginary) entities and that of all possible predicates in any given (practically natural though possibly artificial) language. The characteristic function maps the Cartesian product of these sets (i.e. the set of pairs (u,v) where $u \in U$, $v \in V$) into the set $\{0,1\}$:

$$\mu_v(u) = 1 \leftrightarrow [\, v(u) \text{ holds true} \,].$$

Natural languages, however, are not free of contradictions, people often accept, or make, category statements in spite of being aware of counter examples (Hampton, 1982). Therefore classical (or "naive") set theory based eventually on natural languages cannot be contradiciton-free, either. The most famous contradiction is Russel's antinomy: let H denote the set of all sets which are not elements of themselves. (Evidently, a set may belong to itself, like the set of all sets defined in this paper, or may not do so, like the set of all people living at this moment.) Now, let us consider if H belongs to itself. It is immediately clear that it belongs to H if and only if it does not do so. To illustrate that the use of mathematical ideas (set, set membeship) plays no role here, let us cite another, called Grelling's antinomy: Some English adjectives have the property that they denote, e.g. "English", "polysyllabic", while others do not, e.g. "French", "monosyllabic". Calling the latters heterological, it is evident that "heterological" is heterological if and only if it is not heterological.

A mathematical theory cannot lead to contradictions, therefore the discovery of antinomies in naive set theory resulted in the rejection of axiom (1). In the new version, called axiomatic set theory, axiom (1) is replaced by the following Theorem Schema of Separation (or of Subset):

$$\forall z \exists y \forall x [x \in y \leftrightarrow x \in z \land \phi(x)]. \qquad (4)$$

That is, the statement in Equation 1 does not apply to every entity, only to those belonging to a discretionary set z. What makes a collection of entities a set is uniquely and contradiction-freely defined in axiomatic set theory, see e.g., Fraenkel, Bar-Hillel and Levy (1973). What is important here is only that a set cannot belong to itself (nor to a set which belongs to a set ... which belongs to the original one) any more. The collections which do not satisfy the new axioms thus are not sets in the new, more rigorous sense, and are termed proper classes. (The term "class" was introduced to replace "set" in the original, naive sense. Thus each set, in the sense of axiomatic set theory, is a class as well, and classes which are not sets are proper classes.)

The above U does belong to itself. Thus, in light of axiomatic set theory, it cannot be used to define the characteristic function of sets like it used to in Equation 3. It has to be replaced by a set z:

$$\mu_y(x) : z \otimes V \{0,1\} \qquad (5)$$

Evidently, this is the point where concepts and sets became separated. After this point no operation or generalization possibly leading to contradiciton is allowed for sets. Concepts, however do lead to contradictions, thus if sets are to be drawn closer to concepts, it should be done before Equations 1 an 3 are replaced by Equations 4 and 5.

The definition of fuzzy sets does not do so. It generalizes the values in Euations 4 and 5 by allowing for fractions in the truth of predicate $\phi(x)$ and in the characteristic function $\mu_y(x)$ while, however, it maintains the restricitons on the domain, viz. it rests on sets. This is the formal basis of the widespread use of fuzzy sets in mathematical fields and is of little use in the cognitive sciences. Equation 4 is a fundamental point in ("real", i.e. contradiction-free) mathematics, thus any of its genaralization may have consequences in all fields using mathematics (as noted in Section 1.1). Equation 4 is in fact the definition of subsets, this is the root of the notion that fuzzy sets are generalized subsets (Kaufmann, 1975). On the other hand, since Equation 4 lies within the border of ("real") mathematics, any generalization it allows, including "fuzzification", lies there too. Consequently, it can not bridge the gap between sets and concepts.

The next section attempts to bridge this gap by generalzing not only the values, but also the domain in Equations 4 and 5. Thus, conceptually we turn back to Equatin 1 and formally introduce continuous variability not only in Equations 4 and 5 but also in Equations 1 - 3.

1.3 M-fuzzy sets

As we have seen, no mathematical model is possible for concepts in their genrality since such a mathematical model should be free of contradictions, while concepts are exposed to produce contradictions. Accordingly, it is a reasonable aim to define a model addressing concepts in general and being mathematical to the greates possible extent.

To approach this aim, let us first define a fuzzy rerpesentation R by a generalization of Equation 3:

$$R: W \to [0,1]$$
$$W \subseteq U \otimes V \qquad (6)$$

Here, as above, U and V denote the classes of all (real or imaginary) entities and of all possible predicates (in a given language), repectively. The elements of a fuzzy representation are triplets: $a = (u,v, \mu) \in R$; where $u \in U$, $v \in V$, $\mu \in [0,1]$, such that μ denotes the degree to which the predicate v holds true for the entity u.

At the second step, let us replace the set z by the fuzzy representation R in Equation 4:

$$\forall R \exists A \forall a [a \in A\, a \in R \land \Psi(a)] \qquad (7)$$

That is, if $a = (u,v, \mu) \in R$, where R denotes a fuzzy representation, and Ψ (a) denotes a predicate, then there is a class A such that a A iff Ψ (a) holds true.

Let A^+ denote the subclass, called the core, of A consisting of its elements with full membership:

$$A^+ = \{a = (u,v, \mu) \mid a \in A\ \mu = 1\} \qquad (8)$$

The class A is called a modified fuzzy set (m-fuzzy set, for short) if and only if its core A^+ is a set. (It is not necessary to explicitly prescribe that A^+ should be a set, in a mathematical treatment (Fuhrmann, forthcoming c) it can be obtained as a theorem from a more general definition of m-fuzzy sets.)

In this general way an m-fuzzy set connects a concept (a class) which, in general, does not satisfy the axioms imposed on sets, with a set (its core), which can be the subject of a mathematical (thus necessarily contradiciton-free) theory. Since the core, in a sense, represents the underlying class, such a theory may have inherent relations with concepts in their generality.

The condition that A^+ should be a set connects the intuitive and the mathematical senses of membership. The full membership in the intuitive sense does not necessarily require maximal membership in the mathematical sense. Formally, the former means $\mu_y(x)$ 1 while the latter means exact equality: $\mu_y(x) = 1$. Thus the above restriction is fairly mild in its (mathematical) sense: Let us suppose that an element has practically full membership while its inclusion in the core would ren der the core not to be a set. Then a value of $\mu = 1 - \epsilon$ can be attributed to the degree of its membership such that ϵ should be less than any prescribed positive threshold, however, greater than zero.

It is clear that m-fuzzy sets are a generalization of (traditional) fuzzy sets. Let A_T denote a special m-fuzzy set whose domain of entities:

$$U_T = \{u \mid \exists v, \mu : a = (u, v, \mu) \in A_T\} \qquad (9)$$

is a set, and for which the predicate $\Psi_T(a)$ defined in Equation 7 requires only that its domain of predicates:

$$V_T = \{v \mid \exists u, \mu : a = (u, v, \mu) \in A_T\} \qquad (10)$$

should contain a single element only, say v*:

$$\Psi_T(a) = (u, v, \mu) \leftrightarrow v = v^* \qquad (11)$$

Evidently, A_T is a (traditional) fuzzy set defined on the set U_T by the continuous valued application (μ) of the predicted v*.

1.4 M-fuzzy sets and concepts

The principal target of m-fuzzines is the hierachical organization of concepts. Since the same phenomenon led to prototype theory (Mervis & Rosch, 1981), a comparison of the two approaches may be fruitful. As mentioned in Section 1.1., the central idea in prototype theory is not membership but representativeness and most criticism of this theory rest on a confusion of membership and representativeness (or typicality). Thus, a discussion on the relation between these phenomena appears to be an appropriate starting point.

The tradition of identifying membership with representativeness is extremly heavy among the opponents of prototype theory. For example, Armstrong, Gleitman and Gleitman (1983) insist on it in spite of their own experiments where the subjects responded quite differently when asked about membership and representativeness, respectively. This tradition has some terminological preliminaries. Prototype theory deals principally with representativeness and most key publications in its development (Posner & Keele, 1968; Reed, 1972; Rosch, 1975a,b; Rosch & Lloyd, 1978; Mervis & Rosch, 1981) use strictly this term. Others use the term "typicality" though in a purely technical way defined clearly in each experiment (Rosch, Simpson, & Miller, 1976). In contrast, criticisms of prototype theory (Loftus, 1975; Osherson & Smith, 1981) use "typicality" as a central term without clarifying its meaning.

"Representativeness" and "typicality" seem to have a difference which may be crucial in their relation to "membership". Within the limits of the intuitive approach, where both are used, "representativeness" implies more definitely an underlying existence of membership than "typicality" does. For example, a "typical Swede" is tall and has light hair and skin, blue eyes, an athletic build and Scandinavian attitudes. What make someone a "Swede", however, are the place of birth, nationality of parents, citizenship, and residence. Thus, it may be a true characterization for someone that "he is a typical Swede except he is not Swede at all".

Accordingly, in a mathematical approach "membership", "representativeness", and "typicality" cannot be described with a single variable. The above example also suggests that typicality can be seem as membership though not in the original category, say "A" but in an other one, viz. "typical A". Thus in the application of m-fuzzy sets to concepts we should primarily address "membership" while regarding "typicality" as membership in another category and "representativeness" as a distinct kind of variable.

Assuming this explanation to be valid, m-fuzzy sets appear to be in accordance with prototype theory in that the domain of both typicality and membership is an unrestricted concept (a class in th sense of set theory), and both attribute continuous valued membership to the category elements, which also specifies a core consisting of those elements with maximal membership. (Although "maximal" is meant intuitively in prototype theory and mathematically in m-fuzzy sets, at this point these senses are easy to connect, as noted in Section 1.3)

What has not been connected from prototype theory to m-fuzzy sets are the prototypes themselves. For this, let us consider the third of the above terms, "representativeness". The studies on representativeness that led to prototype theory revealed no n-equivalence not only between the elements in a given category, but vice versa; between the categories comprising a given element. Just like the maximally representative elements for a given category, called prototypes, there is a maximally representative category for a given element, called basic level category (Mervis & Rosch, 1981). This indicates a symmetric aspect of the representativeness between elements and categories. Indeed, considering some prototypes themselves, it can be immediately seen that most of their features can best be characterized by referring to the category in which they are prototypes. For example, the shape, the color, the size, as well as the taste, the touch, and the smell of a prototypical apple can be simply and precisely characterized as "(just) apple-like".

Thus, we consider both the representativeness of an element for a category and that of a category for an element. The former can be identified with membership or, at least, assumed to be proportional with it. The latter can be assumed in the above sense, i.e. how efficient the membership in the given category is in characterizing different features of the given element. In this setting prototypes are defined as the elements holding maximal mutual representativeness, i.e. maximizing both of the above variables. Since these are defined over two arguments each, the element p is a prototype in the category A if and only if two double criteria are simultaneously met: 1. p is maximally representative for A, i.e. α. there is no element more representative for A than p is, and β. there is no category for which p is more representative than for A. 2. A is maximally representative for p, i.e. α. there is no category more representative for p than A is, and β. there is no element for which A is more representative than for p.

Even without a rigorous proof, the collection of prototypes can be assumed to be finite for each category. R regarding the categories like "A" and "typical A" as different (as explained above), we also may assume the collections of prototypes to be contradiciton-free, i.e. to be sets. On this basis, the finite subsets of prototypes denoted by A^p is defined within the core A^+ of an m-fuzzy set A as:

$$A^p = \{a = (u,v,\mu) \mid a \in A^+ \wedge \Pi(a)\} \qquad (12)$$

Here a A^+ guarantees the maximal representativeness of p for A (in the sense explained above) while the predicate $\Pi(a)$ denotes the maximal representativeness of A for p.

The role of A^p in representating the actual m-fuzzy set and, accordingly, also the underlying concept, is fundamental. The core A^+ cannot play this role, in general, since it is not necessarily finite or even enumerable. Furthermore, the elements of the core are not equally representative in the everyday, i.e., the mutual sense (though their membership is equal and maximal, and thus, according to the above assumption, their representativeness is "one-sided"). Therefore a randomly chosen subset of the core may be irrelevant or even misleading in characterizing (or representing) the underlying m-fuzzy set (or concept).

Based on all this, a formal scheme of recursive representation is proposed for modelling the internal hierarchy of concepts. The basic domain is an unrestricted concept which corresponds to a class in the mathematical (more percisely the set theoretical) sense. An m-fuzzy set transforms it into a category furnished with a continuous valued membership function. The core of this m-fuzzy set selects those elements posessing two important properties: holding the maximal degree of membership, and forming a set. Finally, the finite subset of prototypes provides a maximally representative (in the appropriate, mutual sense, as explained above) and conveniently treatable collection within the core. In this enumeration each element is to represent the preceding one, and thus indirectly all of the ones before it.

A fundamental property of the above recursive scheme of representation is that it localizes the possible occurrence of contradictions. As explained in Section 1.2., such a possibility cannot be ruled out in the logic of concepts in general (Hampton, 1982). In the above scheme, however, contradictions may occur in the first two items only (classes and m-fuzzy sets) while they really are ruled out from the other two (sets and finite subsets). Thus, concept cores and prototypes can feasibly be the subject of mathematical theories in which concepts in their generality can also be directly involved as represented by these two ideas. (The relation between prototypes and m-fuzziness is discussed more in details in Fuhrmann, forthcoming b.)

1.5 M-fuzzy grammars and the hierarchy of concepts

As noted in the preceding section, not only the elements in a given category but also the categories comprising a given element have been found unequal. Like the prototypes among the former, the basic level among the latter has a distinguished, viz. the most representative, status (Mervis & Rosch, 1981). Having reviewed the m-fuzzy approach to the intra-categorial hierarchies, let us turn to the inter-categorial ones. Since the latter rest on the relations of a given entity (or element) to different categories, a study on this hierarchy should certainly involve the act of categorization.

"Categorization" or "category identification" as used in the cognitive sciences is very close in meaning to "pattern recognition" as used in Artificial Intelligence. Thus, methods that have proved fruitful in the latter may be utilized in modelling the former. Pattern recognition has basically two apaproaches. Statistical pattern recognition (Chen, 1973) regards the entities (or patterns) as points in a metric space of the underlying features. Here the categories are manifest in clusters within this space. Although the implementation of statistical pattern recognition may require several steps (feature extraction, determination of the optimal metric, analysis of topological relations, learning the categories, etc.), the actual recognition takes place principally in a single step: a given point in the (original or transformed) space of features is attached to one (possibly to a few to none) of the clusters. The methods of statistical pattern recognition have also been applied to the psychic process of categorization (Reed, 1973) yet the other kind of methods where the act of recognition itself is, in general, inherently hierarchical seems more adequate in this field.

The latter approach originates from the idae of formal languages, therefore it is called syntetctic pattern recognition (Fu, 1982). A traditional formal language is based on a generative grammar $G = (V_N, V_T, P, S)$, where V_N and V_T D are the alphabet of nonterminal and terminal symbols, respectively, P is a set of production rules, and S is a distinguished element of V_N called "sentence" symbol. Each production rule connects two strings of symbols by an arrow such that each symbol belongs either to V_N or V_T and the arrow denotes that the string before it can be freely replaced by the one after it. The words of the underlying language are those strings consisting of terminal symbols only and possibly produced by successive applications of the rules in P beginning with S. When applied to pattern recognition, S corresponds to a category, V_T consists of the features, and an entity (or pattern) is regarded as the word (i.e string) of its features. Thus a pattern belongs to a given category if and only if the word representing it belongs to the corresponding language.

This conceptual framework appears suitable for modelling concepts: an entity (or pattern) belongs to a category if and only if it can be "produced" by appropriate rules of a logical nature operating, feasibly in parallel, on confined information (selection of features).

There are, however, also some important differences between the two fields. In formalized pattern recognition, different categories (different grammars and languages) are considered as separate (either serial or parallel) processes. Futhermore the categories are usually preclusive in the practical sense, i.e. an entity (or pattern) belongs to one category only. In the psychic case, however, many categories are inclusive (in the intuitive sense, not confined to any formal operation). Thus the entity under consideration, in general, belongs to several categories and the task is to identify not one (i.e. any one) nor all ones but the optimal one in a sense to be specified (below).

Accordingly, all categories ("sentence" symbols) should appear as words in the same formal language. Such a situation can be achieved by abolishing the traditional distinction between nonterminal and terminal symbols. Grammars and languages where this distinction is not made are known as pure ones (Maurer, Salomaa, & Wood, 1980; Gabrielian, 1981). Here each string generated by the rules belongs to the underlying language thus the production of any string from any other can be considered as principally equal.

The appropriate pure alphabet for the proposed model is a fuzzy representation R. The production rules may connect any pair of strings just like in pure grammars in general. This expresses in our case that any concept (or category) may occur as (a part of) any step in the chain of production, including the beginning and the concluding steps. In other words, any concept can be regarded in itself, or as a constituent of another concept, or as a compound of other concepts. Possible "sentence" symbols are all of those elements in R belonging to an m-fuzzy set, i.e. satisfying a predicate Ψ (a) which defines an m-fuzzy set (according to Equation 7). Thus formally, an m-fuzzy grammar is the unification of as many pure grammars as is the number of such elements in R. The alphabets, as well as the sets of production rules in these "constituent" grammars, are parts or the whole of those in the resulting m-fuzzy grammar. The strength, or reliability, of the recognition (or category identification) is shown by the third element (μ) in the actual "sentence" symbol a = (u,v, μ) $\in$ R.

The production rules decompose the members of the alphabet (individual or in strings) either logically, e.g.

$$(u, woman, \mu) \rightarrow \qquad (13)$$
$$(u, human, \mu_1), (u, female, \mu_2), (u, adult, \mu_3);$$
$$\mu = \min(\mu_1, \mu_2, \mu_3) \; \mu_1, \mu_2, \mu_3 / 3$$

or physically, e.g.

$$(u, triangle, \mu) \rightarrow \qquad (14)$$
$$(3 \text{ staight lines, pairwise intersecting, such as } \mu)$$

It is fundamental in the proposed approach that a fuzzy representation, thus also an m-fuzzy grammar, refer to one, definite mental representation (or a "subjective lexicon"). For example, the production rule:

$$(u, woman, 1) \rightarrow (u, \text{capable of giving birth}, 1) \qquad (15)$$

may be either included or excluded depending on personal factors.

Instead of the organizing principle lost by not distinguishing nonterminal or terminal symbols, another is offered by the structure of subordinate /

superordinate categories. According to this relation the categories within a given m-fuzzy grammar can be ordered into series. It may be useful in organizing the acts of production (or of recognition) to assume that categories related as subordinate/superordinate may occur at different sides of a production rule only if they are adjacent in the corresponding series. Thus the rewriting (i.e. the steps in the production and recognition) becomes regulated, which is familiar in the theory of formal languages (Salomaa, 1973).

It seems necessary to propose explicitely two basic assumptions about the subordinate/superordinate relation, noted above. One is that this relation may vary over different m-fuzzy grammars (reffering to different mental representations). The other is that even within a single m-fuzzy grammar this relation is intuitive (a special case of "conceptual inclusion"). Consequently, this relation, according to the proposed approach, cannot be confined to any of the accustomed formal operations, e.g. inclusion as used in Aristotelian logic and in set theory, or any of its specific generalization to (traditional) fuzzy-set theory. Thus, none of the usual operations (addition, multiplication, "min" and "max" rules, etc.) applies universally to the calculation of strengths (or reliabilities) between the two sides of an m-fuzzy production rule (e.g. Equation 13).

Let us return to the question of "optimal" category, i.e., if a pattern (a sring) can be produced from different "sentence" symbols within an m-fuzzy grammar (it belongs to different categories), as is the case in general, then which of these should be distinguished as "optimal". (Obviously this "optimal" category should also be identified as the result of the corresponding act of recognition.) It is plausible to choose the category into which the given pattern "best fits". Here again we are confronted by the incongruity of intuitive and formal senses. The first obvious idea to formalize the "best fit" is certainly maximal membership. This, however, cannot be accepted since, in general, several categories can be found for a given pattern comprising it at the same, maximal degree of membership. Another choice is the "most informative" category. This can reasonably be identified with the choice maximizing the "mutual representativeness" as defined in Section 1.4. The latter includes "maximal membership" thus it does not contradict the "best fit" choice but renders it more rigorous (maximal membership becomes only a necessary but not sufficient condition). This, more stringent condition also prevents the miscategorization of elements belonging to a category "typical A" only, to the category "A" (as exemplified in Section 1.4. by "typical Swede" and "Swede").

2. POSSIBLE IMPLEMENTATION IN THE BRAIN

An essential property of syntactic pattern recognition, as mentioned in Section 1.5, is that the rules of rewriting do not involve all symbols in the actual string. This is crucial in two respects: One is that the rules operate on confined information only, thus their implementation requires relatively simple devices. The other is that these devices may work largely in parallel on different selections of symbols in the same string.

Although various techniques have been developed to implement syntactic pattern recognition (Fu, 1982), these hardly utilize the above property. This is not surprising since the means of implementation, the traditional computer, is inherently a serial device. Nevertheless, the situation is quite the opposite in the brain where the psychic act of category identification presumably takes place.

The idea that a concept should be assumed as represented by a collective induced state of a group, or an assembly of neurons was proposed by Hebb (1949). From another approach, Piaget (1962) also argued that psychological phenomena should be explained in terms referring to neural activity while he also emphasized the possibly crucial role of abstract, mathematical consideration in such explanations.

The above abstract (viz. m-fuzzy) approach to concepts can feasibly be coupled to some fundamental new results on neural functioning and brain organization. This is explained as follows.

2.1. On the organization and communication in the brain

Hebb (1949) assumed his "assemblies" of neurons to be self-organized by synaptic communication. That is, 1. The assemblies have no well-defined boundaries, individual neurons may either join (e.g. by developing new synapses) or drop (e.g. by misfunction); 2. The sole cohesive agent is classical, i.e. digital, synaptic interaction: each cell in the assembly has postsynaptic sites (i.e. may receive synaptic input) from one or more other cell(s) in the assembly, with a sufficiently low threshold. This also means that the cells have two possible states only: steady state (in the absence of incoming signals) and excited state. Consequently: 3. The assembly as a whole has also two possible states only. Since the occurrence of a concept is attributed to the excited state of a cell-assembly, concepts also obey, according to this approach, classical, bivalued logic, i.e. they either do or do not occur without any intermediate (or fuzzy) possibility.

Some fundamental results of the neurosciences obtained in recent decades suggest a thorough revision of the above ideas. Concerning the functional organization of neurons, the existence of well-defined units, called neuron-modules, has been firmly established (Szentágothai, 1975, 1978; Mountcastle, 1978) in brain sections involved in detailed information procesing (as opposed to general regulatory functions), i.e., where the manifestation of concepts seems most feasible. Accepting Hebb's (1949) idea on the direct relation between concepts and excited states of neural structures, which is certainly justified in any monistic view, it is plausible to identify his cell-assemblies with the experimentally revealed neuron-modules (Fuhrmann, 1985c).

Concerning interneuronal communication, the situation, in a sense, is just the opposite. The possible variety of the functional units is severely reduced by replacing Hebb's (1949) loose cell-assemblies by the much more definite neuron-modules. In contrast, the mode of interaction between neurons has been revealed to be much more variable than assumed earlier. The simple, digital synaptic contact is now generally seen as only one kind of interneuronal communication. It has been firmly established that the synapses can be modulated (Karczmar, 1987), furthermore that neurons can also interact without synapses (Vizi, 1984). Interneuronal connections can also be categorized according to the chemical nature of the involved compounds (Iversen, 1984). Furthermore, even "simple", classical synaptic signals may interact in complex, nonlinear ways (Koch, Poggio, & Torre, 1983).

The organization of information processing in single neurons turned out to be as complex and sophisticated as previously thought about networks of neurons only (Lewis, 1983). This also implies that the neurons, and all the more the structures of neurons, e.g., the modules, have various induced (or excited) states too. Different kinds of induced states in the modules (or assemblies) may refer to different kinds of concepts. It seems plausible that some collective induced states are known to the system (the actually involved modules) from experience, these may represent clearly recognized concepts. Other such states, possibly similar to several known ones may represent ambiguous (or vague, or fuzzy) concepts.

Since the collective induced states develop on the basis of local (interneuronal) communication, various concepts appearing in various parts of the same system (the same brain) may not be logically consistent. Accordingly, the organization of the concepts represented may not be adequately described by any mathematical (i.e. contradicition-free) theory while the setting of m-fuzziness may be useful. (Incidentally, the surge of fuzzification also penetrated into the mathematical modelling of neural activity, viz. a fuzzification of the classical (McCulloch-Pitts type) formal neuron was proposed (Lee & Lee, 1974). In this field, however, the (traditional) fuzzy approach has been even less successful than in the cognitive sciences.)

2.2 Neuronal interactions, states, and encoding

To develop a formal description of the collective induced states of the neuron-modules (the "modular states") which can feasibly be assumed as the manifestation of concepts, let us first consider the neurons, the elementary units in the underlying system. Experimental studies on neuronal functions form a rapidly growing field of research. As noted in the preceding section, the number of revealed types of interneuronal communication is also growing. For the time being it seems justified to divide the set types into two groups: classical and nonclassical, in the following way (Fuhrmann, 1982). The former is assumed as fast and not entirely reliable, which involves an individual threshold function at each cell. The latter is assumed as slower and not only reliable but also capable of correcting errors made by the former. This, latter one. involves a collective function of the cells connected by neighbouring relations. This function operates not only on the actual input but also on previous experience. In terms of information processing, the former, called σ-function, is the encoding while the latter, called φ-function, is the evaluation of the encoded information also making use of memory.

According to this double function the model-neurons in the proposed system have two kinds of input and output, denoted by $I\sigma$, $I\varphi$ and $O\sigma$, $O\varphi$, respectively. Since $I\sigma$ and $O\varphi$ belong to the "first" (σ) kind of communication, they refer to classical synaptic signals, i.e. close sequences of uniform impulses ("spikes"). $O\sigma$ also belongs to this type of classical synaptic signal because it will serve as a constituent of $I\sigma$ at the next stage in information processing. It is $I\varphi$ only whose manifestation has not been specified uniquely. This, as mentioned above, is assumed as a "nonclassical" complex of modulated synaptic, nonsynaptic, and possibly other constituents.

In the model, each neuron has a finite ordered set of (individual) states denoted by $<0>$, $<1>$, ..., $<m-1>$, where the value of m refers to the individual thermodynamic properties of the actual cell. The steady state is denoted by $<0>$, and the larger the number in the bracket,, the more distant the underlying state from the steady one. In formal functioning each neuron is in its steady state when a signal of I may arrive. Since the impulses (spikes) are uniform, the reception of each elevates the cell's state one step farther from the steady state. Thus, having received s spikes of $I\sigma$, a neuron is in its s-th state, i.e. in $<s>$ provided the $s<m$. In general, however, the number of spikes, say n, in a signal of $I\sigma$ exceeds (m-1) for the receiving cells, i.e., the stepwise departure from the steady state cannot continue until all spikes have arrived. (This refers to the thermodynamic constraints on the possible excitation.) If a neuron is in its highest possible excited state, m-1, when another spike in $I\sigma$ arrives, the cell fires a spike of $O\sigma$ and thus resets its steady state $<0>$. Then it continues the reception of the impulses. This act of resetting,

$<m-1> \to <0>$, may take place several times during the reception of $I\sigma$. Accordingly, after all n spikes of $I\sigma$ have arrived, the cell is in its k-th state, $<k>$, such that k is the residue number of the division of n by m as a modulus, i.e.,

$$k = n - [n/m] \cdot m \qquad (16)$$

where [ ] denotes integer part.

Although the above process (σ-function) is accomplished by the actual neuron individually, the cells are not isolated. As mentioned in Section 2.1, the basic functional unit in the model is the (neuron) module. Most cells in a module receive identical σ - input ($I\sigma$). Owing to their different values of modulus (m), referring to their different thermodynamic properties, the reception of the same number of $I\sigma$ spikes (n) brings them into states of different serial numbers in their respective scales of states. These neurons, called receptor cells, send their $O\sigma$ signals to each of the other neurons, called control cells, in the same module. In this way, after all spikes in an actual signal of $I\sigma$ have arrived at the receptor cells, the actual (individual) state of each neuron in the module reflects a part of the information received by the module.

For the recpetor cells, this part of the information is simply the residue of the number of received spikes after division by the cell's modulus. From a collection of such residue numbers the original integer can be uniquely reconstructed provided th at it does not exceed the least common multiple of the moduli (Ore, 1952). Thus the system of residue numbers (as states of the receptor cells) reflects all the information arrived at the module in a distributed form whose constituents have been generated in parallel. (As mentioned above, the spikes are uniform, thus it is evident that only their number conveys information. The role of control cells lies in error correction outlined below.)

The residue numbers can be not only generated but also processed in parallel so far as addition, subtraction, and multiplication are involved only:

$$|X_1 \pm X_2|_{\mod m} = ||X_1|_{\mod m} \pm |X_2|_{\mod m}|_{\mod m}$$

$$|X_1 * X_2|_{\mod m} = ||X_1|_{\mod m} * |X_2|_{\mod m}|_{\mod m} \qquad (17)$$

These are just the operates which can most easily be implemented by chemical concentrations or electrical potentials, i.e., by the most common agents in the nervous system.

Although this parallel, distributed mode of information storage and processing may be fruitful in the computer sciences, too, most studies on residue number systems deal with the very special case of pairwise relative prime moduli only (Szabó & Tanaka, 1967; Fuhrmann, 1986). This restriction is extremely strict not only on the choice of moduli but, consequently, also on possible error correction in the system. Concerning the moduli themselves, their presumable number in a feasible useful residue number system is only a couple of hundred or a few thousand. The requirement that no pair among so many integers should have any common divisor seems impracticable even for artificial systems. It is all the more so for the formalized neuron module outlined above since here, according to the model, the moduli reflect the individual thermodynamic properties of the cells which evolved largely independently of each other (i.e. how could and why would they avoid having any common divisor with any of their fellows in the module).

Concerning error detection, identification, and correction, these faculties originate from redundancy which is inherent in a residue number system only in case of shared divisors: If the moduli are pairwise relative primes, then each possible collection of residue numbers corresponds to an integer, i.e. forms an admitted code-word. Consequently, redundancy can be introduced only by attaching some extra variables to the system. Most classes of error correcting codes defined on residue number system. Most classes of error correcting codes defined on residue number systems follow this line (these are reviewed in Furhmann, 1986). If, however, the moduli share some divisors, then redundancy is inherent in the system, e.g. there is no integer X for which $|X|_{\mod 6} = 2$, while $|X|_{\mod 4} = 1$ (the former renders X to be even while the latter odd). Conditions for error detection, identification, and correction on this basis have also been developed for single errors (Barsi & Maestrini, 1980) and

belong either to the same or to different modules. The evaluation of the actual states of neighbours (Iφ) determines the φ-output (Oφ) of the considered cell. This output is a close sequence of uniform impulses (spikes) such that the number of spikes shows how similar the actual state-configuration is to such configurations known to the considered neuron from experience.

The similarity of state-configurations depends on the difference between the respective neuronal states as well as on the strength of the involved neighbouring relations. Formally, to each cell there belongs an antisymmetric matrix δ such that the absolute value of its element δ_{ij} shows the distance between the i-th and the j-th states (i.e. between $<i>$ and $<j>$) for the actual cell. Thus the cells' individual scales of states are nonlinear ($|\delta_{i,j+k}| = |\delta_{ij}| + |\delta_{jk}|$, in general, does not hold) reflecting that the impact of the same spike on the thermodynamic milieu even in a given cell varies with the distance from the equilibrium (the steady state). The number of elements in δ is evidently m*m, where m is the modulus belonging to the actual neuron. Also to each cell, there belongs a vector γ whose element γ_α shows the strength of coupling between the considered neuron and its α-th neighbour (each cell is its own 0-th neighbour). The numbers of both the states and the neighbours are assumed to be finite for every neuron. Accordingly, the number of state-configurations is also finite for any given neuron, thus they can be ordered discretionarily and referred to by their finite serial number. In such a (fixed) series the distance between the p-th and q-th state-configurations is calculated as:

$$\zeta_{pq} = \Sigma \gamma_\alpha \; \delta_\alpha (p,q) \qquad (18)$$

where the summation covers all neighbours, and $\delta_\alpha {(p,q)}$ denotes the distance between the two states of the α-th neighbour wich belong to the p-th and the q-th state-configuration, respectively.

According to the proposed model, some state-configurations are known, assumed to be learned, others are unknown for the given neuron (as noted in Section 2.1). If the actual state-configuration is known, the φ-output of the cell (i.e. the number of spikes in Oφ) simply identifies it. If it is unknown, Oφ reflects how similar it is to the known ones. In this aspect the proposed model is fairly congruous with prototype theory: Not all cases are presented in the same way but there are distinguished ones, distinguished by their characteristic of being known, while other cases are represented by their similarity to these ones.

Let $<p|$ denote the p-th state-configuration if it is unknown (to the actual neuron) and $|q>$ denote the q-th one if it is known. Then the similarity between $<p|$ and $|q>$ around the cell C is calculated as:

$$<p|C|q> \;=\; 1 - |\,\zeta(p,q)\,| \,/\, \zeta(p) \qquad (19)$$

where $\zeta(p)$ denotes the maximum of $\zeta(p,r)$ over all $|r>$'s.

As immediately seen from Equation 19, the degree of similarity between an unknown and a known state-configuration is confined to the interval [0,1]. Making use of this propoerty an m-fuzzy set P is defined to represent the unknown state-configurati on $<p|$ in the functional sense: P has as many elements as the number of known state-configurations around the given cell C is, and its elements are triplets: $(<p|,\,\text{like}\,|r>,\,<p|C|r>)$. This P belongs to a special type of m-fuzzy sets, in one respect similar to (traditional) fuzzy sets while in another respect contrary to those: As we have seen, in the case of (traditional) fuzzy sets the underlying (m-fuzzy) domain of predicates (Equation 10) consists of a single element only. In the above special case of m-fuzzy sets, e.g. in P, the other domain, that of the entities (Equation 9) is confined to a single element (to $<p|$ in the above example).

Making use of the similarity to all known state-configurations we can characterise how fuzzy (in the intuitive sense) a given unknown state-configuration $<p|$ is. Formally, for the above type of m-fuzzy sets the degree of m-fuzziness is defined as:

$$d_p = \frac{\sum_{|r>} <p|C|r>}{\max_{|r>} <p|C|r>} \qquad (20)$$

Thus, the larger the number of known state-configurations to which $<p|$ is similar to about the same extent, the greater the m-fuzziness of $<p|$, more precisely of the underlying special m-fuzzy set P. (For traditional fuzzy sets there have been various "measures" of fuzziness proposed (Loo, 1977; Ebanks, 1983). Most of these share two properties: They explicitly use some "fuzzy complement", and are not measures in the mathematical sense (Halmos, 1950). Owing to the intricate role of contradictions, accordingly also of "negation" or "complementation" in the logic of concepts (noted in Section 1.2 and also discussed in Section 3.1), no universal operation called "negation" or "complementation" has been proposed on m-fuzzy sets. Formal characterization of m-fuzziness has been defined for the above special type only, and this is called a "degree" to avoid possible confusions with measure theory.)

Equation 20 also offers a characterization of the underlying neural/mental representation. As seen, even the known state-configurations are not free of m-fuzziness except that they are "orthogonal" in the sense that they are not similar to each other according to the definition in Equation 19. Thus, the more "orthogonal" (i.e. the less m-fuzzy) the known state-configuration throughout the network, the clearer the underlying neural/mental representation.

The degrees of similarity and m-fuzziness (Equation 19 and 20) are fundamental in the case of state-configurations, since, as mentioned above, these determinate the φ - output (O_φ) of the considered neuron, i.e. the evaluation of the encoded information at the given stage of processing. What is more, however, these expressions also apply to the modular states which, as also mentioned above, directly correspond to the concepts represented in the system.

The modular states develop on the basis of the state-configurations. A modular state is called known if and only if each neuron in the given module is in a known intramodular state-configuration. The latter involves only those neighbours belonging to the same module. An intramodular state-configuration is called known if and only if there exists (whether actually occurs or not) a known state-configuration (with respect to all neighbours) that comprises it. In this way no direct intermodular condition applies to a modular state for being known, however, intermodular organization still plays a role in determining whether the actual intramodular state-configurations are known. This reflects that the modules, as well as the concepts represented by their states, are self-contained though interconnected units.

The above properties suggest, and formal theorems (Fuhrmann, 1985c) corroborate, that the organization plays the crucial role in the model. That is, the relations (residue, distance, similarity, etc.) are much more important in determinig the emergent (collective) phenomena than the absolute number and values of the variables (number of cells, states, neighbours, etc.). This property is also generally assumed to characterize real neural and mental organization.

2.4 The act of recognition

According to the double function attributed to the neuron, there are two kinds of module-to-module communication in the model: through φ-pathways (bundles of axons of neurons) or φ- pathways (interneuronal neighbouring relations).

The modules connected by δ-pathways work successively (as mentioned above, the network is synchronous, the neurons in a module work in parallel with each other as well as with their intermodular neighbours). The output of a module is the sum of the O φ signals from all of its cells, and this contributes to the input of one or more other module(s), i.e. to the I signals of all receptor cells in the latter(s). The bundles of axons conveying these signals may branch or fuse, thus the output of a module is not necessarily identical with the input of one or more other modules. Accordingly, the modules receiving their input synchronously from one or more module(s) in one definite collection work in parralel. They are, in general, connected by φ-pathways (more percisely they may be connected by φ-pathways while they cannot be connected by φ ones), thus they also work collectively. Such a (φ-connected) collection of modules forms a higher unit called a functional layer. These units correspond to the levels in the underlying mental representation, thus they also constitute the stages in the act of recognition.

Formally, the actual states of modules in a functional layer constitute a word in the underlying m-fuzzy language. Accordingly, the syntactic rules of the corresponding m-fuzzy grammar connect adjacent functional layers.

This also defines how the m-fuzzy grammars should be applied to the considered type of recognition. As explained in Section 1.5, here the task is not to decide whether a given word belongs to a given language as in the case of artificial systems but to select the language to which it "optimally" belongs. In the former case the recognizer (human or machine) chooses from the rules in the given grammar in trying to construct (or reconstruct) a production of the given word. Thus, here the rules are the subject of the basic operation (choosing). In contrast, in the present case the rules are intrinsically manifest by the function (more percisely by the φ-function) of the elements in the recognizer (i.e., the neurons). Accordingly, no algorithm operating on the rules is necessary (Fu, 1982) but the rules themselves determine each step in the act of recognition. One technical operation is, of course, still necessary to apply to the rules: They originally describe the production of words, thus when applied to recognition their two sides must be interchanged.

The underlying syntactic rules can be formulated in various ways corresponding to various levels of abstraction. The lowest level, within the framework of the model, involves the (individual) states of the neurons. In these terms the rules describe how the neuronal states in a functional layer determine those in the next layer.

Let $<r_{ijk}>$ denote the actual state of the k-th neuron in the j-th module of the i-th functional layer (thus the cell is in its r-th induced state, $r = |n|_{\bmod m_{ijk}}$, where n is the number of spikes in its latest δ-input ($l\delta$) and m_{ijk} is its modulus). Then the syntactic rules take the form:

$(<r_{i11}>, <r_{i12}>, ..., <r_{i1z_{i1}}>),$

$(<r_{i21}>, <r_{i22}>, ..., <r_{i2z_{i2}}>), ...,$

$(<r_{iw_i1}>, <r_{i22}>, ..., <r_{i2z_{i2}}>), ...,$

$(<r_{(i+1)11}>, <r_{(i+1)12}>, ..., <r_{(i+1)1z_{(i+1)1}}>),$

$(<r_{(i+1)21}>, <r_{(i+1)22}>, ..., <r_{(i+1)2z_{(i+1)2}}>), ...,$

$(<r_{(i+1)w_{(i+1)}1}>, <r_{(i+1)w_{(i+1)}2}>, ...,$

$(<r_{(i+1)w_{(i+1)}z_{(i+1)w_{(i+1)}}}>).$ (21)

where w_i denotes the number of modules in the i-th functional layer and z_{ij} denotes the number of neurons in the j-th module of the i-th functional layer.

Equation 21 is perfect in the operational sense: Given the states of all cells in a functional layer (say, in the i-th one), those in the next are completely and uniquely determined. However, this expression operating only on neuronal states does not reveal much about the meaning of either the states at the two sides or the connection between them. Somewhat more abstract form can be obtained by making use of the modular states:

$<M_{i1}>_{q_{i1}}, <M_{i2}>_{q_{i2}}, ..., <M_{iz_i}>_{q_{iz_i}} \rightarrow$

$<M_{(i+1)1}>_{q_{(i+1)1}}, <M_{(i+1)2}>_{q_{(i+1)2}}, ...,$

$<M_{(i+1)z_{(i+1)}}>_{q_{(i+1)z_{(i+1)}}}$ (22)

where $<M_{ij}>_{q_{ij}}$ denotes the state of the j-th module in the i-th functional layer, i.e., it is in its q-th state.

Equations 21 and 22 are equivalent in the sense that either uniquely determines the other provided that the underlying system is given. Yet, Equation 22 is called more abstract because its actual determination involves more directly the functional organization, viz. the interneuronal neighbouring relations (φ-pathways), of the system.

Not only can the functional organization be involved in the syntactic rules (by replacing neuronal states by modular ones) but also the experience of the system can be involved. In the simplest case we may note whether the actual modular states are known or not. Thus, the rules take the form, e.g.:

$$| M_{i1} > q_{i1}, M_{i2} > q_{i2}, < M_{i3} | q_{i3}, \ldots, | M_{iz_i} > q_{iz_i} \rightarrow$$

$$< M_{(i+1)1} | q_{(i+1)1}, | M_{(i+1)2} > q_{(i+1)2}, | M_{(i+1)3} > q_{(i+1)3}, \ldots,$$

$$< M_{(i+1)z(i+1)} | q_{(i+1)z(i+1)} \qquad (23)$$

Here, like above, $<\;|$ denotes unknown states, while $|\;>$ denotes known states. Equation 23 shows the rate of known and unknown constituents in both of the connected functional layers, and thus also characterizes, to an extent, the expressed inference. It is possible, for example, that all states at the left hand side are known while none are known on the other side. This reflects that clear-cut conceptual or physical constituents are coupled in a strange way (in the original input). Vice versa, all states may be unknown at the left hand side while none are known on the other side, reflecting that all constituents are somewhat vague, distorted, fuzzy, etc., but on the whole they uniquely identify a (known) more comprehensive pattern (or category, or concept).

The unknown states are represented in the system by their similarity to the corresponding known ones (as mentioned in section 2.3). These similarities as defined by Equation 19 can be utilized to represent the unknown states in the syntactic rules too. Thus we obtaion expressions like:

$$|M_{i1}>q_{i1}, (<q_{i2}|M_{i2}|1>, <q_{i2}|M_{i2}|2>,..., <q_{i2}|M_{i2}|K_{i2}>),...,$$

$$...,(<q_{izi}|M_{izi}|1>, <q_{izi}|M_{izi}|2>,..., <q_{izi}|M_{izi}|K_{izi}>) ---$$

$$\rightarrow (<q_{(i+1)1}|M_{(i+1)1}|1>, <q_{(i+1)1}|M_{(i+1)1}|2>,...,$$

$$<q_{(i+1)1}|M_{(i+1)1}|K_{(i+1)1}), |M_{(i+1)2}>q_{(i+1)2},...,$$

$$|M_{(i+1)z(i+1)}>q_{(i+1)z(i+1)}$$

(24)

where K_{ij} denotes the number of known states of the j-th module in the i-th functional layer.

A further version of the syntactic rules can be obtained by noting the degree of m-fuzziness, as defined by Equation 20, at each unknown (modular) state.

As mentioned above, the known states correspond to well-defined concepts represented in the system. Thus, the known modular states in Equation 24 can be replaced by the corresponding predicates. Accordingly, the degree of similarity between an unknown and a known state can be replaced by the same degree at which the predicate corresponding to the latter applies to the actual situation expressed by the former. By such replacement the syntactic rules take the form of Equations 13, 14, of course, with the two sides interchanged. Evidently, these equations are very simple examples of what may occur at the final steps of an actual process of recognition (or identification). In general, many more terms (predicates or modular states) are contained in the rules. Also the relationship between the strengths or reliabilities (μ's) to which the predicates at the two sides apply (or between the respective similarities of the involved modular states) are, in general, more complex. Regardless of these technical details, however, the functional equivalence of rules operating on the residues of the number of spikes in interneuronal signals (like Equation 21) with rules operating on predicates (like Equations 13 and 14) might be interesting. In this way a syntactic model suited to the intrinsic structure of the system may also account for semantic processes determined by the actual as well as earlier interactions between the system and its enviroment (Fuhrmann, 1984). As noted in Section 2.3, referring to atomic and cristalline structures, this is by no means aimed at a reduction in the philosophical sense. The principal aim of the proposed approach is only to develop a feasible model for the emergence of psychic acts (e.g. the inferences in Equations 13 and 14) from neural activity (e.g. that described by equation 21). The crucial constituents for such an emergence may be the individual organization of the given experience (as expressed by Equation 23) and the relation between the actual situation and these two phenomena (as expressed by Equation 24).

3. ON M-FUZZY RELATIONS

M-fuzzy sets and grammars have been defined for modelling concepts, thus, at least in principle, they also have to account for the relations between concepts. It seems inevitable to discuss explicitly, though briefly, this issue here because most criticisms concerning the application of (traditional) fuzzy sets in the cognitive sciences (Osherson & Smith 1981; Roth & Mervis, 1983; Johnson-Laird, 1983; Smith & Osherson, 1984) stem from considerations on some relations and operations of concepts. (Let us recall that "relation" is a more general idea than "operation". For example, any binary operation $\underline{Q}$ on a domain D, $\underline{Q}:D \otimes D \rightarrow D$, can be substituted by a ternary relation $\underline{R}$ on $D \otimes D \otimes D$ such that $\underline{R}(d_1,d_2,d_3)$ holds if and only if $\underline{Q}(d_1,d_2) = d_3$, where $d_1,d_2,d_3 \in D$. Evidently, this also applies to any n-ary operation with the corresponding $(n+1)$-ary relation.)

Concerning this issue, the standpoint of the proposed approach is that not only are concepts themselves as intricate and versatile as noted by Miller and Johnson-Laird (1976) cited at the beginnig of the present paper, but also the relations between concepts are intricate. For example, Gleitman and Gleitman (1970, p. 67) cite the compound concept: "volume feeding management success formula award" standing for: "the award given for discovering a formula for succeeding at managing the feeding of people in large volumes" in a plaque in a restaurant. Evidently, such complex relations (or operations) can be defined (or modelled) on extremely general domains only. Furthermore, these relations can also lead to contradictions (think, e.g., of contradictory evaluations of the same statement, situation, etc.). On this basis, m-fuzzy variables appear to form the narrowest domain which may be general enough for modelling the relations of concepts.

This generality, however, has two aspects: While it allows for realistically complex relations, it abolishes the universal validity of simple and well-known relations. The latter aspect has been crtiticised sharply by the opponents of (traditional) fuzzy aproaches referred to above. Since m-fuzziness takes a more radical step towards generality than (traditional) fuzziness, this aspect requires some explanation.

3.1. On traditional relations

First, let us consider "contradiction", a fundamental logical relation which appears in various respects in modelling concepts. As it is well-known, (traditional) fuzzy, as well as m-fuzzy, approaches deny the law of the excluded middle, thus they allow for statements which are contradictory in the sense of Aristotelian logic. The validity of Aristotelian logic on concepts is, however, taken a priori for granted by many workers in the cognitive sciences, thus they reject "fuzziness" by arguments like: "We are reluctant to argue at length against the truth of explicit contradictions" - Osherson and Smith (1982, p. 313), or: "a self-contradiction [of the form p and not p] surely merits a truth value of zero" (Johnson-Laird, 1983, p. 199). Nonetheless, a huge amount of psychological experiments that led to prototype theory (Mervis & Rosch, 1981) has proved that concepts do not obey the law of the excluded middle. (The same fact appearing in everyday experience led to (traditional) fuzzy-set theory.)

It is evident that concepts like "an apple which is not an apple" is not "logically empty", we may think, e.g., of a plastic apple (Cohen & Murphy, 1984). Thus, the fact that many-valued, e.g. (traditonal) fuzzy logics allow for this very special, viz. Aristotelian type of contradiction argues obviously for and not againts the adequacy of these logics in modelling concepts. What is more, concepts do produce contradictions in a much deeper logical sense too: As mentioned in Section 1.2., explicitly and unambiguously defined concepts, actually with dichotomous membership, may also lead to contradictions (e.g. Russell's antinomy, Grelling's antinomy). Accordingly, an adequate model for concepts has not only to allow but even to account for contradictions, and not only in the Aristotelian sense as (traditional) fuzziness does, but also in the latter, deeper sense as m-fuzziness does.

The operation called "negation" or "complementation" corresponds to the relation called "contradiction". Consequently, although this operation is unique in the highly specific, Aristotelian logic, it can be generalized in various ways to traditional fuzzy logic and even more ways to m-fuzzy one. Concerning traditional fuzziness, seven approaches to "negation" are cited by Dubois and Prade (1985). Concerning m-fuzziness, the number of possible approaches seems so large that it may not be reasonable to define any universal operation under this name. Indeed, traditional fuzzy sets do not have sharp borders, thus they cannot be uniquely separated from their "complements". In the case of m-fuzzy sets, not only is the actual determination of any border problematic, but so is the principal definition of a border. Accordingly, m-fuzzy sets defined in forms like "not A" are considered in their own rights as different from others including "A". No universal relation is assumed a priori. For example, an element may belong not only to both "A" and "not A" (e.g. plastic apple) but also to both "typical A" and "not A" (e.g. a typical Swede who is not Swede at all).

The last remark above involves not only "negation" but also "inclusion": The category "typical A" is usually assumed to be included by the category "A". "Inclusion", as well as the corresponding logical relation of "implication" is also (like "negation") unique in Aristotelian logic, while it is feasibly generalized in numerous ways to (traditional) fuzzy logic and more so generalized to m-fuzzy logic. Aristotelian assumptions about one of the possibly many traditional "fuzzy inclusions" also resulted in rejections of any fuzzy approaches (Osherson & Smith, 1981; Roth & Mervis, 1983). Here again, the key is that traditional fuzzy sets, just like concepts, do not have sharp borders, thus it is not unique whether the border of one is inside or outside of that of the other. As noted above, the borders of m-fuzzy sets are even less unique. Thus, so are their "inclusion".

The relation between subordinate/superordinate categories, usually assumed as inclusion, is also adopted in m-fuzzy grammars for regulating rewriting (Section 1.5). Nevertheless, this is the case in the intuitive sense only. In the formal sense, the rewriting including the prescribed stages in it is determined by the underlying neuronal connections (Section 2.4). The subordinate/superordinate concepts connected by the rewriting rules appear in this way as emerged from the formalized neuronal activity. Accordingly, no universal "m-fuzzy inclusion" has been defined.

Like the above mentioned operations, set theoretical "union" and "intersection" as well as the corresponding logical operations of AND and OR are unique in an Aristotelian setting (more precisely if the law of the excluded middle is assumed) while can be generalized in various ways to traditional fuzzy setting (Dubois & Prade, 1985). Here again undue assumptions of some properties of the former about one of the obviously multifarious, though equally justified, latters has resulted in rejections of fuzzy approaches in general (Osherson & Smith, 1981; Smith & Osherson, 1984) in spite of the fact that these classical operations have no distinguished status among the numerous kinds of "conceptual combination", the favorite idea of these criticisms (Lees, 1960; Gleitman & Gleitman, 1970). Owing to the inherent properties of different (viz. Aristotelian, traditonal fuzzy, and m-fuzzy) borders, mentioned above, no specific operations under the names of "union" and "intersection" have been proposed on m-fuzzy sets.

3.2. On specifically m-fuzzy relations

Owing to the generality of the domain, the relations between m-fuzzy variables are much more complex than those mentioned above. The m-fuzzy relations have to show mathematical rigor on the sets of concept cores and on the finite subsets of prototypes while they also have to involve the classes of the corresponding concepts in an intuitively appropriate way. Accordingly, relatively few and relatively complex relations have been defined using m-fuzzy variables.

The most important m-fuzzy relations are manifest in the rewriting rules of m-fuzzy grammars. These rules, according to the model, describe the basic steps in category identification, thus feasibly in a variety of cognitive activity, e.g., decision making, opinion forming, aesthetic estimation, too (Fuhrmann, 1985b). Accordingly, these syntactic rules reflecting the inherent complexity of cognition cannot, in general, be decomposed into simple and universal operations. Only one type of partial decomposition is proposed: The rewriting is regulated, i.e. only specified concepts may appear on the two sides of such a rule (as mentioned in Section 1.5). Although this specification refers to the subordinate/superordinate relation of concepts, this is not a reduction of the rewriting rules since this relation is not (uniquely) pre-defined, particularly not as simple Aristotelian inclusion (as noted in the preceding section). Just the other way round, an analysis of the rewriting rules as well as the underlying neuronal structures may feasibly result in the definition of an appropriate m-fuzzy inclusion. A special type of m-fuzzy inclusion is formed by concepts like "A" and "typical A". As mentioned in Section 1.4., classical inclusion does not apply to this type, either, nevertheless this type may be relatively easy to study.

The rewriting rules of m-fuzzy grammars establish relations not only between the variables appearing on their two sides but also between those appearing at the same position in different versions of the given rule (viz. in forms like Equations 13, 14, 21-24). This latter type of relations can be divided into two groups: One group consists of relations between neuronal and modular states (between expression like 21-24). These relations have been uniquely defined by Equations 18-20, open qusetions concern the neurophysiological manifestation of the different states and of the I φ signals. The other group consists of the relations betwen modular states and concepts (i.e., between expressions like 23, 24 on the one hand, and 13, 14 on the other). The epistemological basis, i.e. why these relations are not reductionistic in the philosophical sense, was outlined in Section 2.3. A point of possible misunderstanding, however, must be clarified here: The term "module" is used in the proposed model as defined in the brain sciences (Szentágothai, 1975, 1978; Mountcastle, 1978). These neuronal structures are related to psychic activity in the proposed model in the way introduced by Hebb (1949), viz. collective induced states of such neuronal structures are assumed to represent concepts (Fuhrmann, 1981, 1982). In the cognitive sciences, however, structures called "modules" as well as their relation to psychic functions have become recently well-known in a fundamentally different meaning, viz. in that of Fodor's (1983) theory. Fodor associates his "modules" with fixed neuronal architectures whose functions are fast, mandatory, and informationally encapsulated (among other properties). Accordingly, these modules are not to account for inherently psychic functions, e.g. the learning and recognition of concepts. Such higher functions take place, according to Fodor (1983), elsewhere. (The "modules" are called "input systems" while the higher functions are performed by "central systems".) Although two similar types of function are distinguished in the model explained above too, (as δ-, and φ-functions), here both are performed by the same structures called "modules".

Another idea (in addition to syntactic rule), which manifests some basic m-fuzzy relations is prototype. As explained in Section 1.4., "prototypicality" means, according to the proposed model, not only maximal representativeness of the element for the category but vice versa. Thus the element-to-category relations appear in a new, symmetric aspect. This also involves some new aspects of element-to-element relations, i.e. why and how the prototypes are distinguished within a given category, as well as new aspects of category-to-category relations. The latter is involved in two respects: One is similar to the former, viz. why and how is the basic level category (usually defined as the most representative for the given element) distinguished. The other stems from the observation in Section 1.4. that most features of a prototype in the category "A" can be best characterized as "(just) A-like". Thus, the relation between "A" and its constituent categories (features) appear as symmetric in some sense (viz. in the case of prototype they reciprocally define each other). This also involves some new apects of the subordinate/ superordinate relation since the "features" (or "constituent categories") are certainly "subordinate" in some meaningful sense.

CONCLUDING REMARKS

The largest number of experiments that led, separately, to the development of fuzzy-set theory and prototype theory firmly established that the members of conceptual categories are not equal. Fuzzy-set theory deals with the continuous nature of membership while prototype theory with that of representativeness. Resulting from the confusion of the two, there has been a heavy trend in the cognitive sciences opposing the application of fuzzy sets to modelling concepts (Osherson & Smith, 1981; Roth & Mervis, 1983; Johnson-Laird, 1983; Cohen & Murphy, 1984; Smith & Medin, 1984; Smith & Osherson, 1984). Representatives of this trend argue that fuzzy - set theory takes too big a step from classical Aristotelian logic, and, therefore, although it might be use ful in modelling the very act of categorization, certainly is not useful in representing the inherent logic of concepts. In the present paper the same conclusion was drawn from the diametrically opposite observation, viz. arguing the point that fuzzy-set theory takes too little a step by generalizing the value of the characteristic function from the set $\{0,1\}$ to the real interval $[0,1]$ while, however, using the same domain (viz. sets). Based on the contradictions in the inherent logic of concepts that refuted classical, or naive set theory (e.g., Russell's antinomy, Grelling's antinomy) as well as on those pointed out by psychological experiments (Cohen, 1981; Hampton, 1982) a generalization of the domain was proposed from sets to classes in the sense of axiomatic set theory, i.e. from contradiction-free collection to possibly self-contradictory ones.

The proposed, modifed fuzzy sets (m-fuzzy sets) were applied to model the internal organization of concepts. A hierarchy of concepts as classes, m-fuzy sets as a linkage, concept cores as sets, and prototypes as finite subsets was constructed such th at each item represents the preceding one and thus indirectly all ones before it, and that the inevitable possibility of contradictions is localized to the first two items. Thus a mathematical (i.e. necessarily contradiction-free) theory can be developed for the second pair of items also involving the first pair as represented by the other one.

For modelling the (experimantally proved) inequality of categories that comprise a given element (Mervis & Rosch, 1981) m-fuzzy grammars were proposed. These are pure, i.e., no distinction is made between nonterminal and terminal symbols (Maurer, Salomaa, & Wood, 1981; Gabrielian, 1982), and regulated, i.e. different levels of the rewriting involve different collections of the rules (Salomaa, 1973). Pureness provides that any category may occur at any step in various chains of rewriting (i.e., that each concept can be considered in itself, or as a constituent for another concept, or as a compound of other concepts). Regulatedness provides a direct account for the subordinate/superordinate relation. It was also pointed out, however, that the relations and operations on m-fuzzy sets and grammars are roughly as intricate as those on concepts (Lees, 1960; Gleitman & Gleitman, 1970). Accordingly, definite m-fuzzy sets and grammars refer to definite mental representations (or "subjective lexicons").

This individuality at the psychic level was united with that at the neural level: Psychic acts, e.g., category identification, was approached in an emergentist, i.e. monistic though philosophically not reductionistic way. In this approach concepts appeared as represented by collective induced states of specific neural structures called modules, like Hebb's (1949) cell assemblies. The organization, accordingly also the name, of these structures in the model refer to the neuron-module found as the functional unit in cerebral cortex architecture (Szentágothai, 1975; 1978; Mountcastle, 1978). The formalized modules in a distributed, parallel way. Formally, different versions of a given rewriting rule of an m-fuzzy grammar describe, according to the model, a given step of the information processing at different levels from neuronal to psychic (Equations 13-14, 21-24).

Since the activity of the proposed modules is directly related to the representation, including the learning and recognition, of concepts, they sharply contrast the modules in Fodor's (1983) theory, which is often regarded as the exclusive conception of modular organization in the cognitive sciences.

* address: Pöttyös utca 4. H - 1098 Budapest, Hungary

REFERENCES

Armstrong, S. L., Gelitman, L. R., & Gleitman, H. What some concepts might not be. *Cognition*, 1983, *13*, 263-308.

Barsi, F. & Maestrini, P. Error codes constructed in residue number systems with non-pairwise-prime moduli. *Information and Control*, 1980, *46*, 16-25.

Chen, C. H, *Statistical pattern recognition.* New York: Hayden, 1973.

Cohen, B., & Murhy, G. L. Models of Concepts. *Cognitive Science*, 1984, *8*, 27-58.

Cohen, L. J. Can human irrationality be experimentally demonstrated? *Behavioral and Brain Sciences*, 1981, *4*, 317-370.

Dubois, D., & Prade, H. A review of fuzzy set aggregation connectives. *Information Sciences,* 1985, *36*, 85-121.

Ebanks, B. R. On measures of fuzziness and their representations. Journal of *Mathematical Analysis and Applications,* 1983, *94*, 24-37.

Feldman, J. A., & Ballard, D. H. Connectionist models and their properties. *Cognitive Science,* 1982, *6*, 205-254.

Fodor. J. A. *The modularity of mind*. Cambridge, Mass.: MIT Press, 1983.

Fraenkel, A. A., Bar-Hillel, Y., & Levy, A. *Foundations of set theory.* (2nd.rev.ed.) Amsterdam: North-Holland, 1973.

Fu, K. S. *Syntactic pattern recognition and applications.* Englewood Cliffs, N.J.: Prentice Hall, 1982.

Fuhrmann, Gy. Modelling the visual cortex with "modulo system" concept. *Biological Cybernetics*, 1981, *40*, 39-48.

Fuhrmann, Gy. A recognizing neural network: modulo system for extracting and syntactically processing of the important properties of patterns. In R. Trappl, G. Pask, and L. M. Ricciardi (Eds.) *Progress in cybernetics and system research.* Vol. IX. Washington, D.C.: Hemisphere, 1982.

Fuhrmann, Gy. Syntax as the model of semantics in brain modelling. *Cybernetica,* 1984, *27,* 39-56.

Fuhrmann, Gy. Arithmetic model for the distributed encoding in the neuron module. *International Journal of Neuroscience,* 1985, *28,* 91-110. (a)

Fuhrmann, Gy. Interdisciplinary approach to the brain's pattern recognition. *Cybernetica,* 1985, *28,* 107-145. (b)

Fuhrmann, Gy. Mathematical approach to integrating the "neuron-module" and the "cell-assembly". *International Journal of Neuroscience,* 1985, *29,* 197-203. (c)

Fuhrmann, Gy. Fuzziness of concepts and concepts of fuzziness. *Synthese,* (forthcoming a)

Fuhrmann, Gy. "Prototypes" and "fuzzines" in the logic of concepts. *Synthese,* (forthcoming b)

Fuhrmann, Gy. *Note on the generality of fuzzy sets.* (forthcoming c)

Fuhrmann, Gy. *Note on the integration of prototype theory and fuzzy set theory.* (forthcoming d)

Gabrielian, A. Pure grammars and pure languages. *Intrenational Journal of Computer Mathematics,* 1981, *9,* 3-16.

Gleitman, L. R., & Gleitman, H. *Phrase and paraphase: Some innovative uses of language.* New York: Norton, 1970.

Goguen, J. A. Concept representation in natural and artificial languages: Axioms, extensions, and applications for fuzzy sets. *International Journal of Man-Machine Studies,* 1974, *6,* 513-561.

Halmos, P. *Measure theory.* New York: Van Nostrand, 1950.

Hampton, J. A. A demonstration of intransitivity in natural categories. *Cognition*, 1982, *12*, 151-164.

Hebb, D. O. *The organization of behaviour*. New York: Wiley, 1949.

Iversen, L. L. Amino acids and peptides: fast and slow chemical signals in the nervous system? *Proceedings of the Royal Society of London,* 1984, *B-221*, 245-260.

Johnson-Laird, P. N. *Mental models*. Cambridge, Mass.: Harvard University Press, 1983.

Kandel, A. *Fuzzy mathematical techniques with applications*. Reading, Mass.: Addison-Wesley, 1986.

Karczmar, A. G. Introduction to the session on modulators. *Neuropharmacology,* 1987, *26*, 1019-1026.

Kaufmann, A. *Introduction to the theory of fuzzy subsets* 1. New York: Academic Press, 1975.

Koch, C., Poggio, T., & Torre, V. Nonlinear interactions in a dendritic tree: Localization, timing, and role in information processing. *Proceedings of the National Academy of Science, USA,* 1983, *80*, 2799-2802.

Lee, S. C., & Lee, E. T. Fuzzy sets and neural networks. *Journal of Cybernetics,* 1974, *4*, 83-103.

Lees, R. B. *The grammar of English nominalization*. Bloomington, Ind.: Indiana University Research Center in Anthropology, Folklore, and Linguistics (No. 12), 1960.

Lewis, E. R. The elements of single neurons: A review. *IEEE Transactions Systems, Man, and Cybernetics,* 1983, *SMC-13*, 702-710.

Loftus. E. F. Spreading activation within semantic categories: Comments on Rosch's "Cognitive representations of semantic categories". *Journal of Experimental Psychology: General,* 1975, *104*, 234-240.

Loo, S. G. Measures of fuzziness. *Cybernatica,* 1977, *20*, 201-210.

Maurer, H. A., Salomaa, A., & Wood, D. Pure grammars. *Information and Control,* 1979, *44,* 47-72.

McCloskey, M. E., & Glucksberg, S. Decision processes in verifying category membership statements: Implications for models of semantic memory. *Cognitive Psychology,* 1979, *11,* 1-37.

Medin, D. L., & Smith, E. E. Concepts and concept formation. *Annual Review of Psychology,* 1984, *35,* 113-138.

Mervis, C. B., & Rosch, E. Categorization of natural objects. *Annual Review of Psychology,* 1981, *32,* 89-115.

Miller, G. A., & Johnson-Laird, P. N. *Language and perception.* Cambridge: Cambridge University Press, 1976.

Mountcastle, V. B. An organizing principle for cerebral function. The unit module and the distributed system. In G. M. Edelman & V. B. Mountcastle, *The mindful brain.* Cambridge, Mass.: MIT Press, 1978.

Ore, O. The general Chinese reminder theorem. *American Mathematical Monthly,* 1952, *59,* 365-370.

Osherson, D. N., & Smith, E. E. On the adequacy of prototype theory as a theory of concepts. *Cognition,* 1981, *9,* 35-58.

Osherson, D. N., & Smith, E. E. Gradedness and conceptual combination. *Cognition,* 1982, *12,* 299-318.

Piaget, J. L'explication en psychologie et la parallélisme psychophysiologique. In J. Piaget, P. Fraisse, & M. Reuchlin *Trait de psychologie expérimentale, I. Histoire et méthode.* Paris: Presses Universitaires de France, 1963.

Posner, M. I., & Keele, S. W. On the genesis of abstract ideas. *Journal Experimental Psychology,* 1968, *77,* 353-363.

Reed, S. K. Pattern recognition and categorization. *Cognitive Psychology,* 1972, *3,* 382-407.

Reed, S. K. *Psychological processes in pattern recognition.* New York: Academic, 1973.

Reschner, N. *Many-valued logic.* New York: McGraw-Hill, 1969.

Rosch, E. Cognitive reference points. *Cognitive Psychology,* 1975, *7,* 532-547. (a)

Rosch, E. Cognitive representations of semantic categories. *Journal of Experimental Psychology: General,* 1975, *104,* 192-233.

Rosch, E., & Lloyd, B. B. (Eds.) *Cognition and categorization.* Hillsdale, NJ.: Erlbaum, 1978.

Rosch, E., Simpson, C., & Miller, R. S. Structural basis of typicality effects. *Journal of Experimental Psychology: Human Perception and Performance,* 1976, *2,* 491-502.

Roth, E. M., & Mervis, C. B. Fuzzy set theory and class inclusion relations in semantic categories. *Journal of Verbal Learning and Verbal Behavior,* 1983, *22,* 509-525.

Salomaa, A. *Formal languages.* New York: Academic, 1973.

Simpson, G. G. *Principles of animal taxonomy.* New York: Columbia University, 1961.

Smith, E. E., & Osherson, D. N. Conceptual combination with prototype concepts. *Cognitive Science,* 1984, *8,* 337-361.

Sperry, R. W. Mind-brain interaction: Mentalism, yes; dualism no. *Neuroscience,* 1980, *5,* 195-206.

Szabó, N., & Tanaka, R. *Residue arithmetic and its application to computer technology.* New York: McGraw-Hill, 1967.

Szentágothai, J. The "module concept" in cerebral cortex architecture. *Brain Research,* 1975, *95,* 475-496.

Szentágothai, J. The neuron network of the cerebral cortex. *Proceedings of the Royal Society of London,* 1978, *B-201,* 219-248.

Vizi, E. S. *Non-synaptic interactions among neurons.* New York: Wiley, 1984.

Zadeh, L. A. Fuzzy sets. *Information and Control,* 1965, *8,* 338-353.

THE WEIGHTED FUZZY EXPECTED VALUE AS AN ACTIVATION FUNCTION FOR THE PARALLEL DISTRIBUTED PROCESSING MODELS

David KUNCICKY and Abraham KANDEL

*Department of Computer Science
The Florida State University,
Tallahassee, FL. 32306-4019, USA.*

This research presents an alternate method for viewing the firing rules of neural network models (more generally called parallel distributed processing models, or PDP's). The inputs to the processing unit are considered to be elements of a fuzzy set and the firing rule is viewed as an attempt to find a typical value among the inputs. Two such fuzzy measures are analyzed - the fuzzy expected value (FEV), and the weighted fuzzy expected value (WFEV). Simulations using simple one and two layer networks are performed. The WFEV is shown to compare favorably with the usual weighted sum firing rule when inputs are quasi-orthogonal and the WFEV surpasses the latter when the mean orthogonality of the input patterns decreases.

1. INTRODUCTION

The subject of neural network modeling (or using the more general terminology - modeling of parallel distributed processes or PDP modeling) comprises the study of the emergent properties of collections of neuron-like units. These processing units ope rate asynchronously and in parallel and are bound generally, but not strictly, by neurophysiological principles. Rumelhart and McClelland (1986) list some of the general findings of neuroscience that are useful to neural modelers. These findings are summarized in the paragraphs below. The same authors also point out that not enough is known to seriously restrict theories about how the physiological level and cognitive levels of the brain interact. It is interesting to contrast the brain with the computer, and from this contrast it becomes clear that the architecture of each is very different. They are so different that it is surprising that we would expect a serial Von Neumann machine to exhibit human-like intelligence. It is just as amazing that the serial computer demonstrates the power that it does, surpassing human capabilities in speed, accuracy, and computation in many areas. Below, we contrast some of the basic differences between brains and computers.

Neurons are slow, computers are fast. The processing unit in the brain is the neuron and it operates with a processing speed on the order of 10^{-3} seconds. The upper limit of the neural firing rate is a result of the refractory period that is n ecessary for the neural membrane to recover to a resting potential after it has fired. The processing rate of a single processor in a fast computer, on the other hand, is about 1 gigaflop (1 billion floating point operations per second).

Brains perform distributed computations, computers are centralized. Neurons compute autonomously. Although some neurotransmitters may have a global effect on large numbers of cells, they certainly do not act as a clock or synchronizing device in the short run. From studies of brain damage it can be concluded that no one part of the brain acts as a central processor or homunculus. The processing is distributed across a number of cells. Just how distributed the processing is has been an issue of debate. Even a localized computation in the brain, say involving 1% of the neurons, would involve many millions of neurons (Anderson, 1983).

Computers, however, funnel all computation through a central processor. (Newer parallel architectures such as hypercubes employ distributed processing and are one of the devices on which to implement PDP models.)

The brain is massively parallel, the traditional computer is a serial device. It is the slowness of neurons that is the main argument for massive parallel computation in the brain. There is no other concievable way to account for the complex computations that occur in a few hundred milliseconds. As Feldman and Ballard (1982) point out "entire behaviors are carried out in a few hundred time steps". Imagine trying to write a computer program that could recognize a face in a few hundred operations. Recent advances in computer computational speed have come from the introduction of parallel processors rather than faster processors. This is because single processors are constrained by a final limit of speed, the speed of light.

The abilities of brains degrade gracefully, the abilities of computers are brittle in the face of damage. When faced with progressive ablation the brain continues to function and its functioning gradually degrades with increasing damage. When faced with sudden localized damage the brain shows an ability to recover the lost functions over time. Thousands of brain cells are irretrievably lost every day, yet the computational capabilities of the brain increase with time during much of our lifespan. The brain has an incredibly effective fault tolerance system. Computers on the other hand, may be brought down by the failure of a single element.

Brains store information in the connections between processors, computers store data in memory cells. There is a large fan-in and fan-out of connections between neurons. This is on the order of 10^3 up to 10^5 in the cerebral cortex. Furthermore, learning is presumed to take place by modifying the strengths of the connections. Computers store information as discrete bits that are saved specific addressable locations.

A neural model (which is also called a parallel distributed processing model or PDP model) will adhere more or less to the general principles listed above without paying too much attention to physiological details. The goal is to produce interesting cognitive functions such as learning, memory, and computation from the autonomous interaction of many simple processing units.

The potential for application of PDP models lies primarily in the areas of cognitive psychology, neurophysiology, and computer science. This research is concerned with potential computer science applications which means that adherence to physiological principles is less of a concern than in order disciplines. In computer science, any problem that has a large number of simultaneous, weak constraints and that requires a good but not necessarily optimal solution is a candidate for a PDP solution. These include problems in pattern recognition, scheduling algorithms, robotics, and artificial intelligence. Until recently, massively parallel processes could be simulted only on serial machines. With the increasing availability of massively parallel architectures such as connection machines, the opportunity to implement PDP solutions also increases.

The processing unit in most PDP models performs only one operation - that is to assess constantly the status of its large number of inputs and make a decision whether to fire or not. If the unit represents a feature detector, for instance, then the boolean output of the unit signifies the presence or absence of that feature. If the output value is allowed to be a real value rather than a boolean value, then the output denotes the amount of the feature that is present. If the real output value is limited to the unit interval [0,1], then the output describes the degree of confidence that the feature is present (Williams, 1986). The depiction of the degree of confidence as a conditional probability has been treated thoroughly by Peretto and Niez (1986a,b). It has been suggested in several places that the output of a unit might be perceived as a fuzzy measure of truth (Kohonen, 1984; Williams, 1986; Zadeh, 1965). In several papers, Lee and Lee (1974, 1975) described fuzzy neural networks and applied them to the synthesis of fuzzy finite automata, but their networks were static and did not show the emergent properties described above.

The motivation for this research is that many elements of the subject of cognition are vague. The microstructure of cognition is not well understood. It deals with subjective phenomenon such as perceptions and memories. The hardware of the brain is noisy and the functions driving it are imprecisely defined. For these reasons a fuzzy PDP model is suggested in which the output of each processing unit represents a fuzzy level of confidence. The activated inputs to a processing unit are to be treated as elements of a fuzzy set. The justification for the development of a fuzzy neural model is twofold.

One, the input set of a processing unit does not have sharply defined boundaries. An incoming line to a processing unit is simultaneously an input to many other units. This is consistent with fuzzy pattern recognition schemes where a feature has a grade of membership in more than one class (Hall, 1982). (The fuzzy level of confidence that an incoming line is in any one input set is called the grade of membership of that line. The grade of membership is considered to be a value in the unit interval [0,1]. A grade of membership of '1' denotes complete confidence that membership exists and is equivalent to boolean '1'. A grade of membership of '0' implies that the incoming line is not a member of the input set).

The other reason to use a fuzzy model is that once a neural network is larger than a few tens of units, it becomes exceedingly difficult to analyze completely. Kandel (1986) states that

> " ... fuzzines may be a concomitant of complexity. Systems of high cardinality are rampant in real life and their computer simulations require some kind of mathematical formulation to deal with the imprecise descriptions."

There are many PDP models in the literature. Kohonen (1984) gives a thorough description of the subject. We have chosen to apply the notion of a fuzzy network to the dynamic neural network model of Hopfield (1982). Other models could be used as well but the Hopfield model is widely cited and will serve as a good starting point.

The elements of a PDP model will be defined in the next section. We will describe several fuzzy measures of central tendency and substitute them for the activation function (firing rule), performing simulations with each (Section 3 and 4). In Section 5 several assumptions that limit the stochastic interpretation of PDP's are described. Then, in Section 6, tests are performed to see if the fuzzy model is resistant to one of these assumptions (the orthogonality assumption). Finally, the conclusion and directions for further research are discussed in Section 7.

2. ACTIVATION FUNCTIONS

Our general framework for the description of parallel distributed models is presented by Rumelhart and McClelland (1986). Other frameworks are presented by Kohonen (1984). The common characteristics of a PDP model are:

- one or more sets of processing units
- a set of weighted connections
- an activation function (or firing rule)
- a learning rule.

A set of processing units is usually denoted by a vector, and the value of each element of the vector denotes the state of activation for that unit. In some models the vector is a set of real values, in some the vector is a set of values in the closed unit interval [0,1], and in others it is a set of binary values.

The pattern of weighted connections among units is commonly a complete bipartite graph, and usually expressed as a matrix W where the value of a cell w_{ij} represents the strength of the connection between input unit v_j and out unit u_i. The value of a cell in the weight matrix may also have various ranges in different models. The most common are a set of real values in the closed interval [0,1], a set of real values in the range [-1,1], and a two point set $\{+,-\}$.

The **activation function** for a unit determines the unit's activation value based on the state of activation of the unit, the state of activation of its inputs, and the weight of each input. The activation function combines the output values of the incoming lines (v_j) with the strength of each connection (w_{ij}) to produce a net input for the current processing unit (u_i). This value is frequently expressed as a weighted sum of the products of each input value and its associated weight:

$$u_i = \Sigma\ w_{ij} * v_j$$

Weighted Fuzzy Expected Value

In a linear model the value of ui is passed on as the output. A non-linear function uses a threshold to produce an output in the two-point set {0,1}. Other variants of the activation rule reduce the real-valued output to the closed unit interval [0,1] or some other bounded interval. (The latter are sometimes called squashing functions). In general, if the units have activation values from the set A, and there are n inputs to a particular unit, then the activation function is a function α from A^n to A (Williams, 1986):

$$\alpha : A^n \to A.$$

The **learning rule** is a function that determines the value of the connection weights and is usually based on locally available information. Most learning rules are some variant of the Hebbian rule for cellular learning (Hebb, 1949):

> When an axon of cell A is near enough to excite a cell B and repeatedly or persistently takes part in firing it, some growth process or metabolic change takes place in one or both cells such that A's efficiency as one of the cells firing B is increased.

This often quoted rule states that if a connected pair of units are both highly activated then the strength of the connection between them increases.

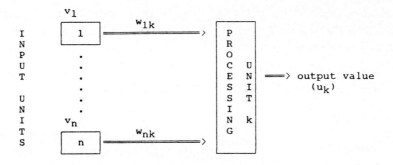

FIGURE 1
Example of one unit and its inputs.

The previous definitions may be understood more clearly by referring to Figure 1 above.

Much of the current PDP reserach focuses on variations of the learning rule. The learning rule determines the strength of the connections between units and generally characterizes the capabilities of a particular model. This paper focuses on a variation of the activation rule. An investigation is made into the implications of this variation on the storage characteritics of a single-layer nonlinear model similar to that of Hoepfield (1982).

Fuzzy Measures as Activation Functions

The activation value of a unit as defined above signifies the amount of some preferred feature that is present. As mentioned before, the activation value may also be viewed as a conditional probability or as a fuzzy measure of truth. It is interesting to note that the weighted sum, which is frequently used in the activation function, is essentially equivalent to the mean of the non-zero inputs. In the same vein, fuzzy measures of central tendency may be used to find a typical value among the inputs to a unit. In what follows, the fuzzy expected value (FEV) and the weighted fuzzy expected value (WFEV) are explored as activation functions for PDP's.

3. THE FUZZY EXPECTED VALUE AS ACTIVATION FUNCTION

The fuzzy expected value (FEV) is a measure of central tendency for fuzzy sets. The FEV depicts a 'typical' or 'representative' value, and is meant to supplant the arithmetic mean and median for fuzzy sets. The following definition of the FEV is drawn from Kandel (1986), and Friedman, Schneider, and Kandel (1987).

Let R be a fuzzy relation defined over the universe Y. For example, the relation 'much less than' may be defined by the matrix

			A1			
	R	1	2	3	4	
	1	0	0	0	0	
A2	2	0.4	0	0	0	
	3	0.8	0	0	0	
	4	1	0.8	0.4	0	

in which the fuzzy sets $A_1 = A_2 = 1 + 2 + 3 + 4$.

Let X_A be the membership function of A where

$$0 \geq X_A(y) \geq 1, \qquad y \in Y$$

In our example X_A is defined by the relation matrix but the relation "x is much less than y" may be defined in many ways.

For example, X_A could also be defined by:

$$X_A(x,y) = \begin{cases} 0, & \text{if } x - y \geq 0 \\ 1 + 0.6(x-y)^{-1} & \text{if } x - y < 0. \end{cases}$$

The fuzzy measure is defined over subsets of Y. The fuzzy measure is a set function that is bounded, nonnegative, monotonic, and continuous.

The fuzzy expected value of X_A, over A, is defined by

$$FEV\{X_A\} = \sup_{0 \leq T \leq 1} \{\min[T, \mu(\sigma_T)]\}$$

where $\sigma_T = \{y \mid X_A(y) \geq T\}$

and T is a threshold ($0 \leq T \leq 1$).

Now $\mu(\sigma_T) = f_A(T)$ is a function of T, so the FEV may be represented geometrically as the intersection of the curves $T = f_A(T)$, which will be at value $T = H$, so that the $FEV\{X_a\} = H \in [0,1]$. Figure 2 depicts this procedure.

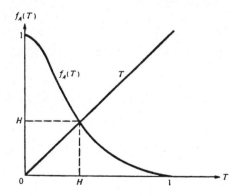

FIGURE 2
A graphic representation of the FEV.

The FEV can also be defined for a finite set of data points, which is more relevant to the application at hand. Let the universe Y be given by the finite union

$$Y = \bigcup_{i=1}^{m} Y_i$$

so that

$$X_A(y) = X_i, \ y \in Y_i; \ 1 \leq i \leq m$$

and the X_i's are ordered

$$0 \leq X_1 \leq X_2 \leq \ldots \leq X_m \leq 1.$$

Essentially, the grades of membership of the data points have been placed in like groups and the groups have been arranged in ascending order.

Define n_i to be the number of data points at level X_i, and N to be the total population. There are m-1 distinct levels of fuzzy measure

$$\mu_i = \frac{\sum_{j=i+1}^{m} n_j}{N}, \ 1 \leq i \leq m\text{-}1.$$

The FEV is the median of the set

$$\{X_1, \ldots, X_m, \mu_1, \ldots, \mu_{m\text{-}1}\}.$$

Let us illustrate the definition with an example. Table 1 below shows the X values of a population of 12 data points grouped into the 5 like categories X_i.

TABLE 1
Example of FEV computation.

i	X_i	n_i
1	0.40	1
2	0.50	3
3	0.55	4
4	0.60	2
5	1.00	2

Using the formula above we calculate the μ_i to be

$\mu_1 = 11/12 = 0.92$
$\mu_2 = 8/12 = 0.67$
$\mu_3 = 4/12 = 0.33$
$\mu_4 = 2/12 = 0.17$

The FEV is the median of the set

$\{0.17, 0.33, 0.40, 0.50, 0.60, 0.67, 0.92, 1.00\}$

which in this example is 0.55. Figure 3 graphically represents the discrete case of the FEV.

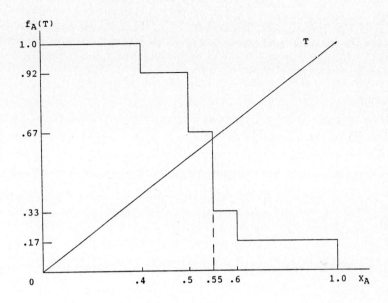

FIGURE 3
A graphic representation of a discrete case of FEV.

First we will describe how the FEV fits into the PDP model, and then give the results of a simulation. If the activated inputs to a processing unit are considered to be a fuzzy set of weak constraints on the unit, then the FEV may be used to replace the weighted sum in the activation function. In the following formulation the weights (w_{ij}) are taken to symbolize the 'grade of membership' in the input set, or the 'degree of confidence' that the input actually represents a particular concept, event, or stimulfus. The FEV of the activated inputs then represents the 'typical value' of the weak constraints impinging on the unit. The output value is a function of the activation value and a threshold. The output is a binary value that signifies whether the unit has autonomously decided to fire or not.

Simulation Using the FEV as an Activation Function

A simulation of the model using the FEV as the activation function was performed using 10 random binary input vectors (n = 100). The inputs were paired with themselves in an auto-associative state array. An auto-associative model pairs an input pattern with itself as output. The model is useful for retrieving whole patterns from partial or distorted input. A hetero-associative model pairs an input pattern with a different output pattern. The Hopfield (1982) learning algorithm was used to assign weights to the connections between the inputs and outputs. The algorithm used is

$$w_{ij} = \sum_k (2v_j-1)(2u_i-1).$$

expect that all weights (w_{ij}) were scaled to [0,1] instead of [-1,1] and the threshold value was set to 0.5 instead of to zero. To test the graceful degradation of the system in the face of partial inputs, the input vectors were progressively distorted. The distortion was measured as distance in Hamming units from the original stored pattern. The recalled vectors were then compared to the desired output vectors and recorded as percent correct.

The average results of many firings as the inputs are gradually distorted is presented in Table 2.

TABLE 2

Average retrieval rate of mean and FEV as activation functions (100 trials for each) Hamming distance).

HAMMING Dx	0	5	10	15	20
Mean	96.5	95.0	92.6	89.7	86.0
FEV	92.3	90.1	87.6	84.4	80.7

It is clear that the FEV's performance is not satisfactory. This is partly due to the FEV's insensitivity to the population density of the data set. Friedman (1987) defines population sensitivity to be:

> Let A be a fuzzy set with populations n_1 and n_2 and membership values X_1 and X_2 respectively. If $n_1 > n_2$, then the typical value should be "much closer" to X_1 than to X_2.

Another reason that the FEV fails in this application is that the decision to fire is often made with results that are very close to the threshold value. From the graphical representation of the FEV in Figure 3 it is clear that if the number of X_i's is small then the FEV is not sensitive to variations close to the threshold. An example showing the calculation of the FEV for one firing from the simulation is persented in the Appendix.

4. THE WEIGHTED FUZZY EXPECTED VALUE AS ACTIVATION FUNCTION

The second quantity that we will use to find a 'typical value' for the inputs to a processing unit is the weighted fuzzy expected value (WFEV). The following definition of the WEFV is from Friedman, Schneider, and Kandel (1987).

Let $w(x)$ be a non-negative monotonically decreasing function defined over the interval [0,1] and τ a real number greater than zero. The solution s^* of

$$s^* = \frac{X_1 w(|X_1-s|) n_1 \tau + \ldots + X_m w(|X_m-s|) n_m \tau}{w(|X_1-s|) n_1 \tau + \ldots + w(|X_m-s|) n_m \tau}$$

is called the weighted fuzzy expected value of order τ. The attached weight function w is generally defined as:

$$w(x) = e^{-\beta x}$$

where $\beta > 0$, and the parameters τ, β are found sufficient for determining the 'most typical value' of X_A. In some instances the values of τ and β are found after the data set is known. It would certainly be desirable to determine the values of the parameters a priori, using information known locally at the processing unit.

In the next Section we will make an **a priori** determination of the τ parameter for each unit. Friedman suggests that s be solved iteratively starting with:

    s = FEV or s = mean
    β = 1, and
    τ = 2.

The iterations are continued until a result of sufficient accuracy is found. In our simulations the first iteration produced satisfactory accuracy.

Simulation Using the WFEV as an Activation Function

The simulation comparing the WFEV and mean as activation functions is a duplication of the simulation in Section 3 except that the WFEV is used in instead of the FEV. After many tests the following conditions for the WFEV were found to produce the most satisfactory results:

    s = mean of activated inputs,
    β = 1, and
    τ = 0.5.

Table 3 presents the results when the inputs are randomly chosen. It can be seen that the retrieval rate for the WFEV compares favorably with that of the mean.

TABLE 3
Average retrieval rate of mean and and WFEV as activation functions
(100 trials for each Hamming distance).

HAMMING Dx	0	5	10	15	20
Mean	99.6%	99.2	98.6	97.5	95.7
WFEV	99.5%	98.8	97.9	96.0	94.0

5. DENSITY AND ORTHOGONALITY ASSUMPTION

There are several assumptions made about the input patterns of a single layer or double layer PDP in the Hopfield model. One is the orthogonality assumption. Non-orthogonal inputs interfere with one another as they are stored in the state array (Stone, 1986). The resulting retrieval rate decreases as the input patterns become less orthogonal (see Table 4). Another assumption made in the Hopfield model is that the binary input vectors have the same ratio of ones to zeros. This is called the density problem. Both of these assumptions and their interaction are discussed below.

Density Assumption

The input patterns must be of similar density if the threshold value is to be constant for all patterns. Density of a binary vector refers to the proportion of ones to the total number of elements. Storing patterns with half ones and half zeros results in an optimum threshold of zero. Other densities result in the optimum threshold varying from zero, and a mixture of densities results in no common threshold value for all patterns. The solution to this problem has been to create patterns with the same density. This is based on the knowledge that any input pattern can readily be mapped to another pattern with a given density (Stiles & Deng, 1987). For example, the two state arrays below are storing a single pattern each. This is calculated using the Hopfield algorithm described earlier. State array A is storing a pattern of density 2/3. State array B is storing a pattern of density 1/3.

OUTPUT

		0	1	1			0	0	1
I									
N	0	-	-1	-1		0	-	1	-1
P	1	-1	-	1		0	1	-	-1
U	1	-1	1	-		1	-1	-1	-
T									

            State Array A   State Array B

By using the firing rule described in Section 2, the following activation values are achieved at the outputs:

	Cell 1	Cell 2	Cell 3
State Array A	-2	1	1
State Array B	-1	-1	0

One can see that a threshold of 0.5 would work for State Array A, but would not be adequate for State Array B, which has an optimal threshold of -0.5. This is a result of the difference in densities of the inputs. When multiple patterns of different densities are stored in the same state array the problem is exacerbated.

Orthogonality Assumption

The pattern stored in a state array must be adequately dissimilar or the recollections are confused. If several patterns are very similar then the minima that represent memories become fused (Hopfield, 1982), and the inputs produce indistinguishable outputs. Some models require that the input patterns be orthogonal to effect maximum recollection. Others have the less restricitive reqirement that the input patterns be linearly independent. Orthogonality means that the input vectors of size n are perpendicular in n-dimensional space or, equivalently, that the dot product of all vector pairs is zero. Linear independence means that the k input vectors span k-dimensional space. This is the same as saying that none of the vectors can be written as a linear combination of the others. The impact of the orthogonality assumption has been minimized in biological models by assuming that preprocessing tends to randomize inputs. The preprocessed material then becomes relatively orthogonal (Kohonen, 1984). Multi-layer models that employ hidden units overcome the assumption by redefining the internal representations to be orthogonal.

TABLE 4
Effect of orthogonality on recall.

Mean dot product	Hamming distance from stored pattern					
	0	5	10	15	20	25
(dissimilar)						
24.9	99.8	99.7	99.2	97.9	97.0	93.4
30.9	89.7	88.8	88.1	87.6	86.5	86.6
38.8	87.9	87.9	87.8	88.0	87.6	87.5
(similar)						

Some learning rules, such as the generalized delta rule, require that the input patterns be only linearly independent (Stone, 1986). However, single layer models that do not assume preprocessed input are severely restrictted in the choice of input patterns. To illustrate the effects of this restriction a simulation was performed that varied the orthogonality of the input patterns. The effect on the ability of the system to recall stored patterns was then measured. The simulation demonstrates the results (see Table 4).

Interaction of Density and Orthogonality

There is an interaction between the density and the orthogonality of a set of input patterns. The mean density of the input patterns affects the maximum number of orthogonal patterns that can be stored. For example, assume a set of binary vectors of size n and density d. There are n possible orthogonal patterns of density $1/n$, but only 2 possible orthogonal patterns of density $1/2$. To illustrate this, consider a binary vector of size 4. If patterns are limited to have a density of 1 then there is a maximum of 4 patterns where all pairs of patterns are orthogonal:

Pattern1	-	1	0	0	0
Pattern2	-	0	1	0	0
Pattern3	-	0	0	1	0
Pattern4	-	0	0	0	1

If the patterns are limited to have a density of 2, however, then there is a maximum of 2 patterns where all pairs of patterns are orthogonal. There is more than one set of these. An example of a possible set is:

Pattern1	-	1	1	0	0
Pattern2	-	0	0	1	1

Effect of Orthogonality on Recollection

The following experiment demonstrates how the stochastic model behaves as the input patterns become less orthogonal. The orthogonality of a set of patterns is measured as the mean dot product of all possible pairs of vectors. One can see from Table 4 that the recall accuracy decreases as the members of a set of input patterns become more similar.

The search for alternations or extensions to the model that would allow a relaxation of the above assumption is based on the genuine robustness of the Hopfield model. Various modifications of the basic Hebbian lerning rule have been proposed that do not eradicate the favorable properties of a distributed, associative system. These research efforts focus on what happens at the synapse. There is less research being done on variations of the activation rule, even though recent studies have shown that classical conditioning may take place by altering the cell body threshold (Tesauro, 1986).

A criterion for biological plausibility of a neural model is that the information needed to execute a particular process is available locally at the site that the process takes place. For example, the information needed for learning by synaptic modification should be available at the synaptic junction. This may include historical or time dependent data. Likewife, the information needed to execute the neural firing should be available at the cell body at the time that the decision to fire is made.

The restriction that information be available locally is important not only for reasons of biological plausibility. As applications are developed using PDP models, the intention is to implement them on machines with autonomous and asynchronous processors such as hypercubes or connection machines. Any algorithm that increases locality of information and thus decreases the quantity of message passing is favorable.

The information about the orthogonality of a set of input patterns is available at the synaptic junction (or in the case of an auto-associative model, at the neuron body). To demonstrate this, recall that the Hopfield memorization algorithm is stated as:

$$w_{ij} = \sum_k (2v_j-1)(2u_i-1).$$

where w_{ij} represents the synaptic junction of input cell v_j and output cell u_i. The weight w_{ij} is summed over the k input patterns. The condition of interest is when $v_j = 1$ and $u_i = 1$, because this is the condition that contributes to the increase in the dot product of any pair of the k input patterns.

For example, consider the vectors u and v:

n	1	2	3	4	5	6	7	8
u =	{0	1	1	0	1	0	1	0}
v =	{1	0	0	1	1	0	0	1}

The only elements that contribute to the dot product u*v are the 5th elements v_5 and u_5. The sum of v and u is:

| t = | {1 | 1 | 1 | 1 | 2 | 0 | 1 | 1} |

It is clear that t_j represents a measure of the contribution of the jth element to the dot product of v and u. The jth element of the sum of the k input vectors is defined as j, the measure of similarity for input unit v_j. This is a local measure of similarity (or, non-orthogonality).

If one allows time dependent information, then knowledge about the orthogonality of the input patterns is available locally for the use by the processing unit. This information is available for use by both the learning function and the activation function. It is debatable whether the computation of is biologically plausible or not. It is certainly realistic to assume that the information is locally available to a processor in a computer application. This is because there are other avenues of communication in the computer, for example, message passing, a central processor, or shared memory. The restriction on locality of information is not as severe as in the biological model.

6. COMPARISONS OF WEIGHTED MEAN AND WFEV

In the following experiment the WFEV is compared to the weighted sum as the mean similarity of inputs is varied. The parameter τ is considered to be a function of δ (the local measure of similarity):

$$\tau = -0.02 * (\delta + 0.6).$$

The firing of a particular unit is then dependent on the local similarity of each unit of the input vector. The results of the preliminary linear function are displayed below in Table 5. Note that in this example the WFEV appears to be resistant to less orthogonal input patterns, but as a trade off it shows a less graceful degradation to altered inputs.

TABLE 5
Effects of orthogonality on recall for mean and WFEV.
The dot product is averaged for all (45) pairs of the stored input patterns.

Mean dot product		Hamming distance from stored patterns					
		0	5	10	15	20	25
25.2	MEAN	100.0	99.9	99.7	99.2	96.5	94.4
	WFEV	99.5	98.9	98.0	95.9	92.1	89.1
31.6	MEAN	88.5	87.9	87.0	88.1	87.6	87.3
	WFEV	98.4	97.4	96.8	92.1	89.8	83.4
38.7	MEAN	87.4	87.4	87.4	87.1	87.5	86.8
	WFEV	91.5	92.1	87.4	83.2	63.2	55.4

When one looks closely at the auto-associative model it appears that using δ as a parameter gives unfair information to the processing unit. The processing unit is supposed to find a typical value among the large number of inputs. The memory is distributed across a large number of connections. If, for example, $\delta_j = 10$ over 10 input patterns then v_j most certainly should fire. Likewise, if $\delta_j = 0$ then v_j should not fire. It appears that δ alone is driving the activation function. This is not the case however. To demonstrate this, a simulation of a hetero-associative model was performed. The input patterns were randomly paired with the output patterns so that information about the similarity of the inputs to a particular unit could not directly influence the output of thet unit. The results of this simulation are presented in Table 6. By comparing the results to those of Table 5, one can see that the fuzzy model is still resistant to the effect of less orthogonal inputs. The results in the two tables vary because the input patterns are chosen randomly and the mean dot products are therefore different. The general results of the hetero-associative simulation are comparable to those of the auto-associative simulation.

TABLE 6

Effect of orthogonality on recall for mean and WFEV, using a hetero-associative network. The dot product is averaged for all (45) pairs of the stored input patterns.

Mean dot product		Hamming distance from stored patterns					
		0	5	10	15	20	25
24.6	MEAN	100.0	99.7	99.5	98.7	96.3	94.6
	WFEV	99.7	99.1	98.7	97.1	93.4	91.0
29.6	MEAN	95.5	94.9	93.3	94.0	92.2	90.2
	WFEV	99.4	98.4	96.2	94.8	90.9	88.5
36.0	MEAN	74.4	74.8	75.3	76.7	77.7	76.8
	WFEV	96.8	96.0	93.9	91.4	88.3	82.2

7. CONCLUSION

The injection of fuzzy techniques into the area of neural network modeling demonstrates that the uncertainty inherent in these complex systems may be interpreted in a non-probabilistic manner. The fuzzy interpretation retains the favorable characteristics of the stochastic model-pattern retrieval from partial inputs, graceful degradation during system damage, and the use of autonomous, asynchronous processors that employ locally available information to perform computations. In addition, the fuzzy model offers a satisfying approach to a subject that is too complex, too vague, and too noisy to be described explicitly. Simulations show that the use of the WFEV as an activation function demonstrates a resistance to interference from non-orthogonal input patterns. The primary disadvantage of using the WFEV in the activation function is the complexity of its computation. Since the algorithm was not implemented on a parallel system, timing constraints are difficult to judge at this point. The other result derived from this study is to reinforce the view that the underlying PDP model is robust with respect to changes in details.

There are many activation functions and many learning rules being proposed for use in PDP models. The fact that they all work to a degree demonstrates the robustness of the underlying model. This is a favorable property of many physical systems and was noted by Hopfield (1982):

> These properties follow from the nature of the ... processing algorithm, which does not appear to be strongly dependent on the precise details of the modeling.

What are some of the weaknesses in the presentation of a fuzzy activation function? One, the choice of the population function

$$\tau = f(\delta)$$

was derived experimentally. This was done by searching for linear functions that produced the most accurate level of output. This approach is not unusual in the history of neural network modeling. An algorithm that works is developed heuristically, and then the mathematical basis for the algorithm is derived. Further research should include the investigation of other population functions, their effect on the model, and some rationale for the use of one function over others.

Another problem is the accessibility of information in the hetero-associative simulations. It is stretching the point a bit to claim that the information needed to compute is available locally at the output units. During learning, the input pattern is presented to the input units and the output pattern is presented to the output units. The strength of the connections between the units are changed according to the learning algorithm.

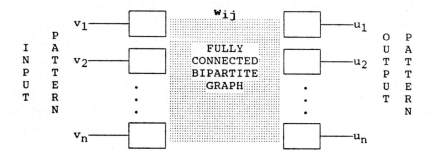

It is reasonable to conjecture that input unit v_i stores some sort of summation of the values presented to it and thus computes δ_i. But, it is difficult to visualize how that information is available locally for use by the corresponding output unit u_i. This is not a fatal error. As was mentioned in the introduction, the physiological constraints on PDP modeling are not very constrictive at this time because of the general lack of knowledge about how the brain actually inplements memory and learning.

There are other popular models (such as back-propagation learning) that contain elements with no obvious physiological basis. In addition to solving these problems, there are several directions for future research in the area of fuzzy parallel distributed processing. These are described below.

One topic for further study is the extension of the fuzzy model to include the learning rule. We have used Hopfield's algorithm for fixing the weights as a starting point but, as we mentioned earlier, the algorithm for the spontaneous learning of proper weights is the most widely varied element of PDP models. Which of the various rules that have been suggested are compatible with the fuzzy model? Does fuzzy set theory have anything to offer in the development of new learning rules?

Other areas of study are not limited to the fuzzy model but are topics for all PDP reserach. One of these is the study of multilayer networks. The cerebral cortex operates with approximately six layers of neurons. What implications does this have for PDP networks? Another interesting topic is the interface of PDP networks with other computational architectures. Can a PDP model act as a front end (or a side end) to an expert system, for example? Of course, as parallel machines become more available, the development of useful PDP applications will become more of an issue for computer scientists. There is potential for useful applications in many areas including robotics, pattern recognition, chip design, meteorological forecasting, fault tolerance, distributed operating systems, as well as cognitive and physiological modeling.

APPENDIX

The following is an example of the calculation of the mean, FEV, and WFEV as activation functions for a single firing.

Correct response is 0 (not firing)
Local similarity (δ) = 2
Threshold (θ) = 0.5
Mean: 0.45
FEV: 0.46
WFEV: 0.45

i	n	μ	X	Min(μ i, X_i)
1	22	1.00	0.38	0.38
2	1	0.56	0.42	0.42
3	5	0.54	0.46	0.46 ⇐ FEV
4	12	0.44	0.50	0.44
5	7	0.20	0.54	0.20
6	2	0.06	0.58	0.06
7	1	0.02	0.62	0.02

Total: 50

There are 50 inputs in the table, representing the active units in the input vector (n = 100). The mean is equivalent to the usual weighted sum of inputs and can be calculated from the table by summing the products of each n_i and X_i:

$$\text{Mean} = \sum_i n_i * X_i$$

The FEV is calculated by finding the maximum of the minimums of each X, μ pair:

$$\text{FEV} = \max_i [\min(X_i, \mu_i)]$$

The WFEV is calculated as described in the Section 4 with the parameter τ defined as a linear function of δ, the local measure of similarity (see section titled "Effect of Orthogonality on Recall" in Section 5):

$$\tau = -0.02 * \delta + 0.6.$$

The threshold is predefined to be 0.5. In this instance, all three measures have compuetd the correct output response of 0 (or not firing) for this unit.

This research has been supported in part by NSF grant IST 8405953.

REFERENCES

Anderson, J. A. Cognitive and psychological computation with neural models. *IEEE Transactions on System, Man, and Cybernetics*, 1983, *SMC-13*, 799-815.

Feldman, J. A., & Ballard, D. H. Connectionist models and their properties. *Cognitive Science*, 1982, *6*, 205-254.

Friedman, M., Schneider, M., & Kandel, A. The use of weighted fuzzy expected vaule (WFEV) in fuzzy expert system. *Fuzzy Sets and Sytems*, 1987, in press.

Hall, L. O. *A possibilistic pattern recognition technique*. Unpublished master's thesis. Tallahassee, FL: Florida State University, 1982.

Hebb, D. O. *The organization of behavior*. New York: Wiley, 1949.

Hoepfield, J. J. Neural networks and physical systems with emergent collective computational abilities. *Proceedings of the National Academy of Sciences USA*, 1982, *79*, 2554-2558.

Kandel, A. *Fuzzy mathematical tecniques with applications*. Reading, MA: Addison-Wesley, 1986.

Kohonen, T. *Self-organization and assocoative memory*. Berlin: Springer, 1984.

Lee, S. C., & Lee, E. T. Fuzzy sets and neural networks. *Journal of Cybernetics*, 1974, *4*, 83-103.

Lee, S. C., & Lee, E. T. Fuzzy neural networks. *Mathematical Biosciences*, 1975, *23*, 151-177.

Peretto, P., & Niez, J. Long term memory storage capacity of multiconnected neural networks. *Biological Cybernetics*, 1986, *54*, 53-63. (a)

Peretto, P., & Niez, J. Stochastic dynamics of neural networks. *IEEE Transactions on Systems, Man, and Cybernetics*, 1986, *SMC-16*, 76-83. (b)

Rumelhart, D. E., & McClelland, J. L. (Eds.) *Parallel distributed processing: Explorations in the microstructure of cognition.* Vol. 1. Cambridge, MA: MIT Press, 1986.

Stiles, G. S., & Deng, D. A quantitative comparison of the performance of three discrete distributed associative memory models. *IEEE Transactions on Computers*, 1987, *C-36*, 257-263.

Stone, G. An analysis of the delta rule and the learning of statistical associations. In D. E. Rumelhart and J. L. McClelland (Eds.) *Parallel distributed processing: Explorations in the microstructure of cognition.* Cambridge, MA: MIT Press, 1986

Teasuro, G. Simple neural models of classical conditioning. *Biological Cybernetics*, 1986, *55*, 187-200.

Williams, R. J. The logic of activation functions. In D. E. Rumelhart and J. L. McClelland (Eds.) *Parallel distributed processing: Explorations in the microstructure of cognition.* Cambridge, MA: MIT Press, 1986.

Zadeh, L. A. Fuzzy sets. *Information and Control*, 1965, *8*, 338-353.

FUZZY LOGIC WITH LINGUISTIC QUANTIFIERS: A TOOL FOR BETTER MODELING OF HUMAN EVIDENCE AGGREGATION PROCESSES?

Janus KACPRZYK

System Research Institute, Polish Academy of Sciences
ul. Newelska 6, 01-447 Warsaw, Poland

Two fuzzy logic based calculi of linguistically quantified propositions are presented. Basically, they are concerned with the calculation of the truth of the linguistically quantified propositions generally written as "QY's are F" and "QBY's are F", where Q is a linguistic quantifier, B is importance, Y is a set of objects and F is a property. The first proposition is exemplified by "most experts are convinced" and the second one by "most of the important experts are convinced". It is then advocated that the above calculi of linguistically quantified propositions can provide means for more adequate and consistent human evidence aggregation processes. This is based on a positive experience from some new multiaspect decision making models in which the aggregation of partial scores (related to the particular aspects) is crucial. In particular, there are considered new multiobjective decision making models seeking an optimal option which best satisfies most (almost all, ... or another linguistic quantifier) of the important objectives, multistage decision making (control) models seeking an optimal sequence of controls which best satisfies constraints and goals at, for example, most of the earlier control stages, and new group decision making models in which new solution concepts have been proposed based on a fuzzy majority given as a linguistic quantifier.

1. INTRODUCTION

Decision making - which is the essence of virtually all human activities at all levels, those of an individual, a group and an organization - is particularly difficult in the present world that is complex, competitive, ill-structured, etc. This clearly calls for some (computer) support to help make (better) decisions. Recent developments in computer technology justify this. At present, and presumably in the foreseeable future, it seems that the most efficient use of decision support systems (DSSs) will be to assist and help humans arrive at a proper decision but by no means to replace humans. The system should therefore carry out the tasks for which it is better suited, such as the processing of well defined and structured information, and provide the user with "good" solution guidelines leaving the user the final choice that involves more "delicate" aspects better handled by the humans.

To effectively and efficiently arrive at a proper and well timed decision within the framework of DSS, some synergy between the human and "machine" should exist. Since any change of human nature is rather difficult to obtain, attempts to make the machine more "human consistent", which parallels what is often termed in ergonomics (human factor engineering) "to fit the task to the man", seem to be more promising.

The human consistency of DSSs have two main aspects. The first, which might be called the input/output human consistency, is related to communication (interface) between the user and system and is most often considered. For instance, user friendly natural language based interfaces have been proposed. The second, which might be termed the algorithmic/procedural human consistency (Kacprzyk, 1986b), is related to the algorithms and procedures used by the system to obtain a solution. Basically, its essence is to try to make mathematical models, algorithms, procedures, etc. somewhat parallel the way the human user perceives their essence, intention and purpose. The rationale is that, from a pragmatic point of view, a "quality criterion" of any DSS is the implementability of its solution proposals. Those procedures which do not depart too much from the user's experience, perception, commonsense, etc., should be easier to implement.

In this paper we are concerned with some aspects of that algorithmic/procedural human consistency of the DSSs. We adopt here the following perspective. A prerequisite for decision making is the availability of a mechanism for evaluating the particular options in question; we assume here for simplicity that this evaluation process yields a numerical value for each option.

Due to the complexity of the present world, any realistic decision making problem should be considered in a multiaspect setting, i.e. accounting for multiple decision makers. This does complicate to a large extent the evaluation process of a particular option. The evaluation with respect to each aspect results therefore in a partial score, and then these partial scores are aggregated (pooled) to obtain the final score which is to be meant as the intended evaluation.

Virtually all attempts to model this process of aggregating the partial scores have been conventionally modeled as some manner of (weighted) averaging which implicitly involves all the partial scores though with possibly different importances. On the other hand, experience clearly indicates that in many practical cases humans do not consider all the aspects, neglecting some of them in the aggregation process. First, if the aspects represent life quality indicators (objectives), then the human perception (evaluation) of life quality can be satisfactory even if some indicators take on too low values (say, below aspiration levels) but the other ones take on satisfactory values. This seems to clearly indicate that some aspects are neglected in the aggregation process. Second, in case of a socioeconomic development planning problem, which is an example of multistage decision making, the human evaluation of a development trajectory is often satisfactory even if at some (sufficiently few and, say, later) stages (say, years) the development indicators take on too low values. This also seems to clearly indicate that some stages are neglected in the aggregation process. Finally, in the case of multiperson decision making, an analyis of human perception of how the individual testimonies (e.g., utilities or preferences) are to be aggregated seems also to clearly indicate that the testimonies of some individuals are neglected.

The above mentioned facts have led to attempts to reformulate some multiaspect (multiobjective, multistage and group) decision making models in which, roughly speaking, that process of neglecting some objectives, stages and individuals be modeled in a possibly human consistent way. Thus, first, new multiobjective decision making models have been proposed (Yager, 1983b; Kacprzyk & Yager, 1984a, 1984b) in which an optimal option is thought which best satisfies, say, most (almost all, much more than 50%, ...) of the important objectives. Second, new multistage decision making (control) models appeared in Kacprzyk (1983c, 1984a) and Kacprzyk and Iwanski (1987) which seek an optimal sequence of controls best satisfying constraints and goals at, say, most of the earlier control stages. Third, new group decision making models have been proposed in Kacprzyk (1984b, 1985b, 1985c, 1986a) where in the determination of solutions a fuzzy majority represented by a linguistic quantifier (e.g., most, almost all, ...) is considered, i.e. the testimonies of, say, most individuals are accounted for.

In the above models the number of aspects to be considered is specified by a fuzzy linguistic quantifier, and fuzzy logic based calculi of linguistic quantified propositions are employed. A positive experience with those models seems to suggest that fuzzy logic with linguistic quantifiers, in our particular case in the form of a calculus of linguistically quantified propositions, can be an adequate and useful tool for modeling the human evidence aggregation processes. This is the main message expressed in this article. Evidently, this is not supported by any psychological investigations, and we hope that this paper could trigger some research efforts in the scientific communities involved in the study of human behavior and mental processes.

Section 2 presents a brief account of two fuzzy logic based calculi of linguistically quantified propositions. Sections 3, 4 and 5 sketch the new multiobjective, multistage and group decision making models. The last section contains the concluding remarks and an extended list of literature.

To make the paper readable to a larger audience the models are sometimes presented in a less formal manner with mathematics and technical details kept at minimum. The interested reader can find details in the cited sources as well as in the author's survey articles (Kacprzyk, 1984a, 1986d, 1987b). Our notation on fuzzy sets is standard. For futher details see e.g. Dubois and Prade (1980), Kacprzyk (1983b, 1986c) or Zimmermann (1985).

2. ON SOME FUZZY LOGIC BASED CALCULI OF LINGUISTICALLY QUANTIFIED PROPOSITIONS

A linguistically quantified proposition is exemplified by "most experts are convinced" and may be generally written as

QY's are F (1)

where Q is a linguistic quantifier (e.g. most), $Y = \{y\}$ is a set of objects (e.g. experts), and F is a property (e.g. convinced).

Importance, B, may also be added to (1) yielding

QBY's are F (2)

that is, say, "most of the important experst are convinced".

Basically, the problem is to find either truth (QY's are F) in case of (1), or truth (QBY's are F) in case of (2), knowing truth (y is F), for each $y \in Y$.

It is easy to see that the conventional logical calculi make it possible to find these truths for crisp (precise) quantifiers only, mainly for "all" and "at least one". To deal with linguistic quantifiers exemplified by "most", "almost all", etc. a fuzzy logic based calculus can be used, and two of them will be presented below.

2.1. The algebraic or consensory method

In the classical method due to Zadeh (1983c) the (fuzzy) linguistic quantifier Q is assumed to be a fuzzy set in $[0,1]$, $Q \supseteq [0,1]$. For instance, Q = "most" may be given as

$$\mu_{\text{"most"}}(x) \begin{array}{l} = 1 \text{ for } x \geq 0.8 \\ = 2x - 0.6 \text{ for } < 0.3 \, x < 0.8 \\ = 0 \text{ for } x \leq 0.3 \end{array} \qquad (3)$$

We only consider the proportional quantifiers (e.g., most, almost all, etc.) as they are more important here than the absolute ones (e.g. about 5). The reasoning for the absolute quantifier is however analogous (see, e.g. Kacprzyk, 1985a).

Property F is defined as a fuzzy set in Y, $F \subseteq Y$. If

$Y = \{y_1, \ldots y_p\}$, then it is assumed that truth(y_i is F) $= \mu_F(y_i)$, $i = 1, \ldots, p$.

Truth (QY's are F) is now calculated using the (nonfuzzy) cardinalities, the so-called Σ Counts, of the respective fuzzy sets in the following two steps (Zadeh, 1983c):

Step 1: $r = \Sigma \text{Count}(F)/\Sigma \text{Count}(Y) = 1/p \sum_{i=1}^{p} \mu_F(y_i)$ (4)

Step 2: $\text{truth}(QY\text{'s are } F) = \mu_Q(r)$ (5)

In case of importance, B = "important" $\subseteq$ Y, and $\mu_B(y_i) \in [0,1]$ is a degree of importance of y_i: from 1 for definitely important to 0 for definitely unimportant, through all intermediate values.

We rewrite first "QBY's are F" as Q(B and F)Y's are B" which implies the following counterparts of (4) and (5):

Step 1 : $r' = \Sigma \text{Count}(B \text{ and } F)/\Sigma \text{Count}(B) =$

$$= \sum_{i=1}^{p}(\mu_B(y_i) \wedge \mu_F(y_i))/ \sum_{i=1}^{p} \mu_B(y_i) \quad (6)$$

Step 2 : $\text{truth}(QBY\text{'s are } F) = \mu_Q(r')$ (7)

The minimum operator "$\wedge$" can be replaced by any other suitable operator as, say, a t-norm (Kacprzyk & Yager, 1984b).

Example 1

Let Y = "experts" = {X,V,Z}, F = "convinced" = 0.1/X + 0.6/V + 0.8/Z, Q = "most" be given by (3) and B = "important" = 0.2/x + 0.5/V + 0.6/Z.

Then $r = 0.5$ and $r' = 0.8$ and
truth ("most experts are convinced") = 0.4
truth ("most of the important experts are convinced") = 1

The method presented may be viewed to provide a consensory like aggregation of evidence. See Yager (1983b) or Kacprzyk and Yager (1984b).

2.2. The substitution or competitive method

The alternative method is developed by Yager (1983a, 1983b). As previously, $Y = \{y_1, ..., y_p\}$ and $F \subseteq Y$. A proposition "y_i is F" is denoted by P_i, and truth P_i = truth(y_i is F) = $\mu_F(y_i)$, $i = 1, ..., p$.

We introduce the set $V = \{v\} = 2^{\{P_1,...,P_p\}}$, where v is "$P_{k1}, ..., P_{km}$" meant as the proposition "P_{k1} and ... and P_{km}"; v and "P_{k1} and ... and P_{km}" are used here interchangeably.

Each v, or its corresponding "P_{k1} and ... and P_{km}", is seen to be true to the degree

$$T(v) = \text{truth}(P_{k1} \text{ and } ... \text{ and } P_{km}) = \text{truth } P_{k1} \wedge ... \wedge \text{truth } P_{km} =$$

$$= \bigwedge_{i=1}^{m} \text{truth } P_{k1} = \mu_F(y_{k1}) \wedge ... \wedge \mu_F(y_{km}) =$$

$$= \bigwedge_{i=1}^{m} \mu_F(y_{k1}) \qquad (8)$$

i.e. we obtain a fuzzy set $T \subseteq V$.

Example 2

For the same data as in Example 1., let Q = "most" be given by (9). Then

truth("most experts are convinced") = 0.6
truth("most of the important experts are convinced") = 0.8

Notice that the method may yield different results than the algebraic one.

The substitution method may be viewed to provide a compatitive like aggregation of evidence (see, e.g., Yager, 1983b or Kacprzyk & Yager, 1984b).

Finally, notice that Zadeh's calculus is conceptually and numerically simpler though some experience indicates that Yager's calculus is more "adequate".

3. MULTIOBJECTIVE DECISION MAKING MODELS BASED ON FUZZY LOGIC WITH LINGUISTIC QUANTIFIERS

For our purposes multiobjective decision making under fuzziness may be formalized as follows (Zimmermann, 1985 or Kacprzyk, 1983b, 1986c). We have a set of possible options (alternatives, variants, decisions, ...), $A = \{a\} = \{a_1, ..., a_n\}$, and a set of objectives (criteria, constraints and/or goals, ...)., $O = \{O_1, ..., O_p\}$. These sets are assumed finite. A degree to which option $a \in A$ satisfies objective $O_i \in O$ is given by truth (O_i is satisfied (by a)) $= \mu\ O_i(a) \in [0.1]$, from $\mu\ O_i(a) = 1$ for full satisfaction, to $\mu O_i(a) = 0$ for full dissatisfaction, through all intermediate values.

It is commonly postulated - what has originated from the seminal (for virtually all fuzzy approaches in decision making, optimization, control!) work of Bellman and Zadeh (1970) - that option $a \in A$ satisfy "O_1 and O_2 and ... and O_p", i.e. all the fuzzy objectives; the degree of this satisfaction is expressed by the so-called fuzzy decision

$$\mu D(a\,"all") = \text{truth}(O_1 \text{ and } \ldots \text{ and } O_p \text{ are satisfied (by a)}) =$$

$$= \text{truth}("all"\,O\text{'s are satisfied}) = \text{truth}(O_1 \text{ is satisfied}) \wedge \ldots$$

$$\ldots \wedge \text{truth}(O_p \text{ is satisfied}) = \mu O_1(a) \wedge \ldots \wedge \mu O_p(a) =$$

$$= \bigwedge_{i=1}^{p} \mu O_i(a) \tag{12}$$

and the problem is to find an optimal option $a^* \in A$ such that

$$\mu_D(a^*\,|\,"all") = \max_{a \in A} \mu_D(a\,|\,"all"). \tag{13}$$

specifically, such that it best satisfies (maximizes the membership function of the fuzzy decision) *all* the fuzzy objectives.

Notice that $\mu_D(.\,|\,"all")$, and related expressions mean only that "all" the elements (objectives) are accounted for; the same convention is used throughout this paper.

As mentioned in Section 1, it is quite natural that in many practical cases the requirement to satisfy "all" the fuzzy objectives may be too rigid, restrictive and even counter-intuitive. More adequate could be to have a mechanism to account for some objectives only, for instance "most", "almost all" of them. An idea of replacing the conventional requirement to satisfy "all" the fuzzy objectives by "most", "almost all", "much more than 50%", etc. (in general, a fuzzy linguisitc quantifier Q that is "milder" than "all") appeared in Yager (1983b) and was then advanced in Kacprzyk & Yager (1984a, 1984b).

In this case the fuzzy decision may be written as

$$\mu_D(a\,|\,Q) = \text{truth}(QO\text{'s are satisfied}) = (\bigwedge_{i=1}^{p} |\,Q)\mu\,O_i(a) \tag{14}$$

which means that in the calculation of $\mu D(..)$ only Q (e.g."most") fuzzy objectives are to be taken into account (but does not indicate how to do this!). We seek therefore an optimal option $a^* \in A$ such that

$$\mu_D(a^* | Q) = \max_{a \in A} \mu_D(a | Q) \qquad (15)$$

specifically, such that it best satisfies QO's.

We can also introduce a varying importance of the particular fuzzy objectives by a fuzzy set B in the set of objectives, $B \subseteq O$, such that $\mu B(Oi) \in [0,1]$ is the importance of objective O_i: from O for definitely unimportant to 1 for definitely important through all intermediate values. In case of importance our multiobjective decision making problem formulation can be written as

$$\mu_D(a\ Q,B) = \text{truth}(QBO\text{'s are satisfied}) = (\bigwedge_{i=1}^{p} | Q,B)\mu\ O_i(a) \qquad (16)$$

which means that only Q of the important objectives are to be taken into account (but once again, does not say how to do this), and we seek an optimal option $a^* \in A$ such that

$$\mu_D(a^*\ Q,B) = \max_{a \in A} D(a\ Q,B) \qquad (17)$$

specifically, such that it best satisfies Q of the important objectives.

The solution of both the above problems will be sketched using the algebraic and substitution methods. A full exposition of the solution would require too many technical details which are beyond the scope of this paper.

Thus, let us introduce a fuzzy set S = "satisfied" $\subseteq$ O such that $\mu S(O_i)$ $\in [0,1]$ is the degree to which objective $O_i \in O$ is satisfied (by option $a \in A$): from 0 for full dissatisfaction to 1 for full satisfaction through all intermediate values; thus $\mu S(O_i) = $ truth (O_i is satisfied)(by a)) $= \mu O_i(a)$, $i = 1, ..., p$; B = "important" $\subseteq$ O is importance and $Q \subseteq [0,1]$ is a fuzzy linguistic quantifier.

Then, if we use the substitution method, we obtain for the case without importance

$$\mu_D(a\,Q) = \text{truth}(QO\text{'s are satisfied}) = \mu_Q(1/p \sum_{i=1}^{p} O_i(a)) \tag{18}$$

and we seek an $a^* \in A$ such that

$$\mu_D(a^* \mid Q) = \max_{a \in A} \mu_D(a \mid Q) \tag{19}$$

In the case with importance we obtain

$$\mu_D(a\,Q,B) = \text{truth}(QBO\text{'s are satisfied}) =$$

$$= \mu_Q(\sum_{i=1}^{p}(\mu B(O_i) \wedge \mu O_i(a))/\sum_{i=1}^{p} \mu_B(O_i)) \tag{20}$$

and we seek an $a^* \in A$ such that

$$\mu_D(a^* \mid Q,B) = \max_{a \in A} \mu_D(a \mid Q,B) \qquad (21)$$

On the other hand, if we use the substitution method, we obtain in the case without importance

$$\mu_D(a \mid Q) = \max_{v(a) \in V} (\mu_Q(v(a)) \wedge \bigwedge_{i=1}^{m} \mu\, O_{ki}(a)) \qquad (22)$$

and we seek an $a^* \in A$ such that

$$m_D(a^* \mid Q) = \max_{a \in V} \mu_D(a \mid Q) \qquad (23)$$

In the case with importance, we obtain

$$\mu D(a\, Q,B) = \text{truth}(QBO\text{'s are satisfied (by } a)) =$$

$$= \max_{v(a)\, \in V} (\mu\, Q(v(a)) \wedge (\bigwedge_{i=1}^{m} (\mu\, B(O_{ki}) \to \mu O_{ki}(a)))) \qquad (24)$$

and we seek an $a^* \in A$ such that

$$\mu_D(a^* \mid Q,B) = \max_{a \in A} \mu_D(a \mid Q,B) \qquad (25)$$

The above problems, in particular (24) and (25), are analytically difficult in the general case. Fortunately enough, and this is also true for virtually all the problems through this paper, under some mild and realistic assumptions the solution becomes much simpler (Kacprzyk, 1985a, 1987b; Kacprzyk & Yager, 1984a, 1984b,; Yager, 1983b).

4. MULTISTAGE DECISION MAKING (CONTROL) MODELS BASED ON FUZZY LOGIC WITH LINGUISTIC QUANTIFIERS

For the purposes of this paper multistage decision making (control) under fuzziness means as follows: A system under control S proceeds from an initial state (output) x_0 under a control u_0 to a new state x_1. Control u_1 is applied and state x_2 is obtained. Finally, under control u_{N-1} from state x_{N-1} state x_N is attained. For each control stage (time) $t = 0, 1, ..., N-1$ its partial (stage) score is evaluated as a cost/benefit relation (related to the control applied and to the state attained). The partial scores are then aggregated to obtain a total score. We seek an optimal sequence of control $u^*_0, ..., u^*_{N-1}$ which gives a best total score.

Formally, at each control stage (time) t, the control $u_t \in U = \{c_1, ..., c_m\}$ is subjected to a fuzzy constraint $\mu_{C^t}(u_t)$, and on the state attained (at $t+1$) $x_{t+1} \in X = \{s_1, ..., s_n\}$ a fuzzy goal $\mu_{G^{t+1}}(x_{t+1})$ is imposed. The state transitions are governed by $x_{t+1} = f(x_t, u_t); x_t, x_{t+1} \in X, u_t \in U; x_0$ is an initial state, N is a termination time (planning horizon), and $t = 0, 1 ..., N-1$.

It is commonly postulated (Bellman & Zadeh, 1970; Kacprzyk, 1983b) that at each t, u_t statisfy the fuzzy constraint C^t and the fuzzy goal G^{t+1}, written as P_{t+1}: "C^t and G^{t+1} are satisfied (by u_t and the resulting x_{t+1}, respectively)"; this satisfaction is to the degree

$$\text{truth } P_{t+1} = \text{truth}(\text{"}C^t \text{ and } G^{t+1} \text{ are satisfied"}) =$$

$$= \mu_{C^t}(u_t) \wedge \mu_{G^{t+1}}(x_{t+1}) \tag{26}$$

Traditionally, a sequence of controls $u_0, \ldots, u_{N-1}$ is required to satisfy the fuzzy constraints and goals at *all* the subsequent control stages; this satisfaction is to the degree expressed by the fuzzy decision

$$\mu_D(u_0, \ldots, u_{N-1} \mid x_0, \text{"all"}) = \text{truth}(P_1 \text{ and } \ldots \text{ and } P_N \mid \text{"all"}) =$$

$$= \bigwedge_{t=0}^{N-1} \mid \text{"all"}) \text{ truth } P_{t+1} =$$

$$= (\bigwedge_{t=0}^{N-1} \mid \text{"all"}) (\mu_{C^t}(u_t) \wedge \mu_{G^{t+1}}(x_{t+1})) =$$

$$= (\bigwedge_{t=0}^{N-1} (\mu_{C^t}(u_t) \wedge \mu_{G^{t+1}}(x_{t+1})) \tag{27}$$

and the problem is to find an optimal sequence of control $u^*_0, \ldots, u^*_{N-1}$ such that

$$\mu_D(u^*_0, \ldots, u^*_{N-1} \mid x_0, \text{"all"}) = \max_{u_0, \ldots, u_{N-1}} \mu_D(u_0, \ldots, u_{N-1} \mid x_0, \text{"all"}) \tag{28}$$

Notice that, as in Section 3, "all" in the above only indicates that all the control stages are taken into account.

For a detailed analysis of this class of problems, and its extensions - mainly with respect to the termination time (fixed and specified in advance, implicitly given through a termination set of states, fuzzy and infinitive) and the system under control (deterministic, stochastic, and fuzzy) - see Kacprzyk (1983b).

Since the new problem formulation to be presented here concerns some more fundamental aspects, the material following concentrates on the basic case, with fixed and specified termination time and the deterministic system under control.

It is easy to see that the reqirement to satisfy the fuzzy constraints and fuzzy goals at all the control stages may often be too rigid and counter-intuitive in practice. An approach of replacing "all" by some milder requirement expressed by a fuzzy linguistic quantifier Q, such as "most", appeared in Kacprzyk (1983c, 1984a) where $u_0, ..., u_{N-1}$ was required to satisfy the fuzzy constraints and fuzzy goals at Q (such as, most) control stages, that is

$$\mu_D(u_0, ..., u_{N-1} \mid x_0, Q) = (\bigwedge_{t=0}^{N-1} \mid Q) \text{ truth } P_{t+1} =$$

$$= (\bigwedge_{t=0}^{N-1} \mid Q)(\mu_{C^t}(u_t) \wedge \mu_{G^{t+1}}(x_{t+1})) \tag{29}$$

and we seek an $u^*_0, ..., u^*_{N-1}$ such that

$$\mu_D(u^*_0, ..., u^*_{N-1} \mid x_0, Q) = \max_{u_0,...,u_{N-1}} \mu_D(u_0,...,u_{N-1} \mid x_0, Q) \tag{30}$$

The importance of the particular control stages can also be introduced.

The only practically relevant approach is to equate importance with discounting that is commonly used in multistage decision making. In such a case B = "earlier", that requires $u_0, ..., u_{N-1}$ to satisfy the fuzzy constraints and goals at Q (say, most) of the earlier control stages, that is

Fuzzy Logic with Linguistic Quantifiers

$$\mu_D(u_0, ..., u_{N-1} \mid x_0, Q\text{"earlier"}) = (\overset{N-1}{\underset{t=0}{\wedge}} \mid Q\text{"earlier"}) \text{truth } P_{t+1} =$$

$$= (\overset{N-1}{\underset{t=0}{\wedge}} \mid Q\text{"earlier"})(\mu_{C^t}(u_t) \wedge \mu_{G^{t+1}}(x_{t+1})) \qquad (31)$$

and we seek an $u^*_0, ..., u^*_{N-1}$ such that

$$\mu_D(u^*_0, ..., u^*_{N-1} \mid x_0, Q\text{"earlier"}) = \max_{u_0,...,u_{N-1}} \mu_D(u_0,...,u_{N-1} \mid x_0, Q\text{"earlier"}) \qquad (32)$$

The solution of these problems will not be discussed since this is too specific and technical, and not particularly illustrative. Basically, using the algebraic method, the solution reduces to some set of dynamic programming recurrence equations (cf. Kacprzyk, 1983c, 1984a) which is, at least conceptually, relatively uncomplicated.

The use of the substitution method, which seems to be much more promising in this case, results in problem formulations which are conceptually and numerically much more difficult. The solution procedures have been proposed by Kacprzyk (1983a) for the problem without importance (discounting) given by (29) and (30), and by Kacprzyk and Iwanski (1987) for the problem with importance (discounting) given by (31) and (32). Basically, in both the papers the problem is solved by an implicit enumeration technique based on neglecting nonpromising partial solutions (sequences of controls from $t=0$ to $t=k<N-1$).

5. GROUP DECISION MAKING MODELS WITH A FUZZY MAJORITY BASED ON FUZZY LOGIC WITH LINGUISTIC QUANTIFIERS

In the context of this paper group decision making means as follows: We have a set of n options, $S = \{s_1,...,s_n\}$, and m individuals. Each individual k, $k = 1,...,m$, provides his or her preferences over S. As these preferences may not be clear-cut, their representation by individual fuzzy preference relations is strongly advocated (Blin and Whinston, 1973; Nurmi, 1981).

An individual fuzzy preference relation of individual k, R_k, is characterized by its membership function

$$\mu_{R_k}: S \times S \to [0,1] \tag{33}$$

If card S is small, R_k may be conveniently represented by an n x n matrix $R_k = [r^k_{ij}]$, $i,j = 1,...,n$, $k = 1,...,m$, whose elements $0 \le r^k_{ij} \le 1$ are meant as the higher $[r^k_{ij}]$ the higher the preference of individual k of s_i over s_j: from 0 for a definite preference of s_j over s_i, to 1 for a definite preference of s_i over s_j, through all intermediate values, with 0.5 for indifference.

Now, given the idividual preference relations of the particular individuals, $R_1, ... R_m$, we seek an option (or a set of options) which is "best" acceptable by the group of individuals as a whole.

Basically, in our context the two lines of reasoning may be followed:

$\{R_1,...,R_m\} \to$ solution

and

$\{R_1,...,R_m\} \to R \to$ solution

where R is a social fuzzy preference relation that is similarly defined as its individual counterpart but concerns the preference of the whole group as to the particular pairs of options (Blin and Whinston, 1973).

Thus, in the first case we derive a solution just from the individual fuzzy preference relations, while in the second case we derive first a fuzzy social preference relation and then use it to derive a solution.

A solution is here not clearly understood (Nurmi, 1981) for diverse solution concepts. Basically, each solution contains options that are "best" acceptable by some majority of individuals.

Usually, a majority (50%) is assumed to be a crisp threshold number of individuals. This is certainly necessary and adequate in political or election situations, however, in less formal settings and smaller groups an "adequate majority", on which decisions are to be made, is often perceived in a less precise manner, such as "most" or "almost all". Evidently, these can be equated with fuzzy linguistic quantifiers. Such an approach has been proposed in Kacprzyk (1984b, 1985b, 1985d, 1986a), where some commonly used solution concepts have been redefined: two of them - the core, and the consensus winner will be considered below.

5.1. Fuzzy cores

We follow the scheme $\{R_1,...R_m\} \rightarrow$ solution, i.e. we determine a solution just on the basis of the individual preference relations. Among many solution concepts proposed in the literature the core is intuitively appealing and often used. Conventionally, the core is defined as a set of undominated options, those not defeated by a required majority $r \leq m$. Specifically

$$C = \{s_i \in S: \nexists s_j \in S \text{ such that } r^k_{ij} > 0.5 \text{ for at least r individuals}\}. \qquad (34)$$

Nurmi (1981) extends the core to the fuzzy α-core defined as

$$C\alpha = \{s_i \in S: \exists s_j \in S \text{ such that } r^k_{ij} > \alpha \geq 0.5 \text{ for at least r individuals}\}. \quad (35)$$

i.e., as a set of options not sufficiently (at least to degree 1-α) defeated by the required majority.

Using a fuzzy majority specified by a fuzzy linguistic quantifier we deal with the above cores.
First, one can use the algebraic method:

$$h^k_{ij} = 1 \quad \text{if} \quad r^k_{ij} > 0.5$$
$$= 0 \text{ otherwise} \quad (36)$$

where, if not otherwise specified, i, j = 1, ...,n; k = 1, ..., m, throughout the paper. Then

$$h^k_{ij} = 1/n-1 \sum_{n=1, i=j}^{n} h^k_{ij} \quad (37)$$

is the extent to which individual k is not against options s_j.

Next
$$h^k_j = 1/m \sum_{k=1}^{m} h^k_j \quad (38)$$

is to what extent all the individuals are not against s_j. And

$$v^i{}_Q = \mu_Q(h_j) \tag{39}$$

is to what extent Q individuals are not against s_j.

The fuzzy Q-core is now defined as a fuzzy set

$$C_Q = v^1{}_Q/s_1 + \ldots + v^n{}_Q/s_n \tag{40}$$

specifically, a fuzzy set of options that are not defeated by Q (for example, "most" individuals. Analogously we can define the fuzzy /Q-core. First, we denote

$$h^k{}_{ij}(\alpha) = 1 \quad \text{if} \quad r^k{}_{ij} < \alpha \le 0.5$$
$$= 0 \text{ otherwise} \tag{41}$$

and then, in an analogous way we define the fuzzy /Q-core as a fuzzy set of options that are not sufficiently (at least to degree 1-) defeated by Q individuals. We can also explicitly introduce the strength of defeat and define the fuzzy s/Q-core. Namely,

$$h^k{}_{ij} = 2(0.5 - r^k{}_{ij}) \quad \text{if} \quad r^k{}_{ij} < 0.5$$
$$= 0 \quad \text{otherwise} \tag{42}$$

and then analogously define the fuzzy s/Q-core as a fuzzy set of options that are not strongly defeated by Q individuals.

Using the substitution method, one can also start with (36), i.e. deriving $h^k{}_{ij}$, and then deriving $h^k{}_j$ which yields the extent to which individual k is not against option s_j.

Introducing the statements $P^k{}_j$: "individual k is not against option s_j" which are true to the extent truth $P^k{}_j = h^k{}_j$. Constructing the set $V_j = \{v_j\} = \{p^{kl}{}_j$ and ... and $P^{kp}{}_j\} = 2^{\{P_{1j}, ..., P_{jm}\}}$ whose generic element $v_j = p^{k1}{}_j$ and ... and $p^{kp}{}_j$ is is meant as "individual k1 is not against option s_j" and ... and "individual kp is not against option s_j".
One can introduce the statements

$P^Q{}_j$: "Q individuals are not against option s_j" (43)

whose degree of truth is

$$\text{truth } P^Q{}_j = \max_{v_j \in V_j} (\mu_T(v_j) \wedge \mu_Q(v_j)) \quad (44)$$

Finally, the fuzzy Q-core is defined as

$$C_Q = \text{truth } P^Q{}_1/s_1 + ... + \text{truth } P^Q{}_n/s_n \quad (45)$$

i.e. as a fuzzy set of options that are not defeated by Q individuals.

And analogously, if one uses (41), then the fuzzy α/Q-core can be defined as a fuzzy set of options that are not sufficiently (at least to degree 1-α) defeated by Q individuals.

Moreover, if one uses (42), then the fuzzy s/Q-core can be defined as a fuzzy set of options that are not strongly defeated by Q individuals.

5.2. Fuzzy consensus winners

Following the scheme $\{R_1, ..., R_m\} \to R \to$ solution, where R is a social fuzzy preference relation, deriving a solution from the social fuzzy preference relation which is in turn derived on the basis of the set of individual preference relations.

Here $\{R_1, ..., R_m\} \to R$ will not be dealt with, and the assumption will be made that $R = [r_{ij}]$ is given by

$$r_{ij} = 1/m \sum_{k=1}^{m} a^k_{ij} \quad \text{if } i \neq j$$

$$= 0 \quad \text{if } i = j \tag{46}$$

where

$$a^k_{ij} = 1 \quad \text{if } r^k_{ij} > 0.5$$

$$= 0 \quad \text{otherwise} \tag{47}$$

For other methods of determinig R, see. e.g., Blin and Whinston (1973).

Below will be discussed the R → solution, how to derive a solution from the social fuzzy preference relation. A solution concept with much intuitive appeal is the consensus winner (Nurmi, 1981), that is a set of options not dominated by other options.

First, employing the algebraic method:

$$g_{ij} = 1 \quad \text{if } r_{ij} > 0.5$$
$$= 0 \quad \text{otherwise} \tag{48}$$

then

$$g_i = 1/n\text{-}1 \sum_{j=1, j=i}^{n} g_{ij} \tag{49}$$

and

$$z^i_Q = \mu_Q(g_i) \tag{50}$$

and finally defining the fuzzy Q-consensus winner as

$$W_Q = z^1_Q/s_1 + \ldots + z^n_Q/s_n \tag{51}$$

specifically, as a fuzzy set of options that are preferred over Q other options.

Analogously,

$$g_{ij}(\alpha) = 1 \quad \text{if} \quad r_{ij} > \alpha \geq 0.5$$
$$= 0 \quad \text{otherwise} \tag{52}$$

can be used and thus define the fuzzy /Q-consensus winner as a fuzzy set of options that are sufficiently (at least to degree) preferred over Q other options.

The strength of preference can also be explicitly introduced into (49) by

$$g_{ij} = 2(r_{ij} - 0.5) \quad \text{if} \quad r_{ij} > 0.5$$
$$= 0 \quad \text{otherwise} \tag{52}$$

and thus define the fuzzy /Q-consensus winner as a fuzzy set of options that are strongly preferred over Q other options.

Using the substitution method, beginning with (48) and then, following the reasoning in Subsection 5.1, obtaining:

P_i^j: "option s_i is preferred over option s_j", $i \neq j$

$$\text{truth } P_i^j = g_{ij} \tag{54}$$

$$V_i = \{v_i\} = \{P_i^{j1} \text{ and ... and } P_i^{jp}\} = 2^{\{P_{i1}, ..., P_{in}\}} - P_i^i \tag{55}$$

$$\mu_T(v_i) = \text{truth} (P_i^{j1} \text{ and...and } p_i^{jp}) = \text{truth } P_i^{j1} \wedge ... \wedge \text{truth } P_i^{jp} \qquad (56)$$

$$P_i^Q: \text{"option } s_i \text{ is preferred over Q options"} \qquad (57)$$

$$\text{truth } P_i^Q = \max_{v_i \in V_i} (\mu_T(v_i) \wedge \mu_Q(v_i)) \qquad (58)$$

Finally, defining the fuzzy Q-consensus winner as

$$W_Q = \text{truth } P_1^Q/s_1 + ... + \text{truth } P_n^Q/s_n \qquad (59)$$

specifically, as a fuzzy set of options that are preferred over Q other options.

Analogously, one can define the fuzzy α/Q-consensus winner as a fuzzy set of options that are sufficiently (at least to degree α) preferred over Q other options.

One can also define the fuzzy s/Q-consensus winner as a fuzzy set of options that are strongly preferred over Q other options.

For details, some properties and examples see Kacprzyk (1984b, 1985b, 1985d, 1986a).

5.3. Remarks on degrees of consensus based on fuzzy logic with linguistic quantifiers

An approach which is similar to that is Subsections 5.2 and 5.3 has been also used to derive a new degree of consensus (Kacprzyk, 1987;

Kacprzyk and Fedrizzi, 1986, 1988). Basically, it expresses the degree to which, say, most of the important individuals agree as to their preferences among almost all the relevant options. Consensus need not therefore be a full and unanimous agreement which is conventionally assumed in spite of the fact that this is often unrealistic, even utopian in practice.

6. CONCLUDING REMARKS

The main intended message in this paper is that fuzzy logic with linguistic quantifiers seems to be a realistic and adequate tool for modeling human evidence aggregation processes. This is supported by a positive experience gained from some nonconventional multiaspect decision making models, most of which have been written by this author, in which fuzzy logic with linguistic quantifiers is used to aggregate partial scores resulting from the accounting of the particular aspects.

This message seems also to be supported by a positive experience from the use of a nonconventional database querying system which makes it possible to retrieve data concerning some vague concepts (e.g., In which towns is there a serious envorinmental pollution?) by allowing queries of the type "Find all records in which most of the important attributes (pollution indicators) considerably exceed some levels". For details, see Kacprzyk and Ziolkowski (1986a, 1986b). It is hoped that the message of this paper will stimulate some research in psychology and related fields to investigate and clarify whether fuzzy logic with linguisitc quantifiers could be uselful and adequate for the formalization of various human evidence aggregation processes.

REFERENCES

Bellman, R. E., & Zadeh, L. A. Decision-making in a fuzzy environment. *Management Science*, 1970, *17*, 151-169.

Blin, J. M., & Whinston, A. P. Fuzzy sets and social choice. *Journal of Cybernetics*, 1973, *3*, 28-33.

Dubois, D., & Prade, H. *Fuzzy sets and systems*. New York: Academic, 1980.

Dubois, D., & Prade, H. Fuzzy cardinality and the modeling of imprecise quantification. *Fuzzy Sets and Systems*, 1985, *16*, 199-230.

Kacprzyk, J. *Multistage decision-making under fuzziness*. Cologne: TÜV Verlag Rheinland, 1983. (a)

Kacprzyk, J. A generelization of fuzzy multistage decision making and control via linguistic quantifiers. *International Journal of Control*, 1983, *38*, 1249-1270. (b)

Kacprzyk, J. A generalized formulation of multistage decision making and control under fuzziness. In E. Sanchez and M. M. Gupta (Eds.) *Proceedings of IFAC Symposium on Fuzzy Information, Knowledge Representation and Decision Analysis*. Oxford: Pergamon, 1984. (a)

Kacprzyk, J. *Collective decision making with a fuzzy majority.* Proceedings of 5th WOGSC Congress. Paris: AFCET, 1984. (b)

Kacprzyk, J. Zadeh's commonsense knowledge and its use in multicriteria, multistage, and multiperson decision making. In M. M. Gupta (Eds.) *Approximate Reasoning in Expert systems*. Amsterdam: North-Holland, 1985. (a)

Kacprzyk, J. Some "commonsense" solution concepts in group decision making using fuzzy linguistic quantifiers. In J. Kacprzyk and R. R. Yager (Eds.) *Management Decision Support System Using Fuzzy Sets and Possibility Theory*. Cologne: TÜV Verlag Rheinland, 1985. (b)

Kacprzyk, J. Group decision making with a fuzzy majority via linguistic quantifiers. Part I: A consensory-like pooling. *Cybernetics and Systems,* 1985, *18,* 135-145. (c)

Kacprzyk, J. Group decision making with a fuzzy majority via linguistic quantifiers. Part II: A competitive-like pooling. *Cybernetics and Systems,* 1985, *18,* 147-160. (d)

Kacprzyk, J. Group decision making with a fuzzy linguistic majority. *Fuzzy Sets and Systems,* 1986, *18,* 105-118. (a)

Kacprzyk, J. Toward "human-consistent" multistage decision making and control models using fuzzy sets and fuzzy logic. *Fuzzy Sets and Systems,* 1986, *18,* 299-314. (b)

Kacprzyk, J. *Fuzzy sets in systems analysis* .Warszawa: PWN, 1986. (c)

Kacprzyk, J. A down-to-earth managerial decision making via a fuzzy logic based representation of commonsense knowledge. In L. F. Pau (Ed.) *Proceedings of IFAC Workshop on Artificial Intelligence in Economics and Management.* Oxford: Pergamon, 1986. (d)

Kacprzyk, J. Towards an algorithmic/procedural "human consistency" of decision support systems: A fuzzy logic approach. In W. Karwowski and A. Mital (Eds.) *Applications of Fuzzy Sets in Human Factors.* Amsterdam: Elsevier, 1986. (e)

Kacprzyk, J. On some fuzzy cores and "soft" consensus measures in group decision making. In J. Bezdek (Ed.) *The analysis of fuzzy information.* Boca Raton, Louis.: CRC Press, 1987. (a)

Kacprzyk, J. Towards "human-consistency" decision support systems through commonsense knowledge based decision and control models: A fuzzy logic approach. *Computers and Artificial Intelligence,* 1987, *6,* 97-122. (b)

Kacprzyk, J., & Fedrizzi, M. "Soft" consensus measure for monitoring real consensus reaching processes under fuzzy preferences. *Control and Cybernetics,* 1986, *15,* 309-322.

Kacprzyk, J., & Fedrizzi, M. A "soft" measure of consensus in the setting of partial (fuzzy) preferences. *Europian Journal of Operational Research*, 1988. (in-press)

Kacprzyk, J., & Iwanski, C. A generalization of discounted multistage decision making and control via fuzzy linguistic quantifiers. *International Journal of Control*, 1987, *45*, 1909-1930.

Kacprzyk, J., & Yager, R. R. "Softer" optimization and control models via fuzzy linguistic quantifiers. *Information Sciences*, 1984, *34*, 157-178. (a)

Kacprzyk, J., & Yager, R. R. Linguistic quantifiers and belief qualification in fuzzy multicriteria and multistage decision making. *Control and Cybernetics*, 1984, *13*, 155-173. (b)

Kacprzyk, J., & Yager, R. R (Eds.) *Management decision support systems using fuzzy sets and possibility theory.* Cologne: TÜV Verlag Rheinland, 1985. (a)

Kacprzyk, J., & Yager, R. R. Emergency-oriented expert systems: A fuzzy approach. *Information Sciences*, 1985, *37*, 147-156. (b)

Kacprzyk, J., & Ziólkowski, A. Database queries with fuzzy linguistic quantifiers. *IEEE Transactions on Systems, Man, and Cybernetics*, 1986, *SMC-16*, 474-479. (a)

Kacprzyk, J., & Ziólkowski, A. retrieval from databases using queries with fuzzy linguistic quantifiers. In H. Prade and C. V. Negoita (Eds.) *Fuzzy logics in knowledge engineering.* Colgone: TÜV Verlag Rheinland, 1986. (b)

Nurmi, H. Approaches to collective decision-making with fuzzy preference relations. *Fuzzy Sets and Systems*, 1981, *6*, 249-259.

Yager, R. R. Quantified propositions in a linguistic logic. *International Journal of Man-Machine Studies*, 1983, *19*, 195-227. (a)

Yager, R. R. Quantifiers in the formulation of multiple objective decision functions. *Information Sciences*, 1983, *31*, 107-139. (b)

Yager, R. R. Reasoning with fuzzy quantified statements. *Kybernetes*, 1985, *14*, 233-240. (a)

Yager, R. R. Aggregating evidence using quatified statements. *Information Sciences,* 1985, *36*, 179-206. (b)

Zadeh, L. A. *A theory of commonsense knowledge.* (Memo UCB/ERL M83/26) Berkely, CA.: University of California, 1983. (a)

Zadeh, L. A. The role of fuzzy logic in the management of uncertainty in expert systems. *Fuzzy Sets and Systems,* 1983, *11,* 199-227. (b)

Zadeh, L. A. A computational approach to fuzzy quantifiers in natural languages. *Computers and Mathematics with Applications,* 1983, *9,* 149-184. (c)

Zadeh, L. A. A theory of commonsense knowledge. In H. J. Skala, S. Termini, and E. Trillas (Eds.) *Aspects of vagueness.* Dodrecht: Reidel, 1984.

Zadeh, L. A. syllogistic reasoning in fuzzy logic and its application to usuality and reasoning with dispositions. *IEEE Transactions on Systems, Man, and Cybernetics,* 1985, *SMC-15,* 754-763.

Zimmermann, H.-J. *Fuzzy set theory and its applications.* Dodrecht: Kluwer, 1985.

ORIGIN, STRUCTURE AND FUNCTION OF FUZZY BELIEFS

Józef CHWEDOROWICZ

Institute of Mathematics
University of Lodz, Poland

The term "belief" has a number of meanings. For a psychologist it is important how the notion of belief denotes subjects' relation with the environment. In one of its current meanings "belief" denotes judgement accepted by the subject, in another "belief" is used as the name of premises of action. The function of belief in human life results from these meanings. Beliefs are determined by the existential dimension of human life. Since beliefs are not isolated elements of reality they always serve some purpose. They have determined functions in determined situations "here and now" as rules and premises of actions. The paper provides an analysis of beliefs about facts and phenomena, with no attention being payed to beliefs about values. In order to examine beliefs from the psychological point of view the following questions should be answered.
1. How are beliefs formed (what is the assertion)?
2. What is the structure of beliefs?
3. What is the measure of dissonance of beliefs?

In order to answer these questions and provide an operational definition we begin with certain cognitive paradigms.

A. Subjects' cognitive representation results from subjective estimates of the perceived world. This proposition is a generalization of the standpoint of many researchers who deal with multidimensional scaling (Abelson & Tukey, 1963; Borg & Lingoes, 1980, 1981; Coombs, 1964, 1976; Guttman, 1968; Lingoes, 1965, 1973; 1981; Lingoes & Borg, 1978, 1980; Kruskal, 1964; Shepard, 1974).

B. Mental phenomena are treated as information processing in human operational system. This direction of analysis is represented by such researchers as Newell & Simon (1972), Lindsay & Norman (1972), Stachowski (1973), Kozielecki (1977, 1986), Bruner (1978) and Nosal (1979), who make use of computer science, computer information processing, structural linguistics and artificial intelligence.

These two approaches complement each other. Information processing through realization of brain programs results in changes in the states of subjects' space of knowledge, in turns the values of these states may be input data to the programs. Subjects' space of knowledge consists of beliefs, values, hypotheses, views, etc.

The subject, surrounded by the environment, forms a rerpesentation of the surrounding reality. Reality is the set of those states and events of the world wich the subject grants with existence independent of self will. This representation may be of a linguistic or numeric nature. Every state of the world can be described by a number of its features, each of whic can assume various values on linguistic or numeric scales (Nowakowska, 1979). The set of possible values does not say much about the given state of the world or event. Pointing only to one particular feature or subset of features is what is generally called the semantic element of a message or the elementary message (Gackowski, 1979). What is a "message"? The definition of message would have to contain a definiton of information. Instead of giving a definiton of these notions, which would in turn contain unprecise notions, we assume that messages and information are undefinable elementary notions, the application of which is clarified by the context in which they have been used (Bauer & Goos, 1974). As Wiener (1971) said -" ... information is information; it's neither matter nor energy". The same message may be interpreted differently and convey different information for different people. Information may then be understood as a result of the interpretation of a message, i.e. a certain transformation of a message into information, which symbolically can be written as

$$M \xrightarrow{\alpha} I \qquad (1)$$

From information which in turn comes from the environment and the subject's own organism, the subject forms a certain space of knowledge.

The considerations in this paper do not concern all the elements of a subject's space of knowledge - they deal only with beliefs. Belief is the judgment (in the logical sense) for which the subject accepts ownership, because of the fulfillment of the contents of judgment in a given reality. Belief can be expressed through utterances and actions.

Earlier works on beliefs concentrated on the relation between the subject and the subject's judgment in the logical sense, irrespective of the act of thinking (Meinong, 1910; Witwicki, 1980). Later works began to examine the relation between acceptance of judgments and readiness to act (Arrow, 1971; Marciszewski, 1972; Shafer, 1976; Hempolinski, 1981; Chwedorowicz, 1984, 1986; Chlewinski, 1984).

The formation of beliefs can be explained on the grounds of social learning. In every situation individuals meet sequences of events in which the two elements: action and reinforcement, always repeat. If an action, being the consequence of an accepted belief, is reinforced, the reinforcement is transferred to the belief. As a belief is subject to reinforcement, reward increases the subject's tendency to use the given belief in a similar situation, whereas punishment decreases it. The very reinforcement may be internal or external, or maybe the fusion of both factors. In effect, a belief may be verified, rejected or a new version of the same belief may be formed.

In the same situation different people may use different beliefs as premises for actions. The theory of social learning explains these differences in beliefs in terms of the different experiences of individuals.

Subjects' space of knowledge contains not only certain or highly probable information, but also much which is not quite certain and which is rather subject to conjectures.

Out of the general class of beliefs only these types, which concern people and situations, i.e., states and events in the world about which only incomplete information is avaliable, are the subject to analysis here.

An event is a set of such states of the world which fulfills a certain given condition. States of the world are subject to conjectures rather than to statements of certainty. The data about the world are represented in a subject's space of knowledge in the form of dimensions. The mechanism of formation of such representations is, as it was said, described as mapping onto linguistic and numeric domains.

Since ranges of categories on linguistic scales are not sharp, expressions formed on their basis are not sharp either. The "fuzzy" approach assumes that decisions about the categorization of states of the world which differ as to the degree of member ship (compatibility) in categories, are partial. Ambiguity of boundary cases results in formation of boundaries between categories which are not sharp. Categories as "fuzzy sets" are sets without sharply defined boundaries (Zadeh, 1965; Kintsch, 1974; Lehrer, 1974). Beliefs which are logical expressions formed on the basis of fuzzy categories have the characteristics of fuzziness.

Fuzzy beliefs differ from both deterministic, probabilistic and from instrumental beliefs, which have been described in literature. As the basis for the acceptance of fuzzy beliefs lies fuzzy acceptance; specifically, the acceptance which takes place only when the motive of assenting to the contents of judgment is in high compatibility with the contents of judgment of determined reality (Chwedorowicz, 1984, 1986). This is in contrast to the motive for accepting a probabilistic belief, which is the high probability of the judgment, and to the motive for accepting a deterministic belief, which is complete compatibility of the contents of judgment with determined reality. Frequently, however, fuzzy and probabilistic beliefs are uttered categorically.

The contents of fuzzy beliefs consists both of external, directly perceived information and of information coming from the subject's own space of knowledge. Both kinds of information are described linguistically.

The present knowledge about beliefs, perception and human information processing allows one to propose a model describing the formation of fuzzy beliefs.

FORMATION OF FUZZY BELIEFS

Every stimulus S reaching the subject can be presented as a pair (M,I) where M denotes message and I the information which it conveys. The same message, via various interpretations, can give various kinds of information (judgments). The difference in interpretation results from the different experience that different people have, and takes the form of the so called "metabeliefs", beliefs about beliefs concerning the world, oneself, and about the relation between the two. Belief is, in the logical sense, a judgment preceded by an assertion (Marciszewski, 1972). Rightness of belief may be expressed by a condition weaker than assertion. The rightness of one's judgment may be expressed by one's expectation that a given judgment will be right in the respective situation. This expectation is a function of two variables: the subjective variable and the perspective variable.

It has been noted by Gestalt psychologists that if the amount of perceived information is not sufficient to build a picture as a whole, the role of the subjective factor increases.

Actions are ascribed to respetive beliefs in the process of coding. Reinforcement of action results in reinforcement of the respective belief. There is a feedback between action and belief B which conveys the reinforcement.

The mechanism of formation of fuzzy beliefs is presented in the following model.

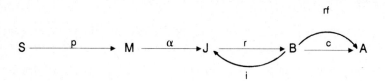

where S-stimulus, p-perception, M-message, α-interpretation of message, J-judgment, r-rating of judgment, i-inference, B-belief, c-coding of belief, rf-reinforcement, A-action.

The given operations depend on metabeliefs about the world, about oneself and about the relations between them.

Rating

Given an emipirical relational system

$$\gamma = <P, S_1, S_2, ... S_n > \qquad (2)$$

where P is a finite set of states of the world, and S_1, S_2, S_n are relations on P (primary, secondary, ... etc.). Also, given a theoretical system

$$\beta = <\rho, R_1, R_2, ... R_n>, \qquad (3)$$

where ρ is the family of all fuzzy subsets of P representing categories on linguistic scales, and where $R_1, R_2, ... R_n$ are relations on (primary, secondary, ..., etc.).

If systems γ and β are homomorphous then there is a mapping on P with values within ρ.

$$\gamma : P \longrightarrow \rho \qquad (4)$$

whereby the name $f \in \rho$ is ascribed to state of the world $x \in P$.

In terms of the theory of measurement (Roberst & Franke, 1976) and the theory of fuzzy sets (Zadeh, 1965, 1971, 1975) the following theorems and definitions can be given to subjects' fuzzy beliefs about states of the world (Chwedorowicz, 1984, 1986).

Definition 1

A fuzzy subset of the set P is any real function mapped upon P such that its values fall within the interval $R_{<0,1>}$. If f is a subset of this kind, then the value f(x) is called degree of compatibility between x and f.

Definition 2

Fuzzy acceptance occurs only when the assent to a judgment is motivated by a high degree of compatibility between the content of the judgment and the reality in question.

Definition 3

The individual threshold of acceptance α varies from person to person, and for any person, may vary from situation to situation over the interval $R_{<0,1>}$.

Theorem 1

A necessary condition for the subject to assent to the content of a judgment is that the degree of compatibility between the judgment and the respective reality exceeds the individual threshold of acceptance.

Compatibility between x and f is checked on a finite set of criteria $I = \{1, 2, \ldots n\}$, some of which have various weights v_i, whereas ratings of beliefs on an i-th criterion are denoted w_i. The degree of compatibility between x and f (the strength of belief) is a function of two variables.

$$f(x) = \max_i(\min(v_i, w_i)), \; i \in I \tag{5}$$

Theorem 2

Two beliefs concerning the same state of the world are dissonant if the difference between them exceeds the individual threshold of acceptance.

If two beliefs are dissonant, then one of them contradicts the other, which can be written as $\varphi_j(x) = f$ and $\varphi_j(x) = f'$. Then the difference between beliefs that are dissonant may be measured by Hamming's linear distance:

$$\delta = 1/n \sum_{i=1}^{n} |f(x_i) - f'(x_i)|, \tag{6}$$

Definition 4

The threshold of tolerance to dissonance is the greates dissonance which does not motivate the person experiencing it to change beliefs. The threshold of tolerance to dissonance α_T varies from person to person and for the same person may vary from situation to situation.

Theorem 3

Fuzzy belief is in the logical sense the judgment to which the subject gives assent if degree of compatibility between the judgment and the respective reality exceeds the individual threshold of acceptance.

The following condition must be met:

$$f(x) > \alpha \text{ and } \alpha > 0.5 \quad \alpha \in <0,1> \tag{7}$$

LANGUAGE OF FUZZY BELIEFS

Belief may be uttered - expressed in words because of its linguistic representation. A set of linguistic representations of beliefs in natural language makes a language of beliefs. The language of beliefs is a subset of all the sentences which concern human beliefs directly or indirectly. By expressing beliefs in linguistic categories it is possible to define operations of negation, union, intersection and semantic implication for beliefs.

Atomic beliefs

The fundamental elements of the language of beliefs are atomic beliefs. An expresion of the kind

$$\varphi_j(x) = f \tag{8}$$

where: j is the considered attribute, is an atomic belief if the person uttering the belief understands its contents in the logical sense, if the mapping φ_j is homomorphism, and if the person uttering the belief has decided to assent to the content of the judgment. The degree of assent given to the judgment is expressed in our model by the funtion of compatibility f(x), while in colloquial language it is expressed by epistemic functors like "I belive", "I reckon", "I am sure", etc. To put it more formally:

$$\varphi_j(x) = f \text{ iff } f(x) > \alpha \text{ and } \alpha > 0.5 \tag{9}$$

where is the individual threshold of acceptance.

Negation of beliefs

The functor of negation $\neg$ p is a symbol of an operation of negation of an uttered judgment. If the person uttering a judgment of the kind $\varphi_j(x) = f$ does not accept it, then

(1) he or she is convinced that the judgment is false and accepts a judgment contrary to it.

$$\neg (j(x)) = f \text{ iff } f'(x) > \alpha \tag{10}$$

(2) he or she has not decided to accept the contents of a given judgment

$$f(x) \leq \alpha \text{ and } f'(x) \leq \alpha \qquad (11)$$

Intersections of beliefs

If two judgments of the form $\varphi_j(x) = f$ and $\varphi_i(x) = g$ are linked by the conjunction "and", there results a compound judgment of the kind

$$\varphi_j(x) = f \text{ and } \varphi_i(x) = g \text{ iff } \min(f(x), g(x)) > \alpha \qquad (12)$$

Union of beliefs

If two judgments of the form $\varphi_j(x) = f$ and $\varphi_i(x) = g$ are linked by the conjunction "or", there results a compound judgment of the form

$$\varphi_j(x) = f \text{ or } \varphi_i(x) = g \text{ iff } \max(f(x), g(x)) > \alpha \qquad (13)$$

Semantic implication

If $\varphi_j(x) = f$ and $\varphi_i(x) = g$ are patterns of judgments, then the relation of the semantic implication

$$\varphi_j(x) = f = \varphi_i(x) = g \qquad (14)$$

is defined by the demand that $\max(1-f(x), g(x)) > \alpha$.

The functor of semantic implication $p \Rightarrow q$ is the pattern of real beliefs formed on the basis of the expression "if ... then ... ". Semantic implication $p \Rightarrow q$ is interpreted as follows:

(a) Judgment p does not contradict judgment q,

(b) Judgment q is the conclusion drawn from judgment p. Semantic implication $p \Rightarrow q$ approximates material implication $p \rightarrow q$ defined by the demand that $\neg p \lor q$.

SYSTEMS OF FUZZY BELIEFS

Beliefs are never isolated in a person's space of knowledge, they rather combine and come together as systems. It is characteristic of those systems that they have mechanisms protecting the cohesion of each system. One of the well-known mechanisms is elimination of contradictory judgments from the system, based on the phenomenon of dissonace. Another, less obvious mechanism is that of sustaining beliefs in spite of the loss of their external sources as comprised by events that had been indirect rea sons for their formation. Yet another mechanism is the censorship of new beliefs introduced into the system. All the mechanisms may be described logically in the form of axioms of a system of beliefs. The axioms are descriptions of certain rules according to which the real system of beliefs works. Accordingly, as these rules are fulfilled, information is accepted and interiorized, or rejected, or may make the system change. The mechanism of eliminating contradictory judgments from a system of fuzzy beliefs may be described as follows:

$$(\varphi_j(x) = f) \Rightarrow \neg (\varphi_j(x) = f') \qquad (15)$$

which reads: If a person uttering a judgment assents that "j-th feature x is f", then he or she cannot assent the "j-th feature x is not f".

Before any state of the world becomes for some reason subject to a particular belief, it has been subject to many earlier formed beliefs. These very beliefs, together with the beliefs concerning the relations between states of the world, are sort of censors on new beliefs and they even determine the shape of possible beliefs.

In case the external source of a belief is no longer present, the very belief may remain unchanged, being internally generated by the system. Both the nature of these relations and the anticipation of possible beliefs is given in the following pattern called the rule of Modus Ponens:

$$((\varphi_i(x) = f) \text{ and } ((\varphi_i(x) = f) \Rightarrow (\varphi_j(x) = g)) \Rightarrow (\varphi_j(x) = g) \qquad (16)$$

Neither of the above mechanisms warrants that a fuzzy belief system will be an approximation of reality perceived by the person uttering the beliefs. In extreme situations a person may generate delusive judgments that make up a coherent system. Therefore it is necessary to adopt an axiom stating that a person's conviction as to some judgment being true is not sufficient evidence of its truth. One may be strongly convinced of the thruth of a false judgment,

$$((\varphi_i(x) = f), \text{ then } x \in f \qquad (17)$$

The three above axioms constitute a belief system based on fuzzy acceptance (Chwedorowicz, 1986), which is parallel to other belief systems based on deterministic or probabilistic acceptance (Marciszewski, 1972; Pap, 1957; Shafer, 1976).

The axioms given above result in a number of new rules such as the rule of simplification:

$$(\varphi_i(x) = f \& g) \Rightarrow \varphi_i(x) = f \qquad (18)$$

$$(\varphi_i(x) = f\&g) \Rightarrow \varphi_i(x) = g \qquad (19)$$

If both partial and non-partial probabilistic assent is used in the description of one and the same reality, there may occur a conflict between these two types of assent. Probabilistic acceptance may, in certain cases, fall short of rationality axioms constructed on the basis of non-partial assent, as in the systems of Pap (1957) or Marciszewski (1972). The contradiction results from the properties of subjective probability (Ramsey, 1931). According to Ramsey subjective probability has the following properties:

(Ps.1) $Ps(p) + Ps(\neg p) = 1$

(Ps.2) $Ps(p/g) + Ps(\neg p/g) = 1$

(Ps.3) $Ps(p\&g) = Ps(p) \times Ps(g/p)$

(Ps.4) $Ps(p\&g) + Ps(p\&\neg g) = Ps(p)$ \qquad (20)

If we assume that $Ps(p)$ and $Ps(\neg p)$ exceed the individual threshold of acceptance k, i.e. that p and $\neg p$ are both accepted at the same time, then the axioms of cohesion in Pap's and Marciszewski's system are not fulfilled. Probabilistic acceptance does not fulfill the axiom of consistency in assent of deterministic systems of beliefs (Pap, 1957; Marciszewski, 1972). This can be shown on the basis of Suppes' theorem (Suppes, 1968).

$$(Ps(p \to q) \geq r \,\&\, (Ps(p) \geq s)) \to Ps(q) \geq r + s - 1 \qquad (21)$$

It is possible to assent to $p \to q$ and assent to p, without assenting to the inference q.

The conflict may be overcome by introducing the notion of fuzzy belief. It results from the definiton of fuzzy belief that fuzzy beliefs fulfill the axiom of cohesion of deterministic system of beliefs. It is ensured by the condition

$$\varphi_j(x) = f \text{ iff } f(x) > \alpha \,\&\, \alpha > 0.5 \qquad (22)$$

The axiom of cohesion of beliefs in Pap's deterministic system of beliefs takes the form

$$B(x, \neg p) \to \neg B(x, p) \qquad (23)$$

where B(x,p) reads: "x believes that p". In terms of fuzzy beliefs the axiom reads as follows:

$$(\varphi_i(x) = f') \to \neg (\varphi_i(x) = f) \qquad (24)$$

Let assume that $f'(x) > \alpha > 0.5$ \qquad (25)

It results from the properties of semantic implication and negation that:

$$\max(f(x), f'(x)) = f'(x) > \alpha \tag{26}$$

This condition means that fuzzy beliefs fulfill the axiom of cohesion of the deterministic system of beliefs.

It is also possible to show that fuzzy beliefs fulfill the axiom of consistency in the deterministic system of beliefs (see Pap, 1957).

$$B(x,p) \,\&\, B(x,p \to q) \to B(x,p) \tag{27}$$

In terms of fuzzy beliefs the axiom reads as follows:

$$((\varphi_i(x) = f) \,\&\, ((\varphi_i(x) = f) \Rightarrow ((\varphi_j(x) = g)) \Rightarrow (\varphi_j(x) = g) \tag{28}$$

Let us assume that $f(x) = a$ $g(x) = b > \alpha$, then

(1) $\max(1-\min(f(x), \max(1-f(x), g(x))), g(x))$

$\quad\quad = \max(1-\min(a, \max(1-a,b)), b) = b > \alpha$

(2) Let assume that $g(x) = a$ $f(x) = b > \alpha$

$\quad\quad \max(1-\min(b, \max(1-b,a)), a) = a > \alpha$

Both of these conditions mean that fuzzy beliefs fulfill the axiom of consistency of the deterministic system of beliefs.

Inference

Being elements of language, fuzzy beliefs form sentences which describe states of the world and relations between them. States of the world which can be observed indirectly may serve as permises concerning those states which cannot be observed, which are subject to conjectures.

Thinking based on premises is called inferencing. According to Ajdukiewicz inferencing is a kind of mental activity in which a judgment, made with some degree of certainty, concerning sentences called premises, results in a judgment, made with an always greater degree of certainty, concerning another sentence called conclusion (Ajdukiewicz, 1965).

Beliefs are formed directly and indirectly. A judgment "it is raining" is assented to directly as compatible with reality if we are getting wet in rain or see drops of rain on a window pane. I assent indirectly to a judgment "My wife needs money for a new car" only because I have assented to judgments: "My wife has been very nice to me lately", "Our neighbour's wife has a new car", "My wife delights in that car","My wife reminds me about her coming birthday". The latter sentences - judgments are assented to indirectly on the basis of my or other people's visual or audial perceptions which have been found compatible with reality. These judgments as well as unuttered judgments about relations between phenomena are premises, while the judgment "My wife needs money for a new car" is the conclusion of presented reasoning. Now I am more strongly convinced of the rightness of the judgment I haven't taken into consideration before. The presented reasoning was necessary for me to believe that my wife needs money.

Distance

Let ρ denote the family of all fuzzy subsets (states of the world). A similarity relation h is mapped on ρ (Zadeh, 1971), its values being

$$\forall\ (f,g),\ (g,r),\ (f,r) \subset \rho \times \rho \quad x$$

there occurs

$$h(f,f) = 1 \qquad (29)$$
$$h(f,g) = h(g,f)$$
$$h(f,r) = \max_{g}\ (h(f,g),\ h(g,r))$$

In terms of similarity relation we may speak of a distance between beliefs

$$\varphi(x) = f \text{ and } \varphi(x) = g$$

Let us denote that distance as $d(f,g)$ $\qquad (30)$
$$d(f,g) = 1-h(f,g) = h(f,g)$$

Distance between beliefs has the following properties:

(1) $d(f,f) = 1-h(f,f) = 0$

(2) $d(f,g) = d(g,f)$

From the transitivity of the similarity relation it follows that

$$h(f,r) \max_{g} \geq (h(f,g), h(g,r)) \tag{31}$$

$$\bar{h}(f,r) \leq \min_{g} (\bar{h}(f,g) + \bar{h}(g,r) - \bar{h}(f,g)\,\bar{h}(g,r))$$

(3) $d(f,r) \leq d(f,g) \;\square\, d(g,r)$

Properties 1-3 are satisfied by Hamming's linear distance (Kaufmann, 1973).

$$d(f,g) = 1/n \sum_{i=1}^{n} |f(x_i) - g(x_i)| \tag{32}$$

If two beliefs are dissonant, then one of them contradicts the other, which can be then written

$$\delta = 1/n - \sum_{i=1}^{n} |f(x_i) - f'(x_i)| \tag{33}$$

The distance between beliefs may be described with non-fuzzy notions. For $f, g, r \in \rho$

$$f =_a g \text{ iff } h(f,g) \geq a \tag{34}$$

which defines non-fuzzy relation $f =_a g$, called a -similarity. The relation satisfies a condition weaker than transitivity.

Theorem 4

If $f =_a g$ and $g =_b r$, then $f =_{a.b} r$

Proof

If $f =_a g$ then $d(f,g) \geq 1-a$

similarly

$$d(g,r) \leq 1-b$$

The thesis results from condition (3) for distance d. The theorem describes the phenomenon of fuzziness of similarity of beliefs in a sequence of similar beliefs.

For $a = \alpha/\alpha$ (-individual threshold of acceptance) relation $=_\alpha$ generates in the set of beliefs subsets of fuzzy beliefs B_α called clusters such that for $f,g \in B_\alpha$

there occurs

$h(f,g) \geq \alpha$

B_α- is the set of beliefs considered by the subject as similar to degree α. The hierarchy of fuzzy beliefs is formed by α clusters with α increasing from zero to one. This concept of hierarchy of fuzzy beliefs coincides with the notion of α-level sets of fuzzy sets.

STRUCTURE OF FUZZY BELIEFS

The structure of fuzzy beliefs may be described by the following formal system

$$B = <P, I, \rho, \varphi>$$

where

P - finite set of states of the world,
I - finite set of scales (attributes) on which the states of the world are rated,
ρ - family of all fuzzy subsets of P, representing categories on linguistic scales

$$\rho = \bigcup_{i \in I} \rho_i$$

φ_i - function ascribing categories to states of the world,

$$\varphi_i: p \to , i \in I$$

The language of fuzzy beliefs based on the system of fuzzy beliefs B is the set of all linguistic representations which directly or indirectly concern human beliefs. The language consists of atomic beliefs of the form $\varphi_i(x) = f$. Atomic beliefs linked with logical functors: negation, intersection, union, and semantic implication etc. form expressions of the language of beliefs, the semantic correctness of which is checked on the basis of axioms of the system of fuzzy beliefs.

CONCLUSION

(1) The theory and definitions presented in this paper are concepts of a theory which seems to explicate the formation and functioning of belief more efficiently than other approaches, and which has certain advantages.

(2) Application of the theory of fuzzy beliefs to the description of belief formation makes it possible to overcome the difficulties which arise if a subject describes the same reality with both detereministic and probabilistic judgments, which results in a contradiction of conclusions reached according to the two different formalisms.

(3) The phenomenon of contradiction within subjects' space of knowledge is, on the grounds of the theory of fuzzy beliefs, explained as contradiction of judgments which belong to his/her space of knowledge but which have not become beliefs because the y have not satisfied the necessary condition that: the degree of compatibility of each judgment with the respective reality exceeds the individual threshold of acceptance and the threshold of acceptance is greater than 0.5.

$f(x) > \alpha$ and $\alpha > 0.5$, $\alpha \in <0,1>$

(4) The theory of fuzzy beliefs explains relations between cluster-like structure and hierarchical structure of beliefs, which, for $\alpha = 0$ and increasing order for $\alpha \in <0,1>$, form a hierarchy of beliefs built of α clusters.

(5) The language of fuzzy beliefs presented in the paper allows one to formulate questions about the system of fuzzy beliefs, and to generate

indirect conclusions which can then be checked using the axioms of fuzzy beliefs.

APPENDIX

The individual threshold of acceptance α varies from person to person and for any person may vary from situation to situation over the interval $R_{<0,1>}$. The individual threshold of acceptance equals the highest value $f(x)$ - degree of compatibility -, for which the subject does not decide to accept the judgment.

External values can be omitted, because for $\alpha = 0$ subjects would accept all the judgments and for $\alpha = 1$ no judgment would be accepted. In both these situations these subject does not receive any information about the world. The following theorem determines the conditions of one's finding the contents of a judgment compatible with the given reality and assenting to it as to his/her own.

Theorem 1

A necessary condition for a subject to assent to the content of a judgment is that the degree of compatibility between the judgment and the respective reality exceeds the individual threshold of acceptance.

Proof

Let there be an empirical relational system

$$\gamma = <P, S_1, S_2, ... S_n>$$

where P is a finite set of states of the world and $S_1, S_2, \ldots S_n$ are relations on P (unary, binary, ..., etc.).

There is also given a theoretical system

$$\beta = <\rho, R_1, R_2, \ldots, R_n>$$

where ρ is the family of all subsets of P some of which represent linguistic categories and $R_1, R_2, \ldots, R_n$ are relations on ρ (unary, binary, ..., etc.). If there is a homomorphism between systems α and β then there is a mapping on ρ with values within (Roberts & Frankl, 1976).

$$\varphi : P \rightarrow \rho$$

whereby $f \in \rho$ is ascribed to the state of the world $x \in P$.

Let g be a category to which a state of the world x really belongs "x is g", while the subject's belief is the judgment "x is f". Let us denote the difference between the belief and the real state of the world as $\wedge$. The measure of distance between the belief "x is f" and the real state of the world "x is g" is the distance between the fuzzy subsets of ρ.

A fuzzy relation of similarity h is mapped on ρ (Zadeh, 1971). In terms of the similarity relation we may speak of the distance between categories f and g. Let us denote that distance as d(f,g).

$$d(f,g) = 1 - h(f,g)$$

This distance between two categories has the following properties:

(1) $d(f,f) = 1-h(f,f) = 0$
(2) $d(f,g) = d(g,f)$
(3) $d(f,r) \leq d(f,g) \square d(g,r)$

It can be shown that properties 1-3 are satisfied by Hamming's linear distance.

$$d(f,g) = 1/n \sum_{i=1}^{n} |f(x_i) - g(x_i)| \qquad (i)$$

Thus the distance between a belief and the reality can be written as

$$\Delta = 1/n \sum_{i=1}^{n} |f(x_i) - g(x_i)| \qquad (ii)$$

where $f(x_i)$ is the degree of compatibility between x and f rated on the i-th criterion, and $g(x_i)$ is the degree of compatibility between x and d rated on the i-th criterion. The fact that "x is g" implies that on any n-th criterion $g(x_i) = 1$ and so equation (ii) assummes the following form:

$$f(x) = 1 - \Delta = 1/n \sum_{i=1}^{n} f(xi) \qquad (iii)$$

Let us denote $1 - \Delta$ as α, the latter being the individual threshold of acceptance. The rating of f(x) is based on a finite set of criteria $I = \{1, 2, ..., n\}$.

If the subject is to accept a judgment, then according to the definition of the individual threshold of acceptance, the degree of compatibility of the contents of a judgment with a given reality should be greater than the value of an individual threshold of acceptance

$$f(x) > \alpha \text{ and } \alpha \in (0,1) \qquad (iv)$$

Applying a modified procedure to rate a belief on a finite set of criteria $I = \{1,2, ... n\}$ we define the following function of compatibility "x is f"

$$f(x) = \max_i(\min(v_i, w_i)) \qquad (v)$$

where w_i is a fuzzy rating of a belief on the i-th criterion v_i is a fuzzy weight of the i-th criterion.

All compatibility functions take values in the closed interval $R_{(0,1)}$.

Assenting the contents of a judgment as one's own is not a sufficient condition for the judgment to become the subject's belief. One may be strongly convinced about the rightness of false judgments. It may happen that in case of a low threshold of acceptance two contradictory judgments are accepted.

This is why an additional condition is needed to warrant that contradictory judgments are not accepted in the process of belief formation.

Lemma

The necessary conditions for a subject not to accept contradictory judgments is that subject's individual threshold of acceptance is greater than or equal than 1/2.

Proof

If a subject accepts two contradictory judgments "x is f" and "x is f'" at the same time, then the functions of compatibility of these judgments are greater than the individual's threshold of acceptance α, $f(x) > \alpha$ and $f'(x) > \alpha$. If so, then $f'(x) = 1-f(x) > \alpha$, Hence $\alpha < 1/2$.

A subject accepts contradictory judgments if the individual's threshold of acceptance is less than 1/2.

In order to satisfy the assumption of the lemma, the value of an individual threshold of acceptance should be equal to or greater than 1/2.

Assumption

Understanding a judgment in the logical sense means that the subject:

(1) Gives the judgment the logical value "true" if the degree of compatibility of the judgment with a respective reality is found to be greater than the individual's threshold of acceptance, or suspends the decision of the acceptance of the judgment.

(2) Respects the rule of non-contradiction, i.e. does not accept two contradictory judgments at the same time.

Notice: The subject describes reality with judgments which are considered true. Thus false judgments are not needed in the subject's system of knowledge and are eliminated.

Theorem 3

Fuzzy belief is in the logical sense the judgment to which the subject gives assent if the degree of compatibility between the judgment and the respective reality exceeds the individual threshold of acceptance $\alpha \geq 1/2$.

$f(x) > \alpha$ and $\alpha \geq 1/2$, $\alpha \in (0,1)$

Proof

The thesis results directly from the assumption, the lemma and theorem 1.

The formation of fuzzy beliefs is a process of successive verifications of compatibility between the contents of a judgment and the respective reality. In case the contents of a judgement is not accepted, the subject may reject the judgment or modify its contensts. New versions of the same judgment may occur, some of which may be accepted by the subject.

REFERENCES

Abelson, R. P., & Tukey, J. W. Efficient utilization of non-numerical infomation in quantitative analysis. *Annales of Mathematical Statistics*, 1963, *34*, 1347-1369.

Ajdukiewicz, K. *Logika pragmatyczna*. Waszawa: PWN, 1965.

Arrow, K. J. *Essays in the theory of risk-bearing*. Chicago, Ill.: Markham, 1971.

Bauer, F. L., & Goos, G. *Infomatik*. Berlin: Springer, 1974.

Borg, I., & Lingoes, J. C. A model and algorithm for multidimensional scaling with external constraints on the distances. *Psychometrika*, 1980, *45*, 25-38.

Borg, I., & Lingoes, J. C. An alternative approach to confirmatory inference and geometric models. *Quality and Quantity*, 1981, *15*, 541-552.

Bruner, J. S. *Poza dostarczone informacje*. Warszawa: PWN, 1978.

Chlewinski, Z. *Geneza i struktura przekonan*. Paper read at the XXV. Polish Psychological Conference. Krakow, 1984.

Chwedorowicz, J. Fuzzy beliefs. *Polish Psychological Bulletin*, 1984, *15*, 51-57.

Chwedorowicz, J. Conditions of fuzzy beliefs change in a dialogue. *Polish Psychological Bulletin*, 1986, *17*, 3-8.

Coombs, C. H. *A theory of data*. New York: Wiley, 1964.

Gackowski, Z. *Projektowanie systemów informacyjnych zarzadzania*. Warszawa: PWN, 1979.

Guttman, L. A general nonmetric technique for finding the smallest coordinate space for a configuration of points. *Psychometrika*, 1968, *33*, 469-506.

Hempolinski, M. *Racjonalnosc przekonan*. Paper read at the Conference on Rational Society and the Sense of History. Warsaw, 1981.

Kaufmann, A. *Introduction a la theorie des sous-ensembles flous*. Paris: Masson, 1973.

Kintsch, W. *The representation of meaning in memory*. Hillsdale, N.J.: Erlbaum, 1974.

Kozielecki, J. *Psychologiczna teoria decyzji.* Warszawa: PNW, 1977.

Kozielecki, J. *Psychologiczna teoria samowiedzy.* Warszawa: PNW, 1986.

Kruskal, J. B. Multidimensional scaling. *Psychometrika,* 1964, *29,* 1-27.

Lehrer, A. *Semantic fields and lexical structure.* Amsterdam: North-Holland, 1974.

Lindsay, P. H., & Norman, D. *Human information processing.* New York: Academic, 1972.

Lingoes, J. C. *The Guttman-Lingoes nonmetric program series.* Ann Arbor, Mich.: Mathesis, 1973.

Lingoes, J. C. Some uses of statistical inference in multidimensional scaling. In I. Borg (Ed.), *Multidimensional data representations.* Ann Arbor, Mich.: Mathesis, 1981.

Lingoes, J. C., & Borg, I. Confirmatory monotone distance analysis - Unconditional. *Journal of Marketing Research,* 1978, *15,* 610-611.

Marciszewski, W. *Podstawy logicznej teorii przekonan.* Warszawa: PWN, 1972.

Meinong, A. *Über annahmen.* Leipzig, 1910.

Nosal, C. *Mechanizmy funkcjonowania intelektu.* Wroclaw: Wydawnictwo Politechniki Wroclawskiej, 1979.

Nowakowska, M. *Teoria dzialania.* Warszawa: PWN, 1979.

Newell, A., & Simon, H. A. *Human problem solving.* Englewood Cliffs, N.J.: Prentice-Hall, 1972.

Pap, A. Belief and propositions. *Philosophy of Science,* 1957, *24.*

Ramsey, E. P. *The foundations of mathematics and other logical essays.* London: K. Paul, Trech, Truber and Co., 1931.

Roberts, F. S., & Franke, C. H. On the theory of uniqueness in measurement. *Journal of Mathematical Psychology,* 1976, *14,* 211-218.

Shafer, G. *A mathematical theory of evidence.* Princeton, N.J.: Princeton University Press, 1976.

Stachowski, R. Modele liniowe przeddecyzyjnego procesu integrowania informacji. *Przeglad Psychologiczny,* 1973, *16,* 335-359.

Suppes, P. Two rules of detachment in inductive logic. In I. Lakatos (Ed.) *The problem of inductive logic.* Amsterdam: Noth-Holland, 1968.

Wiener, N. *Cybernetyka, czyli sterowaniew czlowieku, zwierzeciu imaszynie.* Warszawa: PWN, 1971.

Witwicki, W. *Wiara oswieconych.* Warszawa: Iskry. 1980.

Zadeh, L. A. Fuzzy sets. *Information and Control,* 1965, *8,* 338-353.

Zadeh, L. A. Similarity relations and fuzzy orderings. *Information Sciences,* 1971, *3,* 177-200.

BRAIN ACTIVITY AND FUZZY BELIEF

G. GRECO and A. F. ROCHA

Department of Physiology, Institute of Biology, UNICAMP
13081 Campinas - Brasil

Brain activity was recorded during subjects listening to a text, and then correlated with two different fuzzy measures of belief about the received information. One of these measures (the degree of correlation) expresses the relationship between each decoded piece of information and the structure of the text itself, the other (the degree of confidence) evaluates the individual s confidence in the received information. Both measures depended on the activity of many areas in both cerebral hemispheres. Additionally, the analysis of this dependence confirmed that both measures may be accept as two different psychological constructs.

INTRODUCTION

Human languages, like all social actions, must perform two essential functions simultaneously. They must be directed toward the problem of survival, and they must also be directed toward making sense of existence (Washabaugh, 1980). On the one side, languages must address the everyday technical problems of getting food, shelter, etc. On the other side, they must also contribute to interpreting and transmitting the emotional, cultural, intellectual etc., experiences of the human being (Olson, 1980).

Centering on humans, we speak of universals, contex-free communication, etc., whereas centering on society, we speak of actions, context dependent meanings, and so on (Rocha & Rocha, 1985).

Human languages provide people with a core of common meanings centered on the basic schema, scripts or frames related to survival both in the physical word as well as within society. In this sense, the human utterances form a closed system of self-referred meanings. However, people modify the meanings of the words to speak about individualities, too. In this use, the human utterances become an open system of meanings referred to each individual context. Because of this, human languages are to be treated as partially closed systems, where believes are always evaluated in respect to the language itself, and in respect to the context of the speaker and/or listener (Rocha & Rocha, 1985).

The authors have been using Fuzzy Set Theory to study the commonalities of speech understanding by groups of volunteers, while recording their EEG activities to probe into the individual flavor of each decoding (Rocha & Francozo, 1980; Rocha, 1987). Also, the authors have been using Fuzzy Logic to investigated how people correlated each piece of verbal information to both the speech context and their own contexts (Rocha & Greco, 1987; Theoto, Santos, & Uchiyama, 1987; Rocha, Giorno, & Leao, 1987).

People associate a personal degree of confidence with each piece of information as they pick it up from the speech, but they have to wait until they have at least a grasp of the subject or theme (Sgall, Hajicova, & Benesova,

1973; Oleron, 1980) of the communication in order to assess the correlation between each of these pieces of information and the speech itself (Rocha & Greco, 1987). First of all, they must attempt to construct the scheme of the speech, assembling all the pieces of information into a logical network of meanings, guided by their confidence of the veracity of the received information. Once the initial picture of the speech is thus formed, one may try to verify the speech consistency through the evaluation of the correlation between the accepted pieces of information and the theme or subject of the communication (Rocha, 1982; Rocha & Greco, 1987).

As a consequence of this, the logical networks dealing with both the degrees of correlation and confidence processing during the speech decoding are expected to be the same. However, since correlation - unlike confidence - is sensitive to the structure it helps build its actual values must depend on speech cohesion - which is not the case for confidence (Rocha & Greco, 1987).

The difference between correlation and confidence have been analysed on text recalling, by asking people to build the fuzzy graph of the text decoding (Rocha & Greco, 1987; Theoto, et al., 1987). This graph describes the relations shared by each piece of information used by the subject to conclude for the theme and rhyme of the heard text. The subjects were also requested to assign to each node of the graph either the degree of correlation shared by that node with the theme, or the subject's confidence of the information represented by the node.

The correlation assigned to each node of the graph - unlike confidence - decreased from the left to right nodes and from the root associated with the chosen theme/rhyme to the terminal nodes where each piece of information was represented - whereas confidence was not correlated with the node position. The linear relation between correlation and node position presented also a non-zero linear coefficient, what means that confidence at each node may depend also on the overall text (graph) cohesion.

The logic network of the fuzzy text decoding was obtained by asking the volunteers to assign a logical connective (AND, OR, IF) to each node of the above fuzzy graph. Confidence and correlation were then studied concerning these connectives. Although similar linear functions were established between the truth (confidence or correlation) value of the proposition and the truth of each argument, correlation depended once again on the text cohesion, because the linear coefficients in this case were greater than zero, whereas they were statistically null in the case of confidence. Finally, the truth value of the proposition was equally dependent on each argument if confidence was taken into account, whereas the truth value was mainly dependent on the first argument if correlation was considered.

This paper provides further evidence for the differentiation between confidence and correlation by analysing brain activity recorded during the subjects listening to text and studying the correlation between this activity and the assigned values of confidence and correlation.

THE BRAIN ACTIVITY

The EEG activity evoked by sensory stimuli is considered as a complex variable pattern of several distinct components. It is also believed to correlate with different and sequential stages of message processing. Early or exogenous components have been assumed to reflect stimulus qualities rather than personal evaluation of the sensory environment; whereas late or endogenous components were considered to accompany task relevance (e.g., Donchin, Ritter, & McCallum, 1978; Josiassen, Shagass, Roemer, Ercegovac, & Straumai, 1982; McCallum, Curry, Pocock, & Papakostopoulos, 1983; Ritter, Simson, Vaughan, & Friedman, 1979; Squires, Donchin, Herning, & McCarthy, 1977).

Averaged P300 was initially observed to vary with the reduction of uncertainty about the stimulus, and later with task relevance; stimulus evaluation time; sequential structure; incentive value; etc. This component of the event related activity (ERA) was demonstrated to be concerned with future events rather than to be stimulus bounded (e.g. Beigleter, Porjesz, Chon, & Aunon, 1983); Courchesne, 1978; Ford, Pfefferbaum, & Kapell, 1982; Kutas, McCarthy, & Donchin, 1977; McCallum et al., 1983; Ritter et al., 1979; Ruchkin, Munson, & Sutton, 1982; Squires et al., 1977). Negative components peaking from 200 up to 400 ms (e.g., Na and N2), were observed to be affected by the nature of the classification task and awareness, to vary with the reaction time, and to be inversely dependent on the stimulus probability (e.g. Ford et al., 1982; Josiassen et al., 1982; Ritter et al., 1979; Ritter, Simson, Vaughan, & Macht, 1982; Ritter, Simson, Vaughan, 1983; Squires et al., 1977). Later components in the range of 500 up to 1000 ms were found to correlate with unexpected stimulus attributes (e.g., Kutas & Hillyard, 1980).

Early components (peaking as soon as 15ms and appearing up to 200 ms) were assumed to be stimulus bounded, in the sense that they varied with the physical characteristics of the stimulus. However, it is now observed that they are also associated with the stimulus probability, time interval between stimuli, conceptual information channels (e.g., stimulation of finger vs hand), target vs non-target stimuli, etc. (Ford et al., 1982; Hansen and Hillyard, 1980; Josiassen et al., 1982; McCallum et al., 1983); thus associated to attention requirements, too.

It is apparent that, when a task is relatively simple, important to look for stimulus properties that can be identified - and irrelevant properties filtered out - by the brain substantially earlier than 100 ms. When the task is more difficult, a negative component peaking around 100 ms seems to represent the earliest brain sign of target identification (McCallum et al., 1983). The rapid differentiation of the relevant from the irrelevant is possibly more in keeping with the notion of directed attention, that is to say with the presence of a pre-existing bias or expectation toward a specified class of stimuli (e.g., Rocha, 1985; Rocha & Rocha, 1985).

Along this line of reasoning, it is possible to assume a sensory processing model involving different sequential and parallel stages, such that each stage passes into the subsequent ones whatever information has reached criteria for the encoding properties analysed at that stage (McCallum, et al., 1983; Ritter et al., 1982). Which stages and criteria are to be involved with a particular stimulus at a particular situation, may be a matter of expectation generated by the surrounding environment; previous individual histories; etc. (Rocha, 1985). The greater is the expectation or the smaller is the uncertainty about the stimulus, the earlier may be its recognition, by adjusting information processing at low stages to detect distinguishing properties of the incoming stimulus (Rocha, 1985; Rocha & Rocha, 1985).

Quite similar ERA has been recorded for verbal stimuli presented either through auditory or visual channels (e.g., Brown, Marsh, & Smith, 1973, 1976, 1979; Johnston & Chesney, 1974; Kostandov & Arzumanov, 1977; Kutas and Hillyard, 1980; Matsumyia, Tagliasco, Lombroso, & Goodglass, 1972; Neville, Kutas, & Schmidt, 1982a,b; Roth, Kopell, & Bertozzi, 1970). Components occuring earlier than 100 ms failed to be modified by the use of verbal vs non-verbal material (Matsumyia et al., 1972; Roth et al., 1970); instead they seemed to be linked to the stimulus significance; task involvement; etc. ERA differences reflecting discrimination between the noun vs verb meanings of the same word, are encoded in components as early as N150 and may extend up to P230, N 400, etc. (Brown et al., 1976, 1979; Johnston & Chesney, 1974).

N150 and N400 have been shown to correlate with the semantic processing of verbal material (Brown et al., 1976; Kutas and Hillyard, 1980; Neville et al., 1982a). N400 appears predominantly over left and anterior locations, whereas positive waves predominate over right and posterior sites (Brown et al., 1979; Neville et al., 1982a). This electrophysiological asymmetry was further stressed by its absence in congenitally deaf adults as accurate as hearing subjects in visually identifying words (Neville et al., 1982b).

Verbal N400 that seems similar to non-verbal N2 (see Ritter et al., 1982) may constitute an important step in a semantic analysis performed throughout a controlled process; the guidance at each processing stage provided by both the results attained at the previous levels (e.g., those reflected by N150) and previous learning.

Verbal analysis built up on such an expectancy controlled process may favor good semantic perception (McCallum et al., 1983; Rocha, 1985; Rocha & Rocha, 1985) as soon as necessary to explain speech shadowing at very short latencies (Marslen-Wilson, 1978), as well as to explain the agreement between EEG activity despite visual or auditory presentation of verbal stimuli (compare, e.g., the results of Brown et al., 1979 and Neville et al., 1982). It must be remembered that in this late case, ERA components of important semantic decisions may occur before the word end and that in many other instances, semantic perception may occur even without full word recognition (e.g. Kostandov & Arzumanov, 1977). The fact that late positive components (latence greater than 700ms) are recorded for unexpected long words and late negative shifts in the EEG activity appear for semantically inappropriate words (Kutas & Hillyard, 1980) futher supports this line of reasoning.

The findings of Molfese (1978) about voice onset time (VOT) analysis point also to the hypothesis of a progressive oriented processing of verbal information, from the brainstem to cortical areas, paralleling the analysis from the phonological to the semantical levels. At the one side, Molfese (1978) identified three distinct factors in the EEG activity (latencies ranging from 80 to 485 ms) related with VOT changes. Besides this, he observed an active and balanced involvement of both hemispheres, the right side also participating in the distinction between voiced and unvoiced stimuli. At the other side, it is possible to perform a suitable VOT analysis at subcortical (at least at the thalamic) level. The recorded, well timed, cellular activity were different between voiced and unvoiced sounds at different cells and at different stages of processing. Many of the cells exhibited feature distinguishing patterns, either responding only in case of voiced or unvoiced sound stimuli.

It is assumed that speech perception is the result of a progressive, oriented process begining with a phonological analysis taking place at the brainstem up to bilateral hemispheric areas. The phonological recognition prompts the system to both early and late semantic decisions based on predominantly left hemisferic activity. At low levels phonological analysis is devoted to phonetic pattern recognition leading to word characterization, whereas at high levels, it is related to the aspects of sentence intonation involved with given vs new information decisions (Rocha & Francozo, 1980). At the cortical level, speech processing progresses from words to phrases, and from phrases to text, as analysis moves from central to parietal, and from parietal to frontal areas (Luria, 1974). Thus verbal perception starts with phonological recognition, and continuously moves to semantic analysis of words, sentences and texts, passing at each subsequent stage whatever information has reached chosen criteria; the choice being a matter of strategy and expectation (Rocha, 1985; Rocha & Rocha, 1985) guided by personal belief and the speech cohesion.

Up to now, the electrophysiological approach to such a process reached just the sentence level, because it devoted its attention only to the EEG activity related specifically to the word recognition in the context of the phrase. The next necessary step is the study of the phrase recognition in the context of a text or dialogue (Rocha, 1982; Rocha, 1985), and additionally the building up of the text level. The results presented in this paper are quite innovative in this respect.

RECALLING THE TEXT

Any text or dialogue has a theme or subject, and a rheme or what is said about the theme (Sgall et al., 1973; Rocha, 1983; Rocha & Rocha, 1985). For instance, the theme of the previos section text could be "Brain activity during speech perception", and its rheme accepted as "is founded over an expectation guided process". However, individual expectations flavors speech decoding according to personal histories; conceptions; culture; etc. (e.g.,Oleron, 1980; Olson, 1980). Because of this, the same text or dialogue undergoes, in general, distinct decodings If manipulated by different people. For example, this same text above could be understood as "Electrocortical activity during speech perception" (Theme) "has definite components" (Rheme). The first of the above decodings above could be performed by people involved with expectancy studies, whereas the second one will arise probably from those researchers working with electrophysiology.

Each reader (speaker) organizes the incoming informations according to his (her) expectation in order to get the subject (Theme) of the communication and the message (Rheme) about this subject (Sgall, et al, 1973; Oleron, 1980). This strategy privileges those phrases and words related with the presumed Theme and Rheme. If expectation is initially low, the same strategy may take place as the reading (or listening) and the build up of the text comprehension progress (Luria,1974; Washabaugh, 1980; Rocha, 1985; Rocha & Rocha, 1985).

Although different people may have different understandings of the same speech, their decodings must speek about some commonalities, unless the communication is nonsense. The possible mean understandings of a given speech (text) by a specific population of speakers (readers) may be obtained by calculating the possible mean graphs for the familly of fuzzy graphs of the individual speech or text decodings.

In the case of the present research, a text about the country's dependence on foreign technology was taperecorded, and played to subjects having their EEG recorded. After the recording session, the volunteers recalled the text by writting; pointed out the intended theme and rheme, and organized the recalled text into a fuzzy graph according to the relations share by the recalled phrases in defining the chosen text theme and rheme.

The statistical analysis of these fuzzy recalling graphs revealed 3 possible mean decodings in our population of volunteers (Rocha & Greco, 1987). The dominant decodings (by 65% of the subjects) were those defined by the phrase 1 and 2 speaking about the huge difference between the budget applied on science and the money spent on foreign technology purchase, and by phrases 5 and 6 claiming that science does not receive the backing required to avoid a huge external dependence on technology. The third possible understanding was that embeded in phrases 9 and 10 saying that Brazil faces the problems imposed by a narrowed mind technocracy deciding on how to apply money on science and technology research. Phrase 7 relating the necessity to fund science according to the pressure put by foreign trade was never included in the recalled text. Phrase 4 referring to the foreing technology used by the Brazilian Oil Company in one of its most advertised product was recalled by almost 70% of the people, however it was considered louselly connect with the most frequent decoded themes and rhemes.

The brain activity recorded during the text hearing was highly correlated to these mean decodings because the phrase probability in the recalled texts was linearly correlated to the EEG amplitude and variation recorded at frontal, central and parietal sites (Rocha & Greco, 1988).

The volunteers were asked to assign the degree of correlation (Group A) or confidence (Group B) that they judged the information at each node of the fuzzy recalling graph might share with the decoded theme/rheme or may receive from them. The purpose of the present paper is to investigate the possible correlation between the brain activity during the text decodification and the two fuzzy measure of uncertainty associated to the text comprehension.

METHODS

The text published by a newspaper was recorded by a Brazilian native speaker. Then it was inspected to disclose any stuttering, hesitation, etc. This inspection was based on both a careful listening of tape by different people, as well a visual analysis of the oscillogram display of the text.

The text was played twice within 45 min. in a soundproof room, to university students while their EEGs were recorded. The sound was delivered by two loudspeakers placed two meters in front of the volunteer; volume was adjusted according to the subject's preference. Ag-AgCl electrodes were placed over each hemispheres (at C4, P4, F4 and C3, P3, F3 respectively), and were referred to earlobes. Cut-off frequences were .1 and 50Hz, and the time constant was 0.3s. Impedance was kept below 5Kohm measured at 400Hz. The EEG recordings were fed into the electroencephalograph together with a filtered version of sound (cut-off frequency of 10Hz), used to localize the EEG epochs corresponding to each phrase heard. The EEG and phrase time marks identifying the beginning of each phrase were stored in an Apple compatible computer for further procesing. The storage was accomplished by amplifying the EEG to the +2.5mV range and sampling it at 100Hz and 8 bits.

Between listening sessions, individuals recalled the text and pointed out its theme and rheme. Then they were asked to put the phrase they judged to belong to Theme(T) apart from those judged to relate with Rheme(R). In the sequence, they joined the phrases according with the best intended relations to conclude for T or R. This allowed the construction of graph of the recalled text (RT), describing those associations of information leading to the conclusion of the ideas embodied in the text understanding (Greco, Rocha, & Rocha, 1984; Rocha & Greco, 1987).

The phrases of the recalled text were located at the terminal nodes of the graph according to the importance they were assumed to have defining T and R. The relations between phrases judged as responsible for the deductions of T or R were preserved by the relations between nodes expressed through the arcs. The nodes between the root and the terminals are named 2nd, 3rd, etc. order nodes, and represent the partial deductions leading to T and R. The graph was construed by one of the authors to express the relations disclosed by the volunteer, and received the approval of the volunteer at each step of its construction.

20 volunteers (Group A) assigned to each node the degree of correlation between the information representated at that node and the text T/R. The confidence degree associated with each node was studied using 20 other subjects (Group B). After the second listening session, subjects were permitted to modify any of these early statements.

The statistical analysis of the data (Greco et al., 1984; Rocha & Rocha, 1985) disclosed the mean text decodifications by our population, the most frequent themes and rhemes, as well as the most important associations between phrases and T/R (Rocha & Greco, 1987).

The measure used to study the brain activity during listening the epoch of each phrase was the area under the corresponding EEG trace. This area was calculated by rotating the EEG negative components around the EEG mean value for the entire epoch, in order to allow the areas of the negative components to be added to that of the positive waves. The rotation around the mean value eliminated the influences of any very slow wave on the calculation of the area.

The area under this rotated EEG trace was calculated by summing the area between each two sampled points, and then divided this totally by the epoch duration to furnish a mean EEG area (A) for each phrase. This meas area A measures the mean instantaneous variation of the EEG activity during the entire epoch concerned with a given phrase. The variation coefficient (CV) was calculated as the division of the standard deviation by the above mean.

Two coefficients, Ca and Cv, of asymmetry between the left and right activities were calculated as

Ca = AL/AR

Cv = CVL/CVR

that is, as relations between the mean areas and the coefficients of variations for right and left recordings. These asymmetry coefficients may disclose hemispheric dominances. On the one hand, a Ca greater than 1 may imply a better performance of the left side, and Ca smaller than 1 may signify a right dominance. On the other hand, Cv greater than 1 may imply an EEG amplitude variation at the left side greater than at the right hemiphere. Similar coefficients were calculated between Frontal and Parietal, Frontal and Central, and between Central and Parietal sites, for each hemisphere.

Any possible correlation between the above truth values (confidence and correlation) and the EEG activity was studied using linear regression and correlation analysis of the truth values and all these assimetry coefficients.

RESULTS

The statistical analysis of the data revealed that both the correlation (Cor) as well as confidence (Con) degrees assigned to the recalled phrases were linearly correlated with the brain activity associated with the phrase decoding during the first listening session, but not with the same activity registered during the second listening session. However, the volunteers after the second listening task modified the initial written recall only in some very excepcional instances.

The general picture that has appeared in this kind of research (Greco & Rocha, 1987; Theoto et al., 1987) is that the volunteer decided during the first listening session, such that the second listening is used just as a confirmation of first expectations. Only on rare ocasions did the second listening task cause changes in the volunteers mind, but provoked only minor adjustments of the first decoding. This is true even in the case of very poor initial recalls. The trend is for the volunteer to continue to believe on the first understanding even if the text was badly understood.

Correlating the phrases and the text

The degree of correlation (Cor) between the recalled phrases and the theme/rheme decoding had its dependence on the hemispheric assimetry coefficients calculated for both the Area (A) and Coefficient of Variation (CV) obtained for the Frontal (AF, VF) , Central(AC, VC) and Parietal (AP, VP) records, described by the following multiple regression equations:

$$\text{Cor} = 127 - .25 * AF - .87 * AC - .17 * AP - .8 * VC \quad (1)$$
$$R = 93\% \text{ and } P = 0\%$$

$$\text{Cor} = 132 - .20 * AF - .99 * AC - 0.0 * AP - .8 * VC \quad (2)$$
$$R = 74\% \text{ and } P = 11\%$$

$$\text{Cor} = -89 + .70 * AF + 1.8 * AC - .21 * AP - .9 * VC \quad (3)$$
$$R = 26\% \text{ and } P = 84\%$$

The correlation coefficient change from 26% to 93% and the probability for the null hypothesis ranged from 84% to 0% from Equations 3 to 1. Equation 2 was obtained when the all recalled phrases were taken into account for the caculus. Equation 1 was obtained when phrase 4 - loosely connected to the decoded themes/rhemes, but highly frequent in the written recall due to a massive TV advertising - was disregarded in all the recalls in which it occured. Finally, Equation 3 was calculated the same as Equation 2, but including also a null correlation degree for the never occuring phrase 7, and for all recalled phrases. These 3 different calculations were done to investigate the interdependece between text cohesion, Cor and the brain activity. The stepwise regression analysis confirmed the hypothosis that phrases 4 and 7 were not correlated to the correlation degree, since their values were the only ones encounterd outside the confidence range.

It is possible to conclude from the above results that Cor increased as the dominance of the left hemisphere is reduced when the mean area of the EEG was considered in the frontal, central and parietal areas and when the coefficient of variation was taken into consideration at the central sites. Also, this dependence did not occur if all recalled phrases were considered (Equation 3), but only if those phrases highly correlated with the decoded themes/rhemes were used in the caculations (Equation 1). The Central Area exercised the bigest influence on the final value of Cor (Equations 1 to 3).

Confidence in the received information

The same calculus above were done for the confidence degrees assigned to the recalled phrases by people of the Group B:

$$Con = 326 - 7.4 * AF + 1.8 * AC - 0.0 * AP - 4.0 * VF \quad (4)$$
$$R = .72 \text{ and } P = 13\%$$

$$Con = 246 - 6.5 * AF + 2.5 * AC - 0.0 * AP - 4.2 * VF \quad (5)$$
$$R = .73 \text{ and } P = 12\%$$

$$Con = 292 - 5.9 * AF + 2.3 * AC - 0.0A * P - 3.7 * VF \quad (6)$$
$$R = .74 \text{ and } P = .01\%$$

It is possible to conclude from these results, that confidence in the received information is enhanced when the hemispheric assimetry decreased at the frontal areas and increased at the central sites. This dependence occured -

contrary to the results obtained for Cor - no matter the correlation between the recalled phrases and the decoded themes and rhemes (compare Equations 4,5, and 6). As a matter of fact, the best statistical results appeared if all recalled phrases were used and a null confidence was assigned to the nerver occuring phrase 7 (Equation 6). The Frontal Area exercized the bigest influence on the final value of Con (Equations 4 to 6).

The dependence of confidence on the cerebral activity was fuzzier than the dependence of the correlation degree with the EEG activity because the correlation coefficients calculated for Equations 5 to 6 were smaller than that obtained for Equation 1.

Correlation, confidence and activity in the same hemisphere

Cor and Con were also correlated with the dominance coefficients calculated between Frontal and Central (AF/C, VF/C), Frontal and Parietal (AF/P VF/P), and Central and Parietal (AC/P, VC/P) areas at the same cerebral hemisphere. The most striking difference was observed when Cor and Con were compared; Cor depended on dominances established for areas located at the right cerebral hemisphere - this is in accordance with its dependence on a reduction of the dominance of the left over the right cerebral areas - whereas Con depended on a reduction of the dominance of the left frontal area over the central left sites - this may agree with the fact that confidence augmented as dominance of the left hemisphere decreased over the right frontal areas and increased at the central sites. Besides this, the dependence of Con on the dominance of the left hemispheres was fuzzier than the dependence of Cor on the right dominance if the correlation coefficients in Equations 7 and 8 were compared.

The multiple regressions of these dependences were :
Cor and the right hemisphere:

Cor = 31 -.55 * AF/C + .36 * AF/P + .56AC/P + .10VF/C -.39 * VF/P (7)
R = 94% and P = 1%

Con and the left hemisphere:

Cor = 1586 -12 x AF/C + 8 x AF/P -10 x AC/P (8)
R = 55% and P = 6%

No correlation was found between confidence and correlation degrees assigned to non-terminal nodes and the EEG activity associated to the recalled phrases.

DISCUSSION

The results described in the previous sections pointed to a linear dependence of both the correlation and confidence on the brain activity recorded during the first - but not in the second - text decodification, although the volunteers were instructed about assigning these fuzzy measures to the decoded information just after they had heard the text the first time, recalled it in writting and constructed its fuzzy graph. Besides this, people changed their minds after the second hearing session only on a very few and exceptional occasions. Because of this, it is possible to state that the fuzzy measures of the pertinence of the information to the language (correlation) and individual (confidence) truth sets are real psychological entities having defined correlations with the brain activity during the processing of the information to which they are assigned.

These observations constitute a very innovative electrophysiological demonstration of a neural substrate of a very complex psychological construct. The data presented here indicate that when very abstract concepts such as fuzzy beliefs, are associated with the verbal processing, the well known left hemisphere dominance (see, e.g., Brown et al., 1976, 1978; Luria, 1974; Nevile et al., 1982b; etc.) is reduced by the increase of the participation of the right cerebral areas. This was the case for both correlation and confidence, although correlation seemed to be more dependent on the right hemisphere working than confidence.

Another important finding was that many different areas did participate in the processing of the correlation degree, although central areas seemed to play the most important role in the assessement of this judgement (see Equations 1 and 7). Central areas had always enjoyed a special status in the language processing neural circuits since the pioneer observations made by Wernicke. The involvement of the brain concerning the evaluation of confidence was more restricted to the frontal areas (see Equations 6 and 8), a region well known by its participation in many different types of intellectual processing (Luria, 1974).

Besides this, the correlation degree depended on the activity recorded at the central areas involved the verbal analysis at the word level, on the activity of the parietal circuits involved in the phrase analysis and on frontal areas aggregating phrases on the level of dialogue and/or text (Luria, 1974). Also, the correlation degree depended almost totally on the brain activity related to the phrases that had a high cohesion with the decoded theme and rheme. This was not the case when confidence was considered. The dependence of the confidence degree to the brain activity did not change if different sets of phrases were taken into account in the calculus.

This behavior stresses some important differences between the concepts involved in these two fuzzy measures. This behavior may also be considered as supporting the idea that text decoding may begin with people using their confidence in the information to guide the text deconding, and may then procede, oriented by the correlation between the received pieces of information and the text itself, as soon as its inital structure is captured. This is in line with the proposition that speech understanding is the result of a progressive oriented process, begining with a phonological analysis, taking place at the brainstem, toward a highly abstract semantic processing taking place in both cerebral hemispheres, with guidance reflecting both personal beliefs and speech cohesion (Luria, 1974; Ritter et al., 1982; Rocha, 1985; Rocha & Rocha, 1985).

The fact that no dependence was established between brain activity related to the recalled phrases and the correlation and confidence degrees assigned to non-terminal nodes may be understood as a consequence of this goal oriented processing. In this line of reasoning, both confidence and correlation for these nodes are to be calculated as the fuzzy logic structure of the decoding as it begins to be organised. If this is true, this building-up process cannot be timed during the acquisition of the data because of the EEG segmentation required by the statiscal analysis. This could be the reason for the negative results concerning the non-terminal nodes.

The differences in the dependences established for confidence and correlation stress the idea that these two measures are two different psychological entities as indicated by other researches on language decoding in our group (Greco & Rocha, 1987; Rocha et al., 1984; Theoto et al., 1987). The degree of correlation - but not confidence - depended on the structure of the text decoding, its value increasing from the terminal nodes related to the recalled phrases toward the root associated with the decoded theme and rheme. Also, the correlations decreased from the left toward the right nodes (Rocha & Greco, 1987). In addition, the correlation degree - but not confidence - depended on a non-zero linear coefficient when its value for the proposition was calculated from the values assigned to each argument (Greco et al., 1984; Rocha & Greco, 1987).

The conclusion that two fuzzy measures used on fuzzy logic have a real pschological and physiological existence, has very important consequences for the Fuzzy Set Theory by strengthening the proposition of Zadeh (1964) that fuzzy measures are the theoretical constructs closest to the real way humans reasons. Also the fact that confidence and correlation are two different measures of the uncertainty associated to the processing of any information by humans, have practical consequences for the development of any system trying to simulate the human reasoning as is the case with expert systems.

Curently 'Expert Systems' use in general just one measure of uncertainty (see e.g., Shortlife, 1976, Barr & Feigenbaum, 1982; Rocha et al., 1987) to access the plausibility of the diagnosis, and the measure used by most of them is similar to the correlation degree - correlation between the signal, symptom, etc. and the diagnosis - investigated here. The personal belief in the actual presence of the data is always disregarded in these systems, although this is one of the most important judgements perfomed by the expert when solving any case.

Both fuzzy measures (correlation and confidence) must be taken into account in any calculus of aggregation to be used on decision making systems. We may accept the proposition of Yager (1987) about the use of Ordered Weighted Averaging Aggregation Operators in Multi-criteria Decision Making as an important step in this direction. The main idea here, is that averaging one of these measures - say confidence - during the agregation process (e.g., Rocha & Greco, 1987) may be guided by the strength of the other measure - e.g. correlation.

REFERENCES

Barr, A. & Feigenbaun, E. A. *The Handbook of Artificial Intelligence.* Heuristech Press/Willian Kaufman, 1982

Begleiter, B., Porjesz, B., Chon, L. C., & Aunon, J. I. P3 and stimulus incentive value. *Psychophysiology*, 1983, *20*, 95-101.

Brown, W., Marsh, J. T. & Smith, J. C. Contextucal meaning effect on speech evoked potentials. *Behavior Biology,* 1973, *9,* 755-761.

Brown, W., Marsh, J. T. & Smith, J. C. Evoked potential waveform differences produced by the perception of different meanings of an ambigous phrase. *Electroencephalography and Clinical Neurophysiology,* 1976, *41,* 113-123.

Brown, W., Marsh, J. T. & Smith, J. C. Principal component analysis of ERP differences related to the meaning of an ambiguous word. *Electroencephalography and Clinical Neurophysiology,* 1979, *46,* 709-714.

Courchesne, E. Changes in P30 waves with event repetition: Long term effects on scalp distribution amplitude. *Electroencephalography and Clinical Neurophysiology,* 1978, *45,* 754-766.

Donchin, E., Ritter, W., & McCallum, C. Cognitive psychophysiology: The endogenous components of the ERP. In E. Callaway, P. Tueting and S. H. Koslow (Eds.), *Event Related Brain Potentials in Man.* New York: Academic, 1978.

Ford, J. M., Pfefferbaum, A., & Kapell, B. Event related potentials to a change o pace in visual sequence. *Psychophysiology,* 1982, *19,* 173-177.

Greco, G., Rocha, A. F., & Rocha, M. T. *Fuzzy logical structure of a text decoding.* Proceedings of the 6th International Congress of Cybernetics and Systems, Paris, 1984.

Hansen, J. C, & Hillyard, S. A. Endogenous brain potentials associated with selective auditory attention. *Electroencephalography Clinical Neurophysiology,* 1980, *49,* 277-290.

Josiassen, R. C., Shagass, C., Roemer, R. A., Ercegovac, D. V., & Straumai, J. J. Somatosensory evoked potential changes with a selective task. *Psychophysiology,* 1982, *19,* 146-159.

Johnston, V. S. & Chesney, G. L. Electrophysiological correlates of meaning. *Science,* 1974, *186,* 944-946.

Kostandov, E. & Arzumanov, Y. Averaged cortical evoked potentials to recognized and non-recognized verbal stimuli. *Acta Neurobiologica Experimental,* 1977, *37,* 311-324.

Kutas, M., McCarthy, G., & Donchin, E. Augmenting mental chronometry: P53000 as a measure of stimulus evaluation time. *Science*, 1977, *197*, 792-795.

Kutas, M. & Hillyard, S. A. Reading senseless sentences: Brain potentials reflect semantic incongruity. *Science*, 1980, *207*, 203-208.

Luria, A. R. *Cerebro y Lenguage*. Paidos, 1974.

Marslen-Wilson, W. Linguistic structure and speech shadowing at very short latencies. *Nature*, 1978, *244*, 522-523.

McCallum, W. C., Curry, S. H., Pocock, P. V. & Papakostopoulos, D. Brain event related potentials as indicators of early selective processes in auditory target localization. *Psychophysiology,* 1983, *20,* 1-17.

Matsumyia, Y., Tagliasco, V., Lombroso, C. T., & Goodglass, H. Auditory evoked response: Meaningfulness of stimuli and interhemispheric asymetry. *Science,* 1972, *175,* 790-792.

Molfese, D. L. Neuroelectrical correlates of categorical speech perception in adults. *Brain and Language*, 1978, *5,* 25-35.

Neville, H. J., Kutas, M., & Schmidt, A. Event related potential studies of cerebral specialization during reading: I. Studies of normal adults. *Brain and Language,* 1982, *16,* 300-315. (a)

Neville, H. J., Kutas, M. & Schmidt, A. Event related potential studies of cerebral specialization during reading: II. Studies of congenitally deaf adults. *Brain and Language*, 1982, *16,* 316-327. (b)

Oleron, P. Social intelligence and communication: Introduction. *International Journal of Psycholinguistics,* 1980, *7,* 7-11.

Olson, D. R. On languauge and literacy. *International Journal of Psycholinguistics*, 1980, *7,* 69-83.

Ritter, W., Simson, R., Vaughan, H. G., & Friedman, D. A brain event related to the making of a sensory discrimination. *Sience,* 1979, *203,* 1358-1361.

Ritter, W., Simson, R., Vaughan, H. G., & Macht, M. Manipulation of event related potential manifestations of information processing stages. *Science,* 1982, *218,* 909-911.

Ritter, W., Simson, R., & Vaughan, H. G. Event related potential correlates of two stages of information processing in physical and semantic discrimination tasks. *Psychophysiology,* 1983, *20,* 168-170.

Rocha, A. F. Basic properties of neural circuits. *Fuzzy Sets and Systems,* 1982, *7,* 109-121.

Rocha, A. F. Mobile expert sensory system. In M. Gupta, A. Kandel and B. Kiszka (Eds.), *Approximate reasoning and expert systems.* Amsterdam: North-Holland, 1985.

Rocha, A. F. Toward a theoretical and experimental approach of fuzzy learning. In M. Gupta and E. Sanches (Eds.), *Approximate reasoning in decision analysis.* Amsterdam: North-Holland, 1983.

Rocha, A. F. Brain activity during language perception. In M. Sing (Ed.), *Encyclopedia on system and control.* Oxford: Pergamon, 1987.

Rocha, A. F. & Francozo, E. EEG activity during the speech perception. *Revue Phonetique Appliquee,* 1980, *55/56,* 307-311.

Rocha, A. F. & Greco, G. The fuzzy logic of text understanding. *Fuzzy Sets and Systems,* 187, *23,* 347-360.

Rocha, A. F. & Greco, G. The brain activity during the text decodification. In M. Sing (Ed.), *Encyclopedia on system and control.* Oxford: Pergamon, 1988.

Rocha, A. F. & Rocha, M. T. pecialized speech: The initial prose for language expert systems. *Information Sciences,* 1985, *35,* 215-233.

Rocha, A. F., Giorno, F., & Leao, B. The physiology of the expert system. In A. Kandel (Ed.), *Expected system in the fuzzy age.* Reading, Mass.: Addison Wesley, 1987.

Roth, W. T., Kopell, B. S., & Bertozzi, P. E. The effect of attention on the average evoked response to speech sounds. *Electoencephalography and Clinical Neurophysiology,* 1970, *29,* 38-46.

Ruchkin, D. A., Munson, R., & Sutton, S. P53000 and slow wave in a message consisting of the two events. *Psychophysiology,* 1982, *19,* 629-642.

Sgall, E. P., Hajicova, E., & Benesova, E. *Topic, focus and generative grammar.* Kronberg: Scriptor, 1973.

Shortliffe, E. H. *Computer-based medical consultations: Mycin.* Amsterdam: Elsevier, 1976.

Squires, K. C., Donchin, E., Herning, R., & McCarthy, G. On the influence of task relevance and stimulus probability on event related potential components. *Electroencephalography and Clinical Neurophysiology,* 1977, *42,* 1-14.

Theoto, M. T., Santos, M. R., & Uchiyama, N. The fuzzy decodings of educative texts. *Fuzzy Sets and Systems,* 1987, *23,* 331-345.

Washabaugh, W. The role of speech in the consturction of reality. *Semantics,* 1980, *31,* 197-214.

Yager, R. R. *On ordered weighted averaging aggregation operators in multi-criteria decision making.* Technical Report. New Rochelle, N.Y.: Machine Intelligence Institute, Iona College, 1987.

ACQUISITION OF MEMBERSHIP FUNCTIONS IN MENTAL WORKLOAD EXPERIMENTS

I.B. TURKSEN, N. MORAY and E. KRUSHELNYCKY

Department of Industrial Engineering, University of Toronto
Toronto, Ontario M5S 1A4, Canada

In a mental workload experiment, subjective measurements for linguistic variables such as slight, moderate, and difficult, are recorded to investigate the predictability of the degree of difficulty associated with a task that contained the skill-based, the rule-based, and the knowledge-based components of an operator's behaviour. From the perspective of measurement theory, it was found that the measurement data have the essential properties of weak stochastic transitivity and weak stochastic monotonicity. From the perspective of Rasmussen's taxonomy of human behavior, it was found that the skill- based behavior would dominate while the rule and the knowledge-based behaviours would moderately influence the degree of the difficulty associated with a task.

INTRODUCTION

Measurement of membership values was a controversial topic during the first decade or so of fuzzy set theories. Even as late as 1978, some researchers dared to state that " ... There is no way of determining values of membership functions, either rationally or empirically ... (Watanabe, 1978)". At present, among most fuzzy set researchers it is an accepted fact that membership functions can be determined empirically and rationally (Turksen, 1986; Zysno, 1987; etc.). Based on these previous experimental works, we have launched the measurement experiments to be discussed in this paper for the investigation of the degree of difficulty that would be subjectively associated with a task that has three behavioural components.

Mental workload

Mental workload appears to be an elusive concept. However, whenever we wish to assess the degree to which a person is affected in executing a particular task, we need to determine a measure of the amount of physical effort being exerted as well as the amount of mental effort required by an operator for a task. In current literature one finds that many different measures have been proposed and investigated in terms of physiological, secondary task, and subjective measures (Weirwille, 1979; Ogden, Levine, & Eisner, 1979; Moray, 1982; Gopher & Browne, 1984). In our approach, we attempt to establish the descriptive and predictive power of linguistic (fuzzy) variables for a given task in mental workload experiments.

First, our earlier experiments in the measurement of workload (Moray, King, Waterton, & Turksen, 1987: Moray, Eisen, Money, & Turksen, 1987) clearly indicated that linguistic variables such as "slight", "moderate", and "difficult" used in describing components of a mental workload associated with a task can be measured with an individual's membership rating.

Secondly, in his taxonomy of behaviour, Rasmussen (1983) defined three generic classes of behaviour such as skill-based, rule-based, and knowledge-based. It was suggested that skill-based behaviour would not impose any mental workload since the responses to a task were internalized to the point where no thought would be involved in executing a particular action. Real world examples of this occur in driving a car; e.g., one does not consciously think about maintaining a given level of speed.

For rule-based behaviour, it was suggested that an operator would be required to consult a collection of internalized "IF ... THEN ..." statements in order to react to situations for which the operator had not developed any automatic response patterns ; e.g., "IF fuel low THEN go to gas station".

For knowledge-based behaviour, it was suggested that an operator would undertake a response to a task situation to which he had not been exposed. This would involve an extrapolation of previously encountered situations to this new scenario.

With such background knowledge about (i) the measurement of membership functions, and (ii) Rasmussen's generic classification, we started our investigation of skill-based, rule-based and knowledge-based behaviour componenst of a NASA Hovercraft Simulator (Thornton, 1984). In simulation experiments run with this simulator, it is required that an operator guide the hovercraft along a track defined by its width and turbulence. The width parameter adjusts the track width from one and half to eight hovercraft width. The turbulence parameter causes random disturbances in the movement of the hovercraft. In addition, the track is generated from a series of sine waves. There are nine checkpoints in the course, in which rules may be executed. The rules consist of incrementing the fuel level, oil level, and adjusting the oil pressure, as well as a combinations of these three. These rules may only be excuted at a checkpoint (Thornton, 1984). The rules are executed by a series of key presses. The hovercraft is controlled by an operator via a joystick. The further an operator pushes the joystick forward, the faster the hovercraft travels "up" the track. Side to side motion of the hovercraft is also controlled via a joystick. Each time the hovercraft touches the edge of the track, an audible click is heard and a "wall hit" is registered.

Forty-eight subjects were used in the experiment, none of whom had any previous exposure to the NASA Hovercraft Simulator before the initial training period. Subjects had an appropriate amount of training for the experimentally required combination of skill-rule-knowlegde based behaviour components.

STRUCTURE OF THE EXPERIMENT

We will now briefly describe the skill-based, the rule-based and the knowledge-based components of the tracking task.

(i) The skill-based behaviour is a continuous tracking task and requires an operator to guide the hovercraft between two lines. Operators are asked to do this with the minimum number of "wall hits" in the fastest possible time. The skill-based behaviour is composed of those actions performed by an operator which "evolve ... without conscious attention or control" (Rasmussen, 1983). Three levels of turbulence and three levels of track narrowness were considered in skill-based behaviour.

(ii) The rule-based behaviour consists basically of three rules: (1) control of the fuel level; (2) maintenance of the oil level; and (3) adjustment of the oil pressure. The first two rules are four-step rules while the last is a five-step rule. The rules are further sub-divided by the experimenter, not the subject, into three classes as "slight", "moderate", and "difficult" tasks. The slight, moderate, and difficult tasks consist of one, two, and three rules being imposed on the subject during the course of a simulation run. In total, therefore, there are seven possible rules, i.e., controls: (1) fuel level control, (2) oil level control, (3) oil pressure control, (4) fuel and oil level controls, (5) fuel level and oil pressure controls, (6) oil level and pressure controls, (7) fuel and oil level, and oil pressure controls.

After an initial set of subjects experimented with the simulator, it became apparent that rule 1 (fuel level control) would be sufficiently representative of an easy rule that required a "slight" effort and rule 7 as a representative of a moderate rule, and no rule would represent a difficult task when a rule-based task was combined with a trivially easy skill-based task.

(iii) The knowledge-based runs involve the inclusion of an "abnormality" in the experimental run. The presence of an abnormality is indicated to the subject by a flashing bar at the bottom right corner of the screen during the run. Five knowledge-based runs are included in the measurement of a subject's mental workload. The first type of abnormality is defined by the lack of a malfunction in the system. The second and third types of malfunctions occur at times 80, 200, 320, and 430 from the beginning of a run. The second type is a sudden drop of oil level to 2; the third type is a sudden drop of fuel level to

100; the fourth type is a drop of oil pressure to 0.6 at time 80, a false warning triggered at time 200, and the fuel level drop to 100 at time 320, and finally, another false warning at time 430. For the fifth type of malfunction, the fuel valve and the oil valve are left open near the beginning of a run without the subject's knowing about it. The oil level suddenly drops to 20 at time 15 0; the fuel level suddenly changes to 350 at time 210; and finally, the oil pressure drops to 0.6 at time 330.

Two points should be noted here: (1) with the exception of the false warnings, the abnormalities have to be carefully monitored by a subject in order for a task to be completed successfully; (2) in all cases, the only indication to a subject that some thing is indeed not normal is the presence of the flashing bar. It is up to the subject to determine the nature of the abnormality. It should be recalled that subjects learned to check and control the oil and fuel levels during the rule-based tasks.

TABLE 1

Knowledge-based Runs: Membership Estimates

Levels of Knowledge	Slight			Moderate			Difficult		
	Mean	SD	NM	Mean	SD	NM	Mean	SD	NM
1	0.80	0.09	1.00	0.36	0.20	0.00	0.20	0.15	0.00
2	0.43	0.30	0.00	0.53	0.18	0.43	0.44	0.22	0.86
3	0.43	0.23	0.00	0.76	0.10	1.00	0.48	0.09	1.00
4	0.53	0.27	0.27	0.44	0.10	0.20	0.40	0.26	0.71
5	0.49	0.21	0.16	0.65	0.09	0.73	0.41	0.25	0.75

Explanation: Means are average over six subjects
SD - Standard Deviation,
NM - Normalized Mean

Experimental design

No repetitions are possible for the knowledge-based task. The order of experimental runs is based on three Youden Squares. It consists of (i) 9 conditions by 18 subjects for the skill-knowledge matrix; (ii) 6 conditions by 12 subjects for the rule-knowledge matrix; (iii) 18 conditions by 12 subjects for the skill-rule-knowledge matrix; (iv) 5 conditions by 6 subjects for the basic knowledge-based matrix. The results of the subjective measurement data for these experimental runs are shown in Table 1, 2, 3, and 4.

TABLE 2
Skill-Knowledge Runs: Membership Estimates

Levels of S-K	Slight			Moderate			Difficult		
	Mean	SD	NM	Mean	SD	NM	Mean	SD	NM
SK1	0.85	0.08	1.00	0.26	0.18	0.33	0.06	0.07	0.00
SK2	0.37	0.27	0.41	0.53	0.19	0.93	0.43	0.27	0.42
SK3	0.31	0.27	0.34	0.56	0.17	1.00	0.43	0.29	0.42
SK4	0.42	0.29	0.48	0.53	0.16	0.93	0.39	0.28	0.37
SK5	0.23	0.24	0.24	0.55	0.26	0.98	0.49	0.30	0.48
SK6	0.13	0.08	0.12	0.28	0.14	0.38	0.50	0.31	0.49
SK7	0.08	0.09	0.06	0.28	0.14	0.38	0.87	0.11	0.91
SK8	0.03	0.03	0.00	0.27	0.17	0.36	0.90	0.08	0.94
SK9	0.03	0.04	0.00	0.11	0.08	0.00	0.95	0.05	1.00

Explanation: Means are average over 18 subjects
SD - Standard Deviation, NM - Normalized Mean,
SK1 - Skill1 with KNOW1, SK2 - Skill1 with KNOW3, SK3- Skill1 with KNOW5,
SK4 - Skill4 with KNOW1, SK5 - Skill4 with KNOW3, SK6- Skill4 with KNOW5,
SK7 - Skill5 with KNOW1, SK8 - Skill5 with KNOW3, SK9- Skill5 with KNOW5

SCALE PROPERTIES OF DATA

The scale properties of the measurement data need to be analyzed from the perspective of measurement theory before any other analysis of data could begin.

Every fundamental analysis of measurement theory includes definitions of "weak order" and "algebraic-difference structure" and theorems of "ordinal" and "interval" scales. These essentials are briefly reviewed here in the language of Krantz, Luce, Suppes, and Tversky (1971).

TABLE 3
Rule-Knowledge Runs: Membership Estimates

Levels of R-K	Slight			Moderate			Difficult		
	Mean	SD	NM	Mean	SD	NM	Mean	SD	NM
RK1	0.73	0.31	1.00	0.41	0.21	0.38	0.09	0.04	0.00
RK2	0.63	0.39	0.80	0.34	0.31	0.00	0.30	0.37	0.41
RK3	0.26	0.28	0.06	0.38	0.14	0.21	0.60	0.34	1.00
RK4	0.28	0.23	0.10	0.53	0.25	1.00	0.35	0.28	0.51
RK5	0.38	0.35	0.30	0.44	0.18	0.53	0.34	0.32	0.49
RK6	0.23	0.22	0.00	0.53	0.22	1.00	0.54	0.29	0.88

Explanation: Means are average over 12 subjects
SD - Standard Deviation, NM - Normalized Mean,
RK1 - Rule1 with KNOW1, RK2 - Rule1 with KNOW3,
RK3 - Rule1 with KNOW5, RK4 - Rule123 with KNOW1,
RK5 - Rule123 with KNOW3, RK6 - Rule123 with KNOW5,

TABLE 4
Skill-Rule-Knowledge Runs: Membership Estimates

Levels of S-R-K	Slight			Moderate			Difficult		
	Mean	SD	NM	Mean	SD	NM	Mean	SD	NM
SRK1	0.30	0.10	0.40	0.60	0.00	0.70	0.20	0.10	0.03
SRK2	0.50	0.20	0.71	0.53	0.03	0.60	0.23	0.08	0.29
SRK3	0.23	0.08	0.29	0.45	0.05	0.48	0.33	0.18	0.20
SRK4	0.68	0.08	1.00	0.40	0.00	0.40	0.23	0.08	0.07
SRK5	0.35	0.05	0.48	0.80	0.05	1.00	0.28	0.03	0.07
SRK6	0.55	0.30	0.79	0.50	0.05	0.55	0.33	0.18	0.20
SRK7	0.25	0.10	0.32	0.68	0.18	0.32	0.45	0.00	0.36
SRK8	0.38	0.38	0.52	0.38	0.23	0.37	0.40	0.30	0.29
SRK9	0.13	0.08	0.13	0.35	0.25	0.33	0.58	0.03	0.58
SRK10	0.13	0.13	0.13	0.55	0.25	0.63	0.50	0.30	0.43
SRK11	0.30	0.10	0.39	0.53	0.08	0.60	0.55	0.25	0.49
SRK12	0.25	0.05	0.32	0.55	0.05	0.63	0.18	0.08	0.00
SRK13	0.25	0.15	0.21	0.40	0.20	0.40	0.68	0.23	0.67
SRK14	0.08	0.03	0.05	0.18	0.13	0.07	0.75	0.15	0.76
SRK15	0.05	0.00	0.00	0.13	0.03	0.00	0.75	0.05	0.76
SRK16	0.15	0.05	0.16	0.48	0.18	0.52	0.83	0.03	0.87
SRK17	0.05	0.00	0.00	0.20	0.05	0.10	0.93	0.03	1.00
SRK18	0.08	0.08	0.05	0.23	0.23	0.15	0.85	0.15	0.89

Explanation: Means are average over 12 subjects
SD-Standard Deviation   NM-Normalized Mean
SRK1-Skill1,RULE1,KNOW1   SRK2-Skill1,RULE1,KNOW3,
SRK3-Skill1,RULE1,KNOW5   SRK4-Skill1,RULE123,KNOW1
SRK5-Skill1,RULE123,KNOW3   SRK6-Skill1,RULE123,KNOW5
SRK7-Skill4,RULE123,KNOW5   SRK8-Skill4,RULE1,KNOW3
SRK9-Skill4,RULE1,KNOW5   SRK10-Skill4,RULE123,KNOW1
SRK11-Skill4,RULE123,KNOW3   SRK12-Skill4,RULE123,KNOW5
SRK13-Skill5,RULE1,KNOW1   SRK14-Skill5,RULE1,KNOW3
SRK15-Skill5,RULE1,KNOW5   SRK16-Skill5,RULE123,KNOW1
SRK17-Skill5,RULE123,KNOW3   SRK18-Skill5,RULE123,KNOW5

Definition 1.

Let Θ be a set and be a binary relation on Θ, i.e., is a subset of $\Theta' = \Theta \times \Theta$. The relational structure $<\Theta, \geq>$ is a weak-order iff, for all $\Theta_i, \Theta_j, \Theta_k, \varepsilon \Theta$ the following two axioms are satisfied:

1. Weak connectedness: Either $\Theta_i \geq \Theta_j$ or $\Theta_j \geq \Theta_i$.

2. Weak transitivity: if $\Theta_i \geq \Theta_j$ and $\Theta_j \geq \Theta_k$, then $\Theta_i \geq \Theta_k$.

It is noted that this weak-order is reflexive and asymmetric.

Theorem 1

Suppose that Θ is a finite non-empty set. If $<\Theta, \geq>$ is a weak-order, then there exists a real-valued function on Θ such that for all $\Theta_i, \Theta_j, \varepsilon \Theta, \Theta_i \geq \Theta_j$ iff $\Phi(\Theta_i) \geq \Phi(\Theta_j)$. Moreover, Φ' is another real-valued function on Θ with the same property iff there is a strictly increasing function f, with domain and range equal to Re, such that for all $\Theta_i, \varepsilon \Theta, \Phi'(\Theta_i) = f[\Phi(\Theta_i)]$, i.e., Φ is an ordinal scale.

Definition 2

Suppose that Θ is a non-empty set and $\geq'$ is a quarternary relation on Θ'. The relational structure $<\Theta', \geq'>$ is an algebraic-difference structure iff, $\Theta_i, \Theta_j, \Theta_k, \Theta_l, \Theta'_i, \Theta'_j, \Theta'_k$, and $\Theta'_l \varepsilon \Theta$ the following five axioms are satisfied:

1. $<\Theta', \geq'>$ is a weak order.

2. Sign Reversal: If $\Theta_j \Theta_i \geq' \Theta_l \Theta_k$, then $\Theta_k \Theta_l \geq' \Theta_i \Theta_j$.

3. Weak monotonicity: If $\Theta_j\Theta_i \geq '\Theta'_j\Theta'_i$, and $\Theta_k\Theta_j \geq '\Theta'_k\Theta'_j$, then $\Theta_k\Theta_i \geq '\Theta'_k\Theta'_i$.

4. Solvability: If $\Theta_j\Theta_i \geq '\Theta_l\Theta_k \geq '\Theta_i\Theta_i$, then there exist $\Theta'_l, \Theta_l \, \varepsilon \, \Theta$ such that $\Theta'_l\Theta_i \approx \Theta_l\Theta_k \approx \Theta_j\Theta_l$.

5. Archimedian Condition: If $\Theta_1, \Theta_2, ..., \Theta_i, ...$; is a strictly bounded standard sequence ($\Theta_{i+1}\Theta_i \approx \Theta_2\Theta_1$ for every Θ_i, Θ_{i+1} in the sequence; not $\Theta_2\Theta_1 \approx \Theta_l\Theta_j$, and there exists $\Theta'\Theta'' \, \varepsilon \, \Theta$ such that $\Theta'\Theta'' \geq \Theta_i\Theta_l \geq \Theta''\Theta'$ for all Θ_i in the sequence), then it is finite.

Theorem 2

If $<\Theta', \geq'>$ is an algebraic difference structure, then there exists a real-valued function Φ on Θ such that, for all $\Theta_i, \Theta_j, \Theta_k, \Theta_l \, \varepsilon \, \Theta$, $\Theta_j\Theta_i \geq ', \Theta_l\Theta_k$ iff $\Phi(\Theta_j)-\Phi(\Theta_i) \geq \Phi(\Theta_l)-\Phi(\Theta_k)$.

Moreover, Φ is unique up to a positive linear transformation, i.e., if Φ' has the same properties as Φ, then there are real constants α, β, with $\alpha > 0$, such that $\Phi' = \alpha\Phi + \beta$, i.e., Φ is an interval scale.

A stochastic interpretation

The relational structure $<\Theta, \geq>$ is a weak-order and $<\Theta' \geq'>$ is an algebraic-difference structure if our subjects behave "rationally" and deterministically. We know from our experiments that human subjects behave rather irrationally and nondeterministically (Norwich & Turksen, 1984; Turksen, 1986). Therefore, the "rationality" and deterministic behaviour assumptions must be put aside and all the axioms of measurement theory must be restated in stochastic or possibilistic terms. We will give here a stochastic interpretation.

There are hints in Krantz et al. (1971), in Coombs, Dawes, and Tversky (1970) and in Debrue (1958) for probabilistic treatments. However, our formulation is more general. This is done with the following definitions and conjectures (Turksen, 1983).

Definition 1'

Let Θ be a set and $\geq$ be a binary relation on Θ, i.e., $\geq$ is a subset of Θ'. The relational structure $< \Theta, \geq >_p$ is a weak stochastic order if for all $\Theta_i, \Theta_j, \Theta_k$, ε Θ, the following two axioms are satisfied stochastically above a probability threshold p:

1. Weak Stochastic Connectedness:

Either $P(\Theta_i \geq \Theta_j) \geq p$ or $P(\Theta_j \geq \Theta_i) \geq p$ where $P(\Theta_i \geq \Theta_j)$ is the probability that Θ_i will be in relation $\geq$ to Θ_j. It is defined as:

$$P\{\Theta_i \geq \Theta_j\} = \lim(n_1(\Theta_i \geq \Theta_j)/n), \quad n \to \infty \qquad (1)$$

where n_1 is the number of trials in which a subject states that $\Theta_i \geq \Theta_j$ and n is the total number of trials, i.e.,

$$n = n_1(\Theta_i \geq \Theta_j) - n_2(\Theta_j \geq \Theta_i).$$

At times it is clearly not possible to obtain a large number of observations. In such cases one has to be satisfied with a small number of observations. This is part of the reasons why we might have to resort to a possibilistic as opposed to a probabilistic interpretation.

2. Weak Stochastic Transitivity:

If $P\{\Theta_i \geq \Theta_j\} \geq p$ and $P\{\Theta_j > \Theta_k\} \geq p$ then $P\{\Theta_i \geq \Theta_k\} \geq p$.

Conjecture 1'

Suppose that Θ is a finite nonempty set. If $< \Theta, \geq >_p$ is a weak stochastic order then there exists a random valued function Φ_p on Θ such that for all $\Theta_i, \Theta_j \geq \Theta$: $P\{\Theta_i \geq \Theta_j\} \geq p$ iff $P\{\Phi_p(\Theta_i) \geq \Phi_p(\Theta_j)\} \geq p$. Moreover, Φ'_p is another random valued function on with the same property iff there exists a strictly increasing function f with domain and range equal to Re, such that for all $\Theta \in \Theta$: $P\{\Phi'_p(\Theta) = f[\Phi_p(\Theta)]\} \geq p$, i.e., Φ_p is a stochastic ordinal scale.

Definition 2'

Suppose Θ is a nonempty set and $\geq'$ is a quaternary relation on Θ, i.e., a binary relation on Θ'. The relational structure $< \Theta', \geq' >_p$ is a stochastic algebraic-difference structure iff, $\Theta_i, \Theta_j, \Theta_k, \Theta_l, \Theta'_i, \Theta'_j, \Theta'_k, \Theta'_l \ll \Theta$ and for all sequences $\Theta_1, \Theta_2, ..., \Theta_i, ... \in \Theta$ the following axioms are satisfied stochastically above a threshold probability p:

1. $< \Theta', \geq' >_p$ is a weak stochastic order.

2. Stochastic Sign reversal:
   If $P\{\Theta_j\Theta_i \geq' \Theta_l\Theta_k\} \geq p$
   then $P\{\Theta_k\Theta_l \geq' \Theta_i\Theta_j\} \geq p$.

3. Weak Stochastic Monotonicity:
   If $P\{\Theta_j\Theta_i \geq' \Theta'_j\Theta'_i\} \geq p$ and
   $P\{\Theta_k\Theta_j \geq \Theta'_k\Theta'_j\} \geq p$, then
   $P\{\Theta_k\Theta_i \geq \Theta'_k\Theta'_i\} \geq p$.

4. Stochastic Solvability:
   If $P\{\Theta_j\Theta_i \geq' \Theta_l\Theta_k \geq' \Theta_i\Theta_j\} \geq p$ then there exist $\Theta'_l, \Theta''_l \in \Theta$ such that $P\{\Theta'_l\Theta_i \geq \Theta_l\Theta_k > \Theta_j\Theta''_l\} \geq p$.

5. Stochastic Archimedian Condition:
   This means the probability that "no interval is infinitely large with respect to any smaller one whose jnd, 'just noticeable difference', is

not zero" is greater than or equal to a probability measure p. In other words the imprecise concept in going from Θ_i to Θ_j can only be exceeded by a finite number of successive jnd gains starting at Θ_i is larger than or equal to a probability measure p.

Conjecture 2'

If $< \Theta', \geq >_p$ is a stochastic algebraic difference structure, then there exists a random valued function Φ_p on Θ such that, for all $\Theta_i, \Theta_j, \Theta_k, \Theta_l \in \Theta, P\{\Theta_j \Theta_i \geq' \Theta_l \Theta_k\} \geq p$ iff $P\{\Phi_p(\Theta_j) - \Phi_p(\Theta_i) \geq \Phi_p(\Theta_l) - \Phi_p(\Theta_k)\} \geq p$. Moreover, Φ_p is unique up to a stochastic positive linear transformation, i.e., if Φ'_p has the same properties as Φ_p, then there are real constants α, β, with $\alpha > 0$, such that $P\{\Phi'_p = \alpha\Phi_p + \beta\} \geq p$, i.e., Φ_p is a stochastic interval scale.

It is to be noted that the computations of the left hand sides of all the inequalities in Definitions 1', 2' and Conjectures 1', 2' are to be interpreted in accordance with the probability calculus and the classical rules of inference.

Based on these definitions and conjectures, it could be further argued that

(i) the weak stochastic transitivity, and
(ii) the weak stochastic monotonicity

are the essential minimal requirements for measurement data to have stochastic interval scale property. Under these rather minimal conditions, if the response data passes the weak stochastic transitivity test, then we can be confident, at level p, that the data is at least on a stochastic ordinal scale. It can be observed that the experimental data satisfies the weak stochastic transitivity at the minimal level of $p \geq 0.68 = 562/816$. (See Table 5). As it can be observed, the test is satisfied for the "slight" and "difficult" descriptors of the behaviour components. The "moderate" descriptor is inferred from these.

Next, if the response data passes the weak stochastic monotonicity, then we can be confident, at level p, that the data is at least on a stochastic interval scale. It can be observed that the experimental data satisfies the weak stochastic monotonicity at the minimal level of $p \geq 0.5$ (see Table 6).

TABLE 5
Weak Stochastic Transitivity

Membership Curves	No. of 3-point Ordered Combinations	Proportion of Satisfying Cases
Slight Knowledge	10	10/10
Difficult Knowledge	10	10/10
Slight S-K	83	81/83
Difficult S-K	83	81/83
Slight R-K	20	17/20
Difficult R-K	20	17/20
Slight S-R-K	816	562/816
Difficult S-R-K	816	592/816

TABLE 6
Weak Stochastic Monotonicity

Membership Curves	No. of Test points	Proportion of Satisfying Cases
Slight Knowledge	5	4/6
Difficult Knowledge	5	3/6
Slight S-K	5	12/18
	5	9/18
	5	10/18
	5	10/18
	5	11/18
Difficult S-K	5	13/18
	5	7/18
	5	10/18
	5	10/18
	5	9/18
Slight R-K	5	6/12
	5	7/12
Difficult R-K	5	5/12
	5	7/12

These results, thus, suggest that we have a reasonable ground to proceed with other analyses of data such as curve fitting at the confidence level of 0.5 as a minimum.

SIGNIFICANCE TESTS

Analysis of variance (ANOVA) is used to investigate the significance of the response data with respect to the test conditions, the subjects and the interactions between the two. For example, as for the knowledge-based runs, a one-way ANOVA was evaluated for both the subjects and the experimental conditions with respect to the the three (slight, moderate, and difficult) leves of knowledge-based behaviour.

The results show that the null hypothesis that "there are no differences in responses due to individual subject differences" could not be rejected for the first two levels of knowledge-based behaviour, i.e., for the slight and moderate levels. However, the results of the third level, i.e., difficult knowledge-based behaviour, show that different subjects perceived the difficulty of this knowledge-scenario quite differently ($p < 0.05$). This result may be explained by the use of different strategies in the execution of the difficult knowledge-based task by different subjects.

On the other hand, there seems to be some difference in the experimental conditions with respect to the moderate knowledge-based task while there appears to be no difference for the slight and difficult knowledge-based tasks. This means that all subjects rated the five knowledge-based trials equally in their assessment of slight and difficult knowledge-based tasks.

The interaction effects between the different parameters of the experimental data are also investigated by the use of two-way ANOVA's. The results of the interaction between the skill and knowledge-based tasks indicate that the interaction is higher for the easier levels of skill-knowledge combination tasks and drops as the task gets more and more difficult.

Whereas the interaction effects between the rule and knowledge-based tasks indicate that no interaction really occurs for the rule-knowledge combination. Finally, a three way ANOVA test amongst subjects, tasks and degree of membership types show that neither the subject-task nor the subject-difficulty intercations are significant.

Curve fitting

We have investigated some simple models that are built with the inclusion of either the union, or intersection of the behaviour components. These models take the following forms:

(i) for skill-knowledge based combinations:
$f(s,k) = a_1s + a_2k + a_3U(s,k) + c$
$f(s,k) = a_1s + a_2k + a_3I(s,k) + c$

(ii) for rule-knowledge based combinations:
$f(r,k) = a_1r + a_2k + a_3U(r,k) + c$
$f(r,k) = a_1r + a_2k + a_3I(s,k) + c$; and

(iii) for skill-rule-knowledge based combinations:
$f(s,r,k) = a_1s + a_2r + a_3k + a_4U(s,r,k) + c$
$f(s,r,k) = a_1s + a_2r + a_3k + a_4I(s,r,k) + c$

where s,r,k, refer to the skill, rule, and knowledge based components of the behaviour, U and I refer to union and intersection operators, respectively, and a_1, a_2, a_3, a_4 and c are parameters that are to be estimated by a regression analysis.

All possible regression models were studied using the conjugate pairs of the four well known t-norms and t-conorms, i.e.,

(1) Max(a,b) = a V b,
    Min(a,b) = a ∧ b;

(2) sum(a,b) = a + b,
    Prod(a,b) = ab;

(3) Bold union (a,b) = a V b,
    Bold intersection (a,b) = a ∧ b; and

(4) Drastic union = a V b if a ∧ b = 0
                   1 otherwise

    Drastic intersection = a ∧ b if a V b = 1
                          0 otherwise

The results of the regression analysis can be summarized as follows:

(i) For the skill-knowledge based models, all of the ANOVA results give F-values which indicate that the regression equations are significant beyond the $p = 0.005$ level. The multiple linear correlation coefficients indicate remarkably good fits. The drastic intersection model seems slightly better than the rest of the models.

(ii) For the rule-knowledge based models all of the ANOVA results are insignificant even at the $p > 0.10$ level. Multiple regression coefficients are relatively low. This can be attributed to the number of observations rather than a bad fit. Again, the drastic intersection operator model seem marginally better than the other models.

(iii) For the skill-rule-knowledge models all of the ANOVA results are significant well over the $p = 0.005$ level. All the multiple regression coefficients are remarkably good. In this case there is no clear indication for an operator to be a better fit.

TABLE 7
Linguistic Rules for the Prediction of the Mental Workload

Row #	Component of a task			Mental Workload
	Skill	Rule	Knowledge	Prediction
1	S	-	S	S
2	S	-	M	M
3	S	-	D	M
4	M	-	S	M
5	M	-	M	M
6	M	-	D	D
7	D	-	S	D
8	D	-	M	D
9	D	-	D	D
10	-	S	S	S
11	-	S	M	S
12	-	S	D	D
13	-	M	S	M
14	-	M	M	M
15	-	M	D	D
16	S	S	S	S
17	S	S	M	S
18	S	S	D	M
19	S	M	S	S
20	S	M	M	M
21	S	M	D	M
22	M	S	S	M
23	M	S	M	M
24	M	S	D	D
25	M	M	S	M
26	M	M	M	M
27	M	M	D	M
28	D	S	S	D
29	D	S	M	D
30	D	S	D	D
31	D	M	S	D
32	D	M	M	D
33	D	M	D	D

S = Slight, M = moderate, D = difficult

LINGUISTIC INTERPRETATION

A linguistic interpretation of the task can be based upon the results obtained from the models and the accuracy of these models. One such interpretation of the relationships found in this study is given in Table 7. The entries of Table 7 are based on the means of the raw data, combined over subjects for each of the runs. The results listed agree with our common sense predictions based directly on the observations of the raw data. The linguistic variables such as slight (S), moderate (M), and difficult (D), describe a state of the behaviour components such as skill, rule, and knowledge-based depending on the levels of each as well as describing a state of the mental workload affecting a subject operating under the combined set of task conditions. (Please note the definitions of the terms stated at the end of Table 7.)

Each row of Table 7 should be read as a linguistic rule that predicts the mental workload under a given set of task conditions. For example, rule 16, i.e., row 16, should read as:

IF the predominant state of the skill-based component of a task is described as SLIGHT,
AND the predominant state of the rule-based component of a task is described as SLIGHT,
AND the predominant state of the knowledge-based component of a task is described as SLIGHT,
THEN the predominant state of the mental workload should be predicted as SLIGHT.

For the second example, rule 33, i.e., the last row should be read as:

IF the predominant state of the skill-based component of a task is described as DIFFICULT,
AND the predominant state of the rule-based component of a task is described as MODERATE,
AND the predominant state of the knowledge-based component of a task is described as DIFFICULT,
THEN the predominant state of the mental workload should be predicted as DIFFICULT.

A similar linguistic rule interpretation should be given to all the remaining 31 rules.

In reading these 33 rules, one needs to interpret the linguisitc connective "AND" as the simultaneous occurence of the stated components of behaviour for a given task. A closer look at 26 of these rules (excepting 3, 11, 17, 18, 19, 21, and 27) however, reveals that the prediction of the mental workload can be obtained by taking the linguistic descriptor which has the maximal linguistic variable in the weak ordering of the liguistic variables S,M,D, which are ordered as $S < M < S$. For example, the outcome of rule 24 is:

D(Mental Workload)
= MAX {M(Skill), S(Rule), D(Knowlegde)}

For these 26 rules, in general, we could state that

L(Mental Workload)
= MAX {L(Skill), L(Rule), L(Knowlegde)}

where L stands for the linguistic variable.

For the next 5 rules, i.e., rule 3, 11, 17, 19, and 27, the predicition of the mental workload can be obtained by taking the linguistic descriptor which has the minimal linguistic variable in the weak order $S < M < D$. For example, the prediction of the mental workload for rule 27 can be found as:

M(Mental Workload)
= MIN {L(Skill), L(Rule), L(Knowledge)}.

For these five rules, prediction can be found in general as:

L(Mental Workload)
= MIN {L(Skill), L(Rule), L(Knowledge)}.

Finally, rules 18 and 21 require individual expressions as follows: For rule 18, we can state:

M(Mental Workload)
= NEITHER [S] NOR [D].

For rule 21, we can state:

M(Mental Workload)
= MAX {M(Rule), MIN S(Skill), D(Knowledge)}.

CONCLUSIONS

In summary, therefore, the majority of predictions can be obtained with a MAX selection operator and a few of them can be obtained with a MIN selection operator while some predictors have rather particular and unique selection operators that identify the dominant component in the prediction of the mental workload.

We can also observe that in twenty of the twenty-seven combinations of rules that included the skill-based behaviour, i.e., rules 1, 4, 5, 7, 8, 9, 16, 17,

19, 22, 23, 25, 26, 27, 28, 29, 30, 31, 32, and 33, the linguistic descriptor of the skill-based behaviour also appears as the descriptor of the predicted mental workload. It is in this regard it can be stated that in Rasmussen's taxonomy of human behaviour, the skill-based behaviour would dominate while the rule and knowledge-based behaviour components would moderately influence the degree of difficulty associated with a task.

ACKNOWLEDGEMENT

The authors are grateful for partial supports provided by NASA-AMES, NAGR-429 and NSERC-Operating Grant A-7698. Last but not least the authors are also very appreciative of Ms.H.Santos for her careful typing.

REFERENCES

Coombs, C. H., Dawes, R. M., & Tversky, A. *Mathematical Psychology.* Englewood Cliffs, N.J.: Prentice Hall, 1970.

Debrue, G. Stochastic choice and cardinal utility. *Econometrica,* 1958, *26,* 440-444.

Gopher, D., & Braune, R. On the psychophysics of workload - Why bother with subjective measures. *Human Factors,* 1984, *26,* 519-532.

Kranty, D. H., Luce, R. D., Suppes, P., & Tversky, A. *Foundations of measurement.* (Vol 1.) New York: Academic, 1971.

Moray, N. Subjective mental workload. *Human Factors,* 1982, *24,* 25-40.

Moray, N., King, B., Waterton, K., & Turksen, I. A closed loop causal model of workload based on a comparison of fuzzy and crisp measurement techniques. *Human Factors,* 1987. (in press)

Moray, N., Eisen, P., Money, L., & Turksen, I. Fuzzy analysis of skill and rule-based mental workload. In P. Hancock (Ed.) *Human Mental Workload.* Amsterdam: North-Holland, 1987. (in press)

Norwich, A. M., & Turksen, I. B. A model for the measurement of membership and the consequences of its empirical implementation. *Fuzzy Sets and Systems,* 1984, *12,* 1-25.

Rasmussen, J. Skills, rules and knowledge: Signals, signs, symbols and other distictions in human performance models. *IEEE Transactions on Systems, Man and Cybernetics,* 1983, *SMC-13,* 257-266.

Thornton, C. *Documentation for the NASA Hovercraft Simulator.* (unpublished) Toronto: University of Toronto, 1984.

Turksen, I. B. Measurement of membership functions. In W. Karwowski and A. Mital (Eds.), *Applications of fuzzy set theory in human factors.* Amsterdam: Elsevier, 1986.

Turksen, I. B. *Measurement of fuzziness: An interpretation of the axioms of measurement.* Proceedings of IFAC Symposium. Marseille, France, 1983.

Wierwille, W. Physiological measures of aircrew mental workload. *Human Factors,* 1979, *21,* 575-593.

Williges, R., & Wierwille, W. Behavioural measure of aircrew mental workload. *Human Factors,* 1979, *21,* 549-574.

Zysno, P. *Representing human concepts by fuzzy sets: systematical conditions and empirical results.* Reprints of Second IFSA Congress. Tokyo, Japan, 1987.

A FUZZY SET MODEL OF LEARNING DISABILITY: IDENTIFICATION FROM CLINICAL DATA

J. M. HORVATH

Fort-Hayes State University, Kansas, USA

In the field of learning disability, mathematical modeling of individual theoretical orientations has not been highly successful. Precise modeling of internal structures is difficult because clinicians are more accustomed to practicing their diagnostic and prescriptive skills rather than articulating their thought processes. Until accurate models of learning disability can be developed and debated in the literature, consensus about what constitutes the handicap will be delayed.

Another factor is that traditional mathematical techniques are usually the tools of choice in the attempts to quantify human thought. Zadeh (1973, p. 28) believes this is so because:

> At present, most of the techniques employed for the analysis of humanistic, i.e., human centered, systems are adaptations of the methods that have been developed over a long period of time for dealing with mechanistic systems, i.e., physical systems governed in the main by the laws of mechanics,

electromagnetism, and thermodynamics. The remarkable successes of these methods in unraveling the secrets of nature and enabling us to build better and better machines have inspired a widely held belief that the same or similar techniques can be applied with comparable effectiveness to the analysis of humanistic systems.

Pragmatism is one of the reasons that scientists, mathematicians, and philosophers historically tried to explain human thought through the mathematics of astronomy and physics. There are also philosophical justifications. Rosenfield (1968, p. 205) states:

> In the transition from animal to human automatism, we have endeavored to show how the general mathematical mechanics of the early seventeenth century passed into the Cartesian mechanistic physiology, and how that scientific principle in turn penetrated the general thought of the eighteenth century until it culminated in LaMettrie's total elimination of all non-mechanical forces from his philosophic system. Psychology had become physiology, and such it remains in the behaviorist school of today.

Psychology was not the only discipline affected. George (1973, p. 1) equates cybernetics with behaviorism, states that cybernetics regards human beings and animals as essentially very complicated machines and declares:

> The title of the book, *The Brain as a Computer*, is intended to convey something of the methodology involved; the idea is to regard the brain itself as if it were a computer type of control system, in the belief that by so doing we are making explicit what for some time has been implicit in the biological and behavioral sciences.

Another philosophical justification for mechanistic mathematical systems has to do with the preference for objectivity over subjectivity. This preference is supposed to be more in keeping with the tenets of "science." Given this rationale, it is not hard to see that objectivity in the form of empiricism became paramount for any effort to be considered scholarly and thereby respected. Subjectivism was relegated to virtual nonexistence. Consciousness had become an enigma. Intuition and insight could be explained away immediately as unscientific and not worthy of consideration, reason could be explained as nothing more than an "adding and subtracting of names for that which is sensibly experienced," sense experience could be explained as "simply a motion in the nerves," and the will as "simply the last phase, before overt action, in any conflict of opposing incipient notions" (Hall, 1956). Even consciousness has now been explained in mechanistic fashion. Asmonov (1967) describes consciousness as attention generated by an amplified stimulus in the form of electrical activity moving from one model to another in the cortex. The "mind" construct of Descartes has been increasingly superceded by the "body" entity. Thus, technical activity seemed to be come an end in itself. Mumford (1970, p. 95) describes the climate by stating:

> From Descartes' time on until the present century, to all but the most penetrating minds in science, a "mechanistic" explanation of organic behavior was accepted as a sufficient one. And as machines became more lifelike, Western man taught himself to become in his daily behavior more machine-like. This shift was recorded in the changing meaning of the word "automaton" which was used in English as early as 1611. At first this term was employed to describe autonomous beings with the power to move alone; but it soon came to mean just the opposite: a contrivance that had exchanged autonomy for the powers of motion "under conditions fixed for it, not by it" (New Oxford Dictionary).

A next logical step is to equate quantification with analysis. Bergman (1957, p. 67) states:

> In itself quantification is of course no more a magic key or cure-all than is axiomatization. If the material available warrants it, then its advantages are indeed incomparable. To bring out clearly why this is so is the task of the philosophical analyst. But how much of it the material warrants is a question not of principle but of strategy, on which the philosophical analyst of science cannot advise its practitioners. The matter is one of tact and judgment, the sort of thing one must have at his fingertips. Another cause of all the stridency in favor of quantification lies, I think, in the

social climate, of late the behavior sciences have become the basis, or the alleged basis, of professions whose numerous members are rapidly acquiring managerial power and who, for better or for worse, aspire to even more. Such aspiring groups are in need of prestige symbols. The white coat of the medical man is one; the mathematical formula is another.

DIFFICULTY WITH TRADITIONAL MATHEMATICS

If the human brain mirrored mechanistic mathematics, articulation of human thought would already have been accomplished. Systems which can predict the motions of bodies, deal in nanoseconds, and analyze probabilities with an accuracy and precision which once was considered to be just short of miraculous should be expected to be sophisticated enough to deal with the human brain. So far, mechanistic mathematical systems have proven unsuitable to the task. Zadeh (1973, p. 28) explains it in the following way:

> The conventional quantitative techniques of system analysis are intrinsically unsuited for dealing with humanistic systems or, for that matter, any system whose complexity is comparable to that of humanistic systems. The basis for this contention rests on what may be called the principle of incompatibility. Stated informally, the essence of this principle is that as the complexity of a system increases, our ability to make precise and yet significant statements about its behavior diminishes until a threshold is reached beyond which precision and significance (or relevance) become almost mutually exclusive characteristics. It is in this sense that precise quantitative analysis of the behavior of humanistic systems are not likely to have much relevance to the real world societal, political, economic, and other types of problems which involve humans either as individuals or in groups.

Zadeh (1975) later included human thought in his definition of humanistic systems.

Precision is sacrificed for relevance in humanistic systems because the complexity of the real world forces humans to think not in numbers or sets, but by manipulating classes of objects in which the transition from membership to nonmembership may be gradual rather than abrupt. Degree of membership in the category learning disabled is one example. Instead of manipulating numeric variables by precisely defined, sequential mathematical operations, humans can manipulate both numeric and linguistic variables in parallel to reduce complex phenomena systematically to approximate characterizations. This summarizing ability allows humans to walk, understand distorted speech, read, drive a car, play chess, decipher sloppy handwriting, search for lost objects, exclude irrelevant information in making decisions, and give weightings to data in using it in meaningful ways. So instead of taking a mass of preselected data, sequentially comparing each bit of datum against all the others to make yes/no decisions, and drawing precise, but possibly insignificant answers as mechanistic systems such as computers do, the human system selects the information it will use, reduces it to a trickle, summarizes it, and manipulates it to draw relevant, but not necessarily precise conclusions.

Computer scientists and systems engineers have long recognized that people can understand and operate upon vague, natural language concepts. Computers, however, are extremely rigid and precise information-processing systems. This inherent rigidity severely limits a computer's ability to abstract and generalize fundamental conceptual functions when traditional mathematical systems are used.

Real world considerations also force a reexamination of the objectivity which is the sine qua non of many sciences. While the importance of objectivity cannot be denied, subjectivity is inherent in human thinking. It is what makes us human and should be valued. Huchings (1964, p. 160) criticizes those who would desire to eliminate subjectivity in favor of objectivity in the following way:

> Francis Bacon (1561-1626), Lord Chancellor of England, was one of the most influential writers of his day. He stressed the need

for facts, which he thought could be collected, then passed through a sort of automatic logical mill to give the results. This is an over-simplification of scientific inquiry since it takes no account of what is, after all, the most important and most difficult part of scientific investigation-use of judgment. How does the scientist choose his facts in the first place? Bacon's method simply cannot be applied in practice since facts are infinite in number. Some guidance can be obtained by those who follow in the footsteps of others in observing the success or failure of previous acts of choosing but, in the last resort, the man of science must use his own judgment.

Kaiser (1974, p. 3) puts it succinctly when he says: "The 'objectivity' of science has led investigators away from a valuable source of knowledge and has imposed the single dimension of objective data on man's epistemological investigations."

In sum, traditional mathematical techniques were not developed to capture the vagueness or fuzziness involved in human thought. Human thought is characterized by both precisely defined logic and variables which may or may not be precisely defined. These variables may take a numeric or linguistic form, depending on the complexity and the importance of the task involved.

THE SEARCH FOR MODELS

Two requirements for articulating human thought about learning disability became apparent. The first is that the model had to have the capability of capturing human thought. The second is that the model had to facilitate articulation through an iterative process. Articulation is a process in which a person "talks through" an orientation. Besides fuzzy set theory, Bayesian revision of subjective probabilities and interpretive structural modeling were considered.

Bayesian Methodology

Kaiser (1974, p. 4) describes the method as follows:

> Bayesian methodology provides a means of mathematically incorporating the hunches of specialists into the probability of a given event. The statistics parallel closely the procedures man uses intuitively in dealing with his milieu. Rather than saying one hypothesis is false and another is not false, Bayesian statistics ask what are the probabilities of the alternate hypothesis. In addition to the determination of the probabilities of the alternate hypothesis, Bayesian statistical procedures incorporate the prior information which is available to the specialist because of his professional experience.

Novick (1975, p. 377-378) believes that the Bayesian methodology is superior to traditional statistical methods because two major problems are solved. He states:

> The major problem with both Neyman-Pearson and Fisherian classical statistics is that neither system makes it possible to construct direct probability statements about the parameter of interest. Using classical statistics, we can talk about the probability of a random interval covering the mean, but we cannot talk about the probability of the mean being within a specified interval. This seems very peculiar indeed. We are not to be permitted to make statements about these things that are the subject of our investigation. This is an unsatisfactory state of affairs. What we want is a system that permits us to say: "The probability that the unknown proportion is greater than Φ_1 and less than Φ_2 is 1-d." But to make probability statements about Φ, we must have a probability distribution for Φ. In Bayesian statistics such a distribution is possible and meaningful and indeed is the central objective of a statistical investigation.
>
> A second problem with classical procedures is that they provide no mechanism for incorporating into the statistical processing

prior information we may have about the parameter. Basically we must ask whether it is reasonable to evaluate evidence from an experiment while ignoring other available information extraneous to the particular experiment. Upon reflection, this does not seem unreasonable.

It is the second problem which is of concern in this consideration of human thought or opinion. The Bayesian method may be thought of as a rule for modifying or revising opinion expressed as probabilities in light of new evidence. The process begins by having an expert witness estimate what he considers to be the probabilities of a given set of events. These are known as prior probabilities, i.e., the state of knowledge prior to observations of a specific event or condition. The next step is to make an experimental observation and to estimate its conditional probabilities. These conditional probabilities are then combined with the prior probabilities to yield a posterior probability, which may be thought of as a revised estimate after the observations have been made. In light of new evidence, the posterior probability becomes the prior probability and the process is repeated if the observations are independent. This process may be repeated as long as new information is available.

An example may be illustrative. Assume that one wants to know the probability that the condition of learning disability (LD) exists in a given child. An expert believes that the prevalence of LD is five percent (5%). This five percent is a prior probability. In his experience with the school-aged population, the expert finds that fifty percent (50%) of those labeled LD and ten percent (10) of those labeled nonlearning disabled (NLD) exhibit a characteristic called "A." These are the experimental observations or conditional probabilities. What if the child under observation exhibits "A?" Is that child LD? The Bayesian technique can give a probability estimate with simple calculations. The formula is as follows:

$$P(LD|A) = P(A|LD) P(LD) / \{P(A|LD) P(LD)\} + \{P(A|NLD) P(NLD)\}$$

Where
- P(LDA) is the posterior probability, i.e., the probability that the child is LD when he exhibits "A"
- P(LD) is the prior probability, i.e., the prevalence of LD in the school-aged population
- P(NLD) is another prior probability, i.e., the prevalence of NLD in the school population
- P(ALD) is the conditional probability, i.e., the probability that "A" will be exhibited if a child is, in fact, LD
- P(ANLD) is another conditional probability, i.e., the probability that "A" will be exhibited, given the fact that the child is NLD

$$P(LD\ A) = (.5)(.05) / \{(.5)(.05)\} + \{(.1)(.95)\} = .21$$

Now it can be stated that if a child exhibits "A," the probability that the child is LD is .21 and the probability that the child is NLD is 1.00 minus .21 (1.00 - .21 = .79).

The expert next observes that forty percent (40%) of those labeled LD exhibit characteristic "B" and ten percent (10%) of those labeled NLD exhibit this same characteristic. The previous posterior probability (assuming that "A" and "B" are independent) obtained from the equation (.21) now becomes the prior probability and the data is applied to the formula:

$$P(LD|A\&B) = (.4)(.21) / \{(.4)(.21) + (.1)(.79)\}$$

It can now be stated that if a child exhibits both "A" and "B," the probability that the child is LD is .52 and the probability that the child is NLD is 1.00 minus .52 (1.00 - .52 = .48).

This process could go on indefinitely. As many characteristics as a theoritician desired could be considered if the characteristics were independent of each other. There is even some evidence (Lichtenstein, 1972) to show that the Bayesian methodology is robust enough to withstand nonindependence.

It is reasonable to assume that the Bayesian method would be even more useful if measures of the characteristics could be incorporated. Fortunately, Bayesian methodology can be used with continuous (e.g., test score) data as well as discrete data. In the example cited above, the method assumes discrete data and makes no provision for degree of severity of each characteristic. If each characteristic is continuous rather than discrete, consider Figure 1.

Characteristics "C" and "D" in Figure 1 have been measured by separate test instruments. Each abscissa represents the scores on a measure of one characteristic and each ordinate represents the frequencies of scores. On the first graph in Figure 1, the distributions of scores for the hypothetical characteristic "C" are plotted for both the LD and NLD sample. The shaded area represents the area of overlap or commonality. The second graph is a similar plot for the hypothetical characteristic "D." The question concerns a child whose score places him in the shaded area. Is that child LD or NLD?

If characteristics which an expert believes to be indicative of LD were measured and their distributions for LD and NLD children were obtained, then for each score of each characteristic, a likelihood ratio (LR) could be established. This would be the ordinate of the score distribution for LD children divided by the ordinate of the distribution for NLD at the abscissa value for the score in question. Figure 2 presents a graphic representation of the likelihood ratio.

At the score of fifty (50) for characteristic "D," the likelihood ratio would be calculated as follows:

LR = RT / ST

For any score, a likelihood ratio greater than 1.00 indicates that LD children have that score more often than NLD children. If the likelihood ratio is less than 1.00, NLD children have that score more often than LD children. To obtain

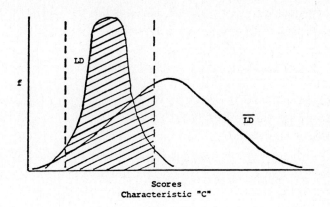

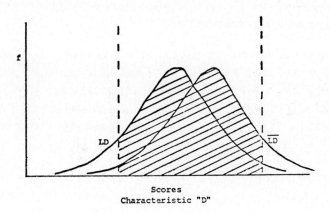

FIGURE 1

Representative plots of continuous data for LD and $\overline{\text{LD}}$ on two different measures

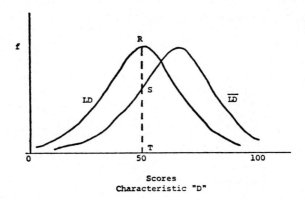

FIGURE 2
Graphic representation of likelihood ratio.

a probability estimate for LD, the Bayesian technique dictates that one or more scores be obtained for a child in question, the likelihood ratio for each score be calculated, and these likelihood ratios be subjected to the following formula:

$$P(LD|C,D,E,F) = LR_0 \Pi LR / (LR_0 \Pi LR) + 1$$

Where $P(LD|C,D,E,F)$ is the probability of LD given the characteristics C,D,E,F,

- LR_0 is the prior odds ratio (in this example 5% of the school-aged population is presumed to be LD and 95% is NLD, hence $LR_0 = .05/.95$)
- ΠLR the product of the likelihood ratios--the likelihood ratio for the set of four data (C,D,E,F) observed (if the observations are statistically independent)

Then, if $LR_C = .07$, $LR_D = 1.5$, $LR_E = 10.2$, and $LR_F = 4.5$, the calculation of the probability of LD for the child is as follows:

$$P(LD|C,D,E,F) = (.05/.95)(.07)(1.5)(10.2)(4.5)/$$
$$/(.05/.95)(.07)(1.5)(10.2)(4.5) + 1 =$$
$$= 2.54/3.54 = 0.72$$

It can be said that a child who earned the scores depicted above on instruments designed to measure hypothetical characteristics "C," "D," "E," and "F" had a probability of .72 of being LD.

The above discussion illustrates the concept of applying Bayes' rule to a group of test scores. But before that can be done, certain mathematical requirements must be satisfied. The distributions must be standardized so that comparisons are compatible. One approach is to model the test score distributions by the beta distribution. The steps are as follows:

1. Rescale the scores into the range 0--1

2. Use those transformed scores to obtain the beta distributions, each of which is assumed to have a beta density of:

$$f(x) = 1/\beta(m,n) \, (x^{m-1})(1-x)^{n-1}$$

where $n = (1-\bar{x})/s_x^2 \, \{\bar{x}(1-\bar{x}) - s_x^2\}$
$m = \bar{x}n/(1-\bar{x})$
and $\beta(m,n)$ is the beta function

3. Calculate the likelihood ratios

$$LR(x) = f(x)_{LD}/f(x)_{NLD}$$

4. Apply the formula to obtain a probability

$$P(LD|C) = LR_0 \Pi LR / (LR_0 \Pi LR) + 1$$

When this standardization has been completed, one need only look in a table to find the likelihood ratio for a score and then apply it in the formula shown in step four above.

While the Bayesian technique is useful for identification of learning disabled children, a median estimate does not elicit the intuitive insights, wisdom, and knowledge of each clinician. The method is not designed for articulation of mental processes. Another problem with applying Bayesian methodology or traditional mathematical methods to the identification of children with learning disability is that each child is presumed to be learning disabled or nonlearning disabled. The Bayesian theorem yields a probability estimate. What if learning disability were thought of as an entity in which the change from membership to nonmembership was gradual rather than abrupt? In that case, any characteristics and severity of those characteristics would interact to determine the degree of handicap of each child. For purposes of illustration, consider a continuum in which there are degrees of severity of neurological dysfunction (LD is thought to be included in this group). One end of the continuum could be labeled "classic cases" and the other end could be labeled "perfectly neurologically functional". Each child would then be placed at some point on this continuum. Of course this example is only illustrative and simplistic because other handicapping conditions are not taken into account. A continuum model may be more indicative of the understanding of learning disability since clinicians appear to prefer different cut-off point would be those called learning disabled; on the other side of the cut-off point would be those who would be called normal, or something else. Because clinicians have specified prevalence figures for learning disability from one percent to over thirty percent (Lerner, 1985), credence is lent to the notion that learning disability results from an interaction between the number of characteristics and the degree of severity of those characteristics. If learning disability is thought of as a range of degrees of dysfunction, then probabilities become less useful than a deception of a point on the continuum where an individual falls.

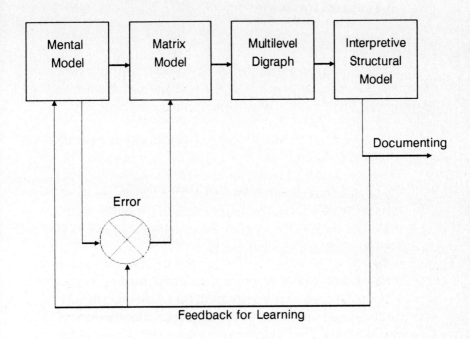

FIGURE 3
Block diagram of interpretive structural modeling process, adopted from Warfield (1974)

Interpretive Structural Modeling

Another promising approach to the question of interaction between number of characteristics and degree of severity is interpretive structural modeling. Farris and Sage (1975, p. 153) describe the technique as follows:

> Interpretive structural modeling is a tool which permits identification of structure within a system. The system may be technical, social, medical, or any system which contains identifiable elements which are related to one another in some fashion. The system may be large or small in terms of elements; and it is in the larger, complex systems which benefit the most from interpretive structural modeling.
>
> The characteristic which these complex systems have in common is that each has a structure associated with it. That structure may be obvious as in the managerial organization of a corporation, or it may be much less obvious as in the value structure of the decision maker. In the physical sciences, structure is most often articulated in the form of mathematical equations. In the behavioral and social sciences, structure is usually not articulated in such a clear fashion. Whether articulated or not, the structure of a system must be dealt with by individuals and groups. The interpretive structural modeling process transforms unclear, poorly articulated mental models of systems into visible well-defined models.

Because the human brain is severely limited in dealing with complexity (problems involving a significant number of elements and relations among the elements), mental models of phenomena tend to be muddy and piecemeal (Waller, 1975). In interpretive structural modeling, the term "mental model" means the mental picture that a person holds regarding the relationships among the elements of a system. Examples of interpretive structural models (Warfield, 1974) are: block diagrams, flow charts, and the organization of books such as *Organizations* (March and Simon, 1958). Because interactive computer programs have been developed, it is possible to structure complex relationships so that mental models can be transformed into clear, precisely defined, well-organized documentation.

The process is represented graphically in Figure 3. In this method, a complete knowledge of content is not necessary. If a specialist knows the relationships among the elements, structuring may take place.

Waller (1975) applied this process to the analysis of the learning disabled child. The first step was to replace a mental model with a matrix model. A specialist describes or is given the elements (content) of a system. The specialist in the Waller study was a learning disability teacher. She was given as a system the thirteen categories (elements) of the Kirk and Kirk (1971) psycholinguistic diagnostic paradigm. The thirteen categories were: (1) auditory reception, (2) visual reception, (3) auditory association, (4) visual association, (5) verbal expression, (6) manual expression, (7) grammatic closure, (8) auditory closure, (9) sound blending, (10) visual closure, (11) auditory sequential memory, (12) visual sequential memory, and (13) visual-motor coordination.

The number assigned to each was used in the construction of a binary matrix. This was done by asking the teacher to consider a particular child and answer a series of questions until the relationships concerning the psycholinguistic process of that child were completely specified. Examples of questions were the following: Is the child weaker in auditory reception (1) than visual reception (2)?; Is the child weaker in auditory reception (1) than auditory association (3)?; Is the child weaker in auditory reception (1) than visual association (4)?

Questions such as these were asked until the matrix was completed. An affirmative answer on the part of the teacher was coded as 1 and a negative answer was coded as 0 (hence the term binary matrix). The matrix for one child is presented in Figure 4.

From this matrix, a digraph (directed graph) was constructed. This is shown in Figure 5. Since elements 1, 2, 4, 5, 10, and 12 had an equal number of negative responses, and more negative responses than any other elements, it was concluded that the teacher believed that this particular child demonstrated strengths to the same degree in elements 1, 2, 4, 5, 10, and 12.

	1	2	3	4	5	6	7	8	9	10	11	12	13
1	0	0	0	0	0	0	0	0	0	0	0	0	0
2	0	0	0	0	0	0	0	0	0	0	0	0	0
3	1	1	0	1	1	0	1	1	0	1	1	1	1
4	0	0	0	0	0	0	0	0	0	0	0	0	0
5	0	0	0	0	0	0	0	0	0	0	0	0	0
6	1	1	1	1	1	0	1	0	1	1	1	1	1
7	1	1	0	1	1	0	0	0	0	1	1	1	0
8	1	1	0	1	1	0	1	0	0	1	1	1	1
9	1	1	1	1	1	0	1	1	0	1	1	1	1
10	0	0	0	0	0	0	0	0	0	0	0	0	0
11	1	1	0	1	1	0	0	0	0	1	0	1	0
12	0	0	0	0	0	0	0	0	0	0	0	0	0
13	1	1	0	1	1	0	1	0	0	1	1	1	0

FIGURE 4
The matrix for one child.

Similarly, element 11 had the next highest number of negative responses, element 7 was below that, and so on. The process is much more complicated than this when the relationships among elements are more complex (Warfield, 1973). An example of what could happen, given complex relationships, is shown in Figure 6.

The interpretive structural model is simply the substitution of written statements or labels for the number codes. This has been done in Figure 7 for the Waller (1975) study.

This method can be used to establish a profile of each child's strengths and weaknesses according to a particular theoretical framework. Data tempered by experiential judgment may be included in the model, leaving those who use

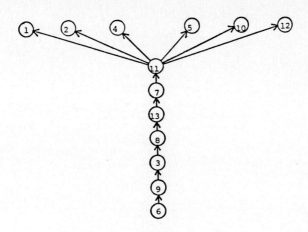

FIGURE 5
Digraph of LD child.

the method to formulate a more precise personal mental model, i.e., establish a more definite and complete set of relationships among the elements.

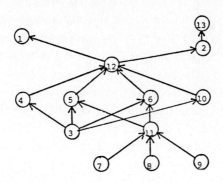

FIGURE 6
Example of a complex Digraph.

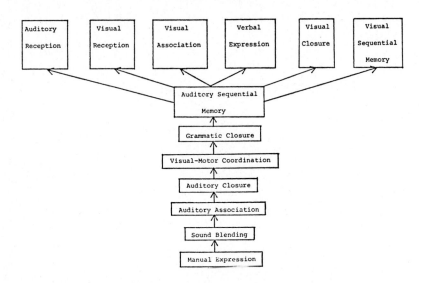

FIGURE 7
An interpretive structural model from the Waller (1975) study.

Interpretive structural modeling seems to have value as a way to structure relationships among theoretical elements for purposes of making general diagnoses. In the Waller (1975) study, however, a glance at the test results alone might have precluded the use of the interpretive structural modeling technique. It is a time consuming technique which is not designed for gaining an understanding of the meaning of the content of a theory. In complex systems, computer assistance is necessary. The method would be more useful if diagnosis could be done directly instead of using the categories of a diagnostic instrument. The diagnosis by interpretive structural modeling may be less useful, however, than that by the instrument itself because the digraph can not specify the quantitative relationships of the instrument. Interpretive structural modeling allows one to structure complex relationships into a computer program, but does not aid understanding of these relationships. Nonclinicians without extensive theoretical background can successfully use this technique in dealing with individual children, but the results are not generalizable.

Fuzzy Set Theory

Human thought processes can be captured and quantified, as stated by Korner (1966, p. 35):

> Classical elementary logic, i.e., the logic of unanalyzed propositions of quantification, rests on the requirement that all classes definable within it be exact, i.e., do not admit of neutral candidates for membership. This does not mean that inexact classes are, therefore, incapable of precise logical treatment. An imprecise logic would be a contradiction in terms. A precise logic of inexact classes is not. It is obvious that once we extend elementary logic by allowing inexact classes, not only the algebra of classes, but also its theory of quantification will be affected.
>
> For we then have to allow the quantifiers to range over inexact extensions; and allow propositions which express that an object is a neutral candidate for membership in a class.

Fuzzy set theory is a way to capture that uniquely human process. More complete descriptions of the model may be found in Kaufmann (1975), Bellman and Giertz (1973), Zadeh (1971), and Nowakowska (1982).

Fuzzy sets may be thought of as classes in which the boundaries are not sharp. An example of such a class is age. At what age is one old rather than young? If age is thought of as consisting of those two distinct periods in life, one could picture these in the classical dualistic sense as shown in Figure 8.

The problem is that humans don't operate under such precision. Sharp boundaries do not exist between young and old. A person who is called old by someone younger than himself may not be willing to apply that label to himself. The transition between "young" and "old," then, can be said to be a fuzzy boundary and could more appropriately be pictured as gradual as in Figure 9, rather than abrupt as is pictured in Figure 8.

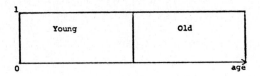

FIGURE 8
Young-old dichotomy.

The ordinate represents the degree of membership in a class. The abscissa is the age in question. A person af age "A" would be said to have .2 membership in the class "young." Linguistically, this might be described as being not very young and not very old.

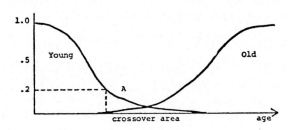

FIGURE 9
Age as a fuzzy set.

In fuzzy set theory, both numerical variables and linguistic variables are taken into consideration. Linguistic variables are summarized descriptions of fuzzy sets. Examples are the following terms (called hedges by Zadeh, 1975): sort of, a little, not very, extremely, a lot of, very very, a few, quite, and more or less.

Humans manipulate both linguistic and numerical variables in conditional statements and algorithms. Three examples of conditional statements are:

If $x = .2$, then $y = .7$
If x is small, then y is sort of large
If x is very large, $y = .4$.

When the complexity of the relationships increase, algorithms are used to define the functions. Zadeh (1973, p. 30) speaks of this when he says:

> Essentially, a fuzzy algorithm is an ordered sequence of instructions (like a computer program) in which some of the instructions may contain labels or fuzzy sets, e.g.:
>
> > Reduce x slightly if y is very large
> > Increase x very slightly if y is not very large and not very small
> > If x is small then stop; otherwise increase x by 2.

Some other examples of fuzzy sets are baldness, height, size, and the earth's atmosphere. How many hairs does a man have to lose before he becomes bald? How tall is tall in a person? By adding one stone at a time, when does the size of a pile change from small to large? At what point is something out, rather than in, the earth's atmosphere? These questions may be thought of as an extremely small sample of the problems that humans encounter in trying to conceptualize the world. Zadeh (1965, p. 338) summarized the dilemma by stating:

> More often than not, the classes of objects encountered in the real physical world do not have precisely defined criteria of membership. For example, the class of animals clearly includes dogs, horses, birds, etc., as its members, and clearly excludes such objects as rocks, fluids, plants, etc. However, such objects as starfish, bacteria, etc. have an ambiguous status with respect to the class of animals. The same kind of ambiguity arises in the case of a number such as 10 in relation to the "class" of all real numbers which are greater than 1.

Clearly, the "class of all real numbers which are much greater than 1," or "the class of beautiful women," or "the class of tall men," do not constitute classes or sets in the usual mathematical sense of these terms. Yet the fact remains that such imprecisely defined "classes" play an important role in human thinking, particularly in the domains of pattern recognition, communication of information, and abstraction.

Inexact classes may be quantified through use of fuzzy set theory (Goguen, 1969), a generalization of the traditional theory of sets (Hersh and Caramazza, 1976).

LEARNING DISABILITY - A FUZZY SET

Learning disability may be thought of as a phenomenon or construct which does not have precisely defined criteria for membership. The heterogeneity mentioned in the literature indicates a degree of complexity which defies precise description. Instead of defining learning disability in opposition to nonlearning disability as had been done in previous research (Kass, 1977), the condition can be defined according to theorized boundaries, with membership within those boundaries ranging from mild to severe. A portrayal of learning disability as a fuzzy set is given in Figure 10.

The universe of discourse is the school-aged population, represented by the rectangle. The concentric rings represent the condition of learning disability, with the outer rings depicting the milder range and the inner rings depicting the more severe range. Fuzziness is shown by the increasing number of breaks in each circle as one moves from the center outward. The innermost circle may be thought of as hard-core learning disabled children or "classic" cases. Children with mild learning problems would be placed in the outermost circle.

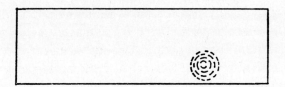

FIGURE 10
Fuzzy set portrayal of learning disability.

If mental models of the construct called learning disability could be accurately quantified, a basis for comparison and ultimate consensus about what constitutes the handicap and how to identify it will exist. There would be three advantages: (1) learning disabled children could be accurately identified by persons other than clinicians, (2) resources could be allocated proportionately to those most in need of services, and (3) scarce resources would be conserved through increased efficiency.

To illustrate the advantage of quantifying a theoretical model, a clinician was asked to articulate her model. The Kass (1977) model was quantified because of its value in the field of learning disability. If one model could be accurately depicted, the value of using fuzzy set theory to quantify human thought in this and other educational and psychological applications could be established.

THE PROCESS OF QUANTIFICATION

The process of quantification can be effectively accomplished with a clinician and two modelers. First, the clinician verbalizes her model by "talking through" which data are needed for identification, the relationships among these data, and the importance of each datum to the model. The modelers record this information and then ask the clinician to weight the importance of the various data. Formulae are constructed according to the specifications laid down by the clinician. The data and formulae are manipulated until the results agree with the clinician's verbal statements. These formulae are then placed in several algorithms. Data from "live children" are used. When the output from the formulae agree with the clinician's intuitive judgments, the value of the membership function for the fuzzy set "learning disability" for each child whose data were used is confirmed. Learning disabled students can then be identified by someone other than the clinician by entering the pertinent data into the formulae. Finally, a quantified model of the theory in the form of a computer program is constructed. The complete model is available in Horvath (1978) and further explanation is offered in Horvath, Kass, and Ferrell (1980).

Five specific procedures were followed in quantifying the theory:

1. A verbal statement of the theory was developed.
2. A quantified model of the verbal statment was established.
3. A questionnaire based on the quantified model was constructed.
4. A computer program was developed.
5. The quantified model was tested using cases.

Verbal Statement

The Kass model is a multidimensional age-related model of deviance in learning which indicates learning disability. Kass (1977) defines learning disability or dyssymbolia as being:

> characterized by extreme deviance in the acquisition and use of symbols in reading, writing, computing, listening, or talking; which deviance is due to an interaction between significant deficits in develop- mental functions and environmental conditions which make the individual vulnerable to those dysfunctions.

Four criteria are necessary for learning disabilty: The first is that the student in question should have normal achievement potential as indicated by a normal or nearly normal score on a standardized test of intelligence.

Secondly, the student must demonstrate behavior characteristics indicative of learning disability. Five types of behaviors indicate learning disability as follows: (1) an attending deficit is noted in confusion about the direction of letters and numbers, in lack of eye-hand coordination, and in not focusing on the task at hand; (2) a labeling deficit is noted in not remembering symbols, remembering one day and not another, exhibiting confusion in what things are called, and in seeming not to have a technique for rehearsing what is to be memorized; (3) an understanding deficit is noted in not separating figure from background, not learning more than one meaning per word, inaccurate guessing from context, and not following directions; (4) an integrating deficit is noted in inaccurate performance in skills such as spelling, writing, syntax, and sometimes in word-calling to such a degree that remediation requires getting rid of bad habits; and (5) an expressing deficit is noted in difficulty in expressing ideas through reading, writing, and talking, not because of a lack of intelligence, but because of inadequate basic skills which are ordinarily acquired in the first few grades.

The third criterion for learning disability is low achievement. A student who demonstrates low achievement earns a score at or below the sixteenth percentile on at least one subtest of an achievement test which has been standardized on a sample of the same age or grade placement. The fourth criterion for learning disability is evidence of component deficits within age-related functions. It is hypothesized that children pass through five stages of development called functions. Each function is qualitatively different from th e others. This is similar to the stage theory posited by Piaget or Erickson. The functions are as follows:

1. Sensory Orientation. This is the physiological or functional readiness stage in which the child's senses are activated. Dysfunctions in this period occur through deficits in body balance, visual pursuit, and auditory discrimination.

2. Memory. Occuring from the end of Sensory Orientation through approximately age eight, memory involves the ability to reproduce sensory images when these are no longer externally present. The young child imitates models and later recalls those imitations. To illustrate this point, consider the fact that children can readily learn more than one language before seven years of age. Some of the deficits which may be noted during the development of the Memory Function are hyper- or hypo-excitability, difficulty with rehearsal, and poor short-term

3. Re-Cognition. This is the understanding of semantic meaning and structural meanings. Sensory impressions are colored by concepts, thus changing earlier cognition of the world. Children from approximately age eight through eleven engage in word play, reflecting more flexibility in thinking than during the memory function, which is highly literal in nature. Deficits include difficulty in haptic discrimination, visual figure ground distinctions, and in visualization.

4. Synthesis. This involves the habituation or automatization of previously learned modes of response to the environment. Observations made by the individual during the previous functions become compacted into internalized representations. Whereas sensory discriminations were learned previously, sensory associations must be made in the age range of eleven to fourteen. Deficits may appear as impairments in temporal sense, monitoring, and auditory-visual-haptic coordination.

5. Communication. This is the process by which meanings are received from others and expressed to others, either consciously or unconsciously. Synthesized skills of speaking, writing, gestures, and reading now take on a personalized style, and personal responsibility is taken for the consequences of what is communicated. This is the most complex of the human functions and involves processing what is received through one's senses and awareness of what one is transmitting to others. Deficits are manifested in mathematical comprehension, reading comprehension, and in writing.

From this base, statements of conditions and logical relations were obtained in a series of interviews with the clinician. Throughout the procedure, it was necessary for the clinician to take an active part so that her model could be accurately captured. Sources of information and cut-off points were included.

Through discussion and revision, these statements were reduced to linguistic variables and logical relations in the context of fuzzy set theory. These relationships were depicted through use of the terms AND and OR in the verbal statement. Between two characteristics or criteria, an AND indicated that both charactreristics must be present for learning disability to be indicated and an OR meant that only one characteristic was sufficient to indicate learning disability. The process was iterative, intense, and depended on interaction between the clinician and modelers. Modeling one's thought is a challenging intellectual exercise.

An example of a portion of the process will be given to illustrate the demands involved. The clinician stated that one characteristic of learning disability in the sensory orientation function included symptoms suggestive of neurological disorder. After further discussion, the specific symptoms were stated as: excessive crying as reported by a parent, supersensitivity to stimuli as reported by a physician or parent, convulsions as reported by a parent or physician or both, regidity of body as reported by a parent, screeching as reported by a parent, lethargy as reported by a parent, premature birth as reported by a physician, high fever for several days as reported by a physician, and reverse swallowing as reported by a physician. Any one of these symptoms suggests a neurological disorder. For that reason, an OR was used between each symptom to denote a suspected neurological disorder. These symptoms were then grouped according to the structure of the clinician's theoretical orientation.

Quantified Model

This step involved having the clinician assign membership values for each characteristic expressed in the verbal statement. Class membership for each characteristic was assigned a value on the [0,1] continuum, which represents the degree to which a child would be said to display the characteristic. All the input data were either judgments or specified test data, and the output was a value of the membership for the fuzzy set of learning disability. For example, in the sensory orientation function example previously cited, the symptom of convulsions reported by a parent was assigned a value of .3, convulsions reported by a physician was .5, convulsions reported by both was .6, premature birth was .1, and so on.

In those cases where age influenced the membership value, a graph was drawn which reflected that influence. For example, delay in developmental milestones must be evaluated month by month. The age at which the milestone occured determined the membership value for learning disability. Other values which depended on age or other functions also became apparent. It was found that these values could be graphed as sigmoid curves or three dimensional unit square functions. Equations were fitted to the graphs, which made it easier to determine the membership function when given the score or month or other appropriate variable. Functions other than sigmoid curves or unit squares could also be used to depict representations of the clinician. The quantified model for re-cognition appears as Table 1 (Horvath, Kass, and Ferrell, 1980).

TABLE 1
The Quantified Model for Re-cognition

A child in the re-cognition function would be said to have indicators of learning disability if he

ALL OF   A. has the trait of normal achievement potential, shown by the symptoms of

        1. adequate intelligence, indicated by

a. IQ score of 80 or above	.8	OR
b. IQ score of 75 to 79	.5	OR
c. IQ score of 50 to 74	.1	OR
d. IQ score below 50	.0	OR

A Fuzzy Set Model of Learning Disability

          2. adequate verbal behavior, indicated through
             a. percentile (p) on a vocabulary test
$$1/(1 + e^{-2(p-25)})$$ OR
             b. percentile (p) on a verbal opposites test
$$1/(1 + e^{-2(p-25)})$$ OR
             c. percentile (p) on a verbal absurdities test
$$1/(1 + e^{-2(p-25)})$$ OR
             d. percentile (p) on a verbal analogies test
$$1/(1 + e^{-2(p-25)})$$

AND   B. has behavior characteristics of LD, shown by

1. difficulty paying attention, indicated in a clinical report	.6	OR
2. difficulty sitting still or lack of perseverance, indicated in a teacher/expert report	.4	OR
3. jerky, uncoordinated motor movements, indicated in a report	.3	OR
4. difficulty working indicated in a teacher report	.4	OR
5. not following directions, indicated in a teacher report	.6	OR
6. difficulty with friendships, indicated in a report	.4	OR
7. immaturity, indicated in a report	.3	OR
8. requiring continuing medication/special diet, indicated by a medical report	.7	

OR   C. shows evidence of component deficits within age-related functions, by

      1. visualization problems, indicated in a

a. score at or below the 16th percentile on the Raven's Coloured Progressive Matrices Test	.8	OR
b. clinical report	.8	OR

		c.	score at or below the 16th percentile on the WISC-R Mazes subtest .8	OR
	2.	haptic discrimination problems, indicated by		
		a.	score at or below the 16th percentile on the Ayres Tactile Discrimination Test .7	OR
		b.	report .8	OR
		c.	score at or below the 16th percentile on the Benton Finger Agnosia Test .8	OR
	3.	a lack of connecting visual and auditory symbols, indicated in a		
		a.	report by an expert of a reading block .6	OR
		b.	report by an expert of excessive reliance on phonetic spelling .4	OR
		c.	report of excessive reliance on auditory memory for sequence of letter names .3	OR
	4.	report of a lack of visual / auditory / /haptic connections .4		
AND	D.	has the trait of significantly low achievement, shown by the symptoms of		
	1.	difficulty with arithmetic, indicated in		
		a.	score at or below the 16th percentile on a standardized arithmetic computation test .6	OR
		b.	report of difficulty with arithmetic computation .5	OR
		c.	score at or below the 16th percentile on standardized test of story problems .6	OR
		d.	report of difficulty with story problems .4	OR
	2.	difficulty with reading, indicated in		
		a.	score at or below the 16th percentile on a standardized reading test .5	OR
		b.	report of difficulty in oral reading .5	OR
		c.	score at or below the 16th percentile on a standardized word discrimination test .6	OR
		d.	teacher report of poor word discrimination .3	OR

A Fuzzy Set Model of Learning Disability 377

      e. score at or below the 16th percentile on a standardized test of reading comprehension   .3 OR
      f. teacher report of difficulty with word meaning   .2 OR
      g. clinical report of difficulty with outlining skills   .7 OR
3. difficulty with penmanship, indicated in a
      a. teacher report   .5 OR
4. difficulty with spelling, indicated by
      a. score at or below the 16th percentile on a standardized spelling test   .6 OR
      b. teacher report of difficulty in spelling   .6

With this quantified model, the membership value for re-cognition can be determined. Each of the other functions was quantified in similar fashion. The model had to be adjusted to the clinician's theoretical viewpoint that learning disability could not be diagnosed before the end of the sensory orientation function. Another adjustment had to be made because of the fact that in the memory function and re-cognition function, lack of data in case reports make it practically impossible to differentiate behavioral characteristics and component deficits. Because of this, OR instead of AND was placed between these two criteria in the memory and re-cognition functions.

The clinician stated that learning disability is a lifelong handicap and information from previous age-related functions must be included in the final membership value. It was necessary, therefore, for her to specify a system which mandated that if the areas of writing, reading, arithmetic, spelling, communication, imagery, reading block, interpersonal relations, or immaturity were answered in the affirmative in a later age-related function, the area or areas would automatically be affirmed in prior age-related functions. This procedure was called cross-indexing. Another specification was that prior information had to be included in the determination of the final overall membership value. This was done by having the clinician specify a weighted system which took the membership values of the prior age-related functions into account.

Questionnaire

A questionnaire was developed which fitted diagnostic indicators into the four operational criteria of achievement potential, behavioral characteristics, age-related deficits, and achievement. For example, in the sensory orientation function, convulsions, premature birth, and high fever were grouped under achievement potential; excessive crying, screeching, lethargy, and reverse swallowing were grouped under behavioral characteristics; and so on. This made it easier for the data collector to fill out the form. Although the assumption is that all four criteria must be present for an at risk condition for learning disability to exist, it was not necessary to place OR's between each characteristic and AND's between each set of criteria. By keeping the questionnaire as simple as possible, data could be collected by relatively unsophisticated personnel, coded, and fed into a computer program.

Computer Program

A computer program was written from the verbal model, questionnaire, and specifications elicited from the clinician. FORTRAN was the language of choice, but any of several languages, including BASIC, would have sufficed. The program had to be able to accommodate assumptions about missing data.

A membership value for each age-related function is given by the program. These values represent the degree to which a child could be classified as learning disabled within each function. A single value is then generated which is a summary indicator of the child's membership in the class called learning disabled.

Testing The Model

The clinician was asked to look at either completed questionnaires or raw data from a number of cases and assign a membership value for learning disability to each. Working independently, one of the modelers took data from each case and fed those into the computer program, which calculated a membership value for each case. A comparison of the membership value assigned by the clinician and that assigned by the computer revealed that the quantified model matched the thought of the clinician to a high degree, i.e., beyond the .001 level of significance. The statistical technique used to test the match was the Pearson product-moment correlation. These analyses confirmed that it is possible to predict clinical classification of children who are learning disabled with a computer program when fuzzy set theory is applied as the methodology. Many cases had missing information, yet accuracy was still maintained.

CONCLUSIONS

Fuzzy set theory is a useful tool which can be used and understood at several levels. Because clinicians and modelers are less likely to have great general mathematical expertise and be well grounded in fuzzy set theory, the use of elementary principles of fuzzy set theory to model theoretical orientations in education and psychology is recommended. Although deeper understanding of the methodology may be more likely to lead to more efficient use of the technique and more accurate models, the increase in time may not warrant the slight gain in accuracy. The process is iterative, active, intense, and time-consuming.

It is possible for learning disability experts to quantify their theories through use of fuzzy set theory. As these theories are quantified, comparison and further refining would occur which could lead to the development of a consensus definition about what constitutes the handicap of learning disability. An identification procedure acceptable to a majority of the profession might follow. Given this, the potential for developing remediation programs is much higher. The whole approach to learning disability would become more efficient and economical. Fuzzy set theory is not exclusive to education and psychology. By demonstrating the utility of using fuzzy set theory, it is hoped that researchers in the other areas will also be encouraged to employ fuzzy set theory in their work.

REFERENCES

Asmonov, N. M. *Modeling of Thinking and the Mind.* New York: Spartan Books, 1967.

Bellman, R. and Giertz, M. On the analytic formalism of the theory of fuzzy sets. *Information Sciences,* 1973, 5, 149-156.

Bergman, G. *Philosophy of Science.* Madison: University of Wisconsin Press, 1957.

Farris, D. R. and Sage, A. P. On the issue of interpretive structural modeling to obtain models for worth assessment. In Baldwin, M. M. (Ed.), *Portraits of Complexity.* Monograph No. 9. Columbus: Battelle Memorial Institute, 1975.

George, F. H. *The Brain as a Computer.* Oxford: Pergamon Press, 1973.

Goguen, J. A. The logic of inexact concepts. *Synthese,* 1969, *19,* 325-373.

Hall, E. W. *Modern Science and Human Values.* Princeton: Van Nostrand, 1956.

Hersh, H. M. and Caramazza, A. A fuzzy set approach to modifiers and vagueness in natural language. *Journal of Experimental Psychology: General,* 1976, *105,* 254-276.

Horvath, M. J. *An Identification Procedure in Learning Disability Through Fuzzy Set Modeling of a Verbal Theory.* Unpublished doctoral dissertation. Tucson: The University of Arizona, 1978.

Horvath, M. J., Kass, C. E., and Ferrell, W. R. An example of the use of fuzzy set concepts in modeling learning disability. *American Educational Research Journal,* 1980, *17,* 309-324.

Huchings, D. W. Science in the sixteenth and seventeenth centuries. In Brierley, J. (Ed.). *Science in its Context.* London: Heinemann, 1964.

Kaiser, C. J. *Theoretical Positions and Bayesian Estimations of Learning Disability Specialists.* Unpublished doctoral dissertation. Tucson: The University of Arizona, 1974.

Kass, C. E. Identification of learning disability (dyssymbolia). *Journal of Learning Disabilities,* 1977, *10,* 425-432.

Kaufmann, A. *Introduction to the Theory of Fuzzy Subsets,* Volume 1. New York: Academic Press, 1975.

Kirk, S. D. and Kirk, W. D. *Psycholinguistic Learning Disabilities: Diagnosis and Remediation.* Urbana, IL: University of Illinois Press, 1971.

Korner, S. *Experience and Theory.* London: Routledge and Kegan Paul, 1966.

Lerner, J. W. *Children With Learning Disabilities* (4th Ed.). Boston: Houghton Mifflin, 1985.

Lichtenstein, S. Conditional non-independence of data in a practical Bayesian decision task. *Organizational Behavior and Human Performance,* 1972, *8,* 21-25.

March, J. G. and Simon, H. A. *Organizations.* New York: Wiley, 1958.

Mumford, L. *The Myth of the Machine: The Pentagon of Power.* New York: Harcourt Brace Jovanovich, 1970.

Novick, M. R. Introduction to Bayesian Inference. In Blommers, P. J. and Forsyth, R. A. (Eds.). *Elementary Statistical Methods in Psychology and Education.* (2nd Ed.). Boston: Houghton Mifflin, 1975.

Nowakowska, M. (Ed.) *Some Problems in the Foundations of Fuzzy Set Theory.* Amsterdam: North Holland, 1982.

Rosenfield, L. C. *From Beast-Machine to Man-Machine.* New York: Octagon Books, 1968.

Waller, R. J. Applications of interpretive structural modeling in management of the learning disabled. In Baldwin, M. M. (Ed.). *Portraits of Complexity.* Monograph No. 9. Columbus: Battelle Memorial Institute, 1975.

Warfield, J. N. On arranging elements of a hierarchy in graphic form. *IEEE Transactions on Systems, Man, and Cybernetics,* 1973, *SMC-3,* 121-140.

Warfield, J. N. Toward interpretation of complex structural models. *IEEE Transactions on Systems, Man, and Cybernetics,* 1974, *SMC-4,* 405-417.

Zadeh, L. A. Fuzzy sets. *Information Science,* 1965, *8,* 338-353.

Zadeh, L. A. Toward a theory of fuzzy systems. In Kalman, R. E. and DeClaris, N. (Eds.). *Aspects of Network and Systems Theory.* New York: Holt, Rinehart, and Winston, 1971.

Zadeh, L. A. Outline of a new approach to the analysis of complex systems and decision processes. *IEEE Transactions on Systems, Man, and Cybernetics,* 1973, *SMC-3,* 28-44.

Zadeh, L. A. The concept of a linguistic variable and its application to approximate reasoning--I. *Information Sciences,* 1975, *8,* 199-249.

TOWARDS A FUZZY THEORY OF BEHAVIOUR MANAGEMENT

Vladimir B. CERVIN and Joan C. CERVIN,

Cervin Associates 80, Quebec Ave., No. 609,
Toronto, Ontario, Canada.M6P 4B7

Behaviour management (BM) is identified as a set of procedures used by managers, experimenters, therapists, businessmen, etc. modifying or maintaining behaviour under given conditions. Modification is considered to be achieved by learning processes in the targeted individuals generated by BM . BM procedures are fuzzy in the sense that the particular actions to be taken are specified only approximately and the results are quantified in terms of fuzzy numbers. The theoretical conception of BM consists of fuzzy algorithmic specifications of the main procedures to be executed by behaviour managers to fuzzy criteria of behaviour change.

The main procedures of BM described in terms of fuzzy algorithms are: 1. Elicitation of a desired responses (generalized from classical conditioning procedures): Learning and maintenance of responses under given conditions with the help of promts, modeling, instructions, teaching aids, or aversive stimuli (which generate defensive responses). 2. (a) Selection of a desired response (generalized from operant procedures): Learning and maintenance by reinforcement of responses from among those spontaneously emitted by *the responding individual, or (b) elimination, by associating aversive, or withdrawing positive stimulation, of a spontaneously occuring undesirable response, usually combined with reinforcement of another, desirable response.*

The formalized procedures are quite general and are expected to be applicable, at primary and secondary levels of learning with suitable physical or social means, respectively.

INTRODUCTION

Behavioural methods are now applied in an extraordinary variety of human activities: animal training (Breland & Breland, 1966), business (Cervin, Bonner, Rae, & Kozeny, 1971), engineering psychology (Cervin, 1973; Pritchard, Leonard, Von Bergen, & Krik, 1976), health (Bandura, 1969; Ladoucer, Bouchard, & Granger, 1977; Wolpe & Lazarus, 1966), hypnotic memory retrieval (Cervin, 1987), instruction and education (Plax & Lacks, 1976; Skinner, 1968; Walters & Grusec, 1977), management and personnel (Hinton & Barrow, 1975; Luthans & Kreitner, 1975; Watson & Tharp, 1972), sex therapy (Kolodny, Masters & Johnson, 1979), sleep disorders, (Rathus & Nevid, 1977), social learning (Bandura, 1977), sports (Suinn, 1979) etc. The purpose of these applications is to achieve a change in human behaviour by developing new desirable responses in familiar situations, eliminating undesirable responses, or maintaining lagging behaviour. These general goals of behavioural intervention are better subsumed under the heading Behaviour Management (BM), rather than under the more restrictive term "behaviour modification".

BM methods include the use not only of "primary" physical or physiological means (stimuli), but also "secondary" social, linguistic and cognitive ones. Behaviour (responding) comprises not only physical performance but also imagery and symbols (Ladoucer et al., 1977), overt (speaking) or covert (imagining). For example, symbolic planning of an action is as much behaviour as is its physical execution. The use of imagery supplements person-environment interaction by a sort of auto-interaction (Cervin, 1987; Ladoucer et al., 1977; Watson & Tharp, 1972). In other words, BM is conceived in very general terms. This paper is an attempt to subsume the variety of BM procedures under one theoretical roof of fuzzy algorithms in order to increase the effectiveness and "portability" of BM procedures. Two large classes of BM methods can be distinguished.

Elicitation

These methods that started as classical conditioning using unconditioned stimuli to elicit the to-be-conditioned responses to conditioned stimuli, have developed by analogy and generalization into general elicitation methods that subsume a variety of techniques, e.g., social modeling (Bandura, 1969), verbal prompting, or even algorithms involving complex cognitive-behavioural operations (e.g., in sport), and of course, aversive techniques. Aversive techniques (Ladoucer et al., 1977, p. 91) are used to elicit, not to punish, target responses that may be desirable because they compete with undesirable responses, e.g. drinking alcohol. The function of all these interventions is to elicit and establish, under specified conditions, a target response, be it at a physical or a symbolic level.

Selection

The second class of behavioural methods that started as operant conditioning is selection of responses for establishment or elimination, by means of critical stimuli, such as rewards-punishments or friendly-unfriendly interventions. Unlike in elicitation, these stimuli are administered after the target response to encourage or discourage its recurrence. The target response either occurs spontaneously or, if desirable but unavailable, is first elicited by one of the elicitation methods, e.g., prompting. The response to a reward is of interest only from a motivational point of view, it must be compatible with and contingent upon the target response, for example, playing well and receiving a prize. The response to punishment is also contingent upon the target response and may be incompatible with it, for example, quitting smoking and smoking. (However, playing well and losing does happen too; this fact is a reason for establishing handicaps). The target response that is selected for elimination is either not rewarded or is punished and an alternative (reciprocal) response is rewarded. Both classes of BM methods - elicitation and selection - can be used in the establishment and maintenance of responses and in keeping out eliminated responses.

Motivation

BM interventions use the momentary motivation or arousal of the narrowly defined as momentary readiness to respond to a particular stimulus. The practitioner should know how and when his client gets motivated or how to make him so. Arousal may be based on the physiology of the organism and thus occur periodically without outside intervention: a person becomes hungry and responds positively to food (approach, ingestion); most people would at all times respond negatively to a pain-producing stimulus (withdrawal, anxiety). Primary positive stimuli induce approach in aroused people, but eventually continued stimulation will reduce arousal, e.g., in sex. Primary negative stimuli induce arousal and defense reactions, but their termination reduces both. Thus primary or unconditional stimuli can produce or reduce arousal and should be handled accordingly.

Arousal may also result in response to previously learned (secondary) stimuli called incentives and disincentives, which may be reproduced, with the practitioner's help, in the environment or in the client's imagination. Arousal and positive reactions can be elicited by the presentation of incentives, even without the physical presence of the associated primary stimuli, e.g., the announcement that dinner is ready. Arousal and negative responses can be elicited by disincentives, such as threats of pain or signals that pain may be caused, e.g., the sight of a needle previously associated directly or vicariously with a painful injection. Incentives and disincentives can thus be used to induce arousal and positive or negative responses, respectively. In BM, incentives and discincentives are often easier to handle than the primary stimuli they stand for. What function an incentive or primary stimulus will perform depends on the circumstances. For example, the promise of money is a motivator, actual payment is a reward; the threat of assault is a motivator, assault is punishment; an invitation to dinner is a motivator, dinner itself a reward; the promise of a kickback is an elicitor, the kickback itself is a reward; a bribe is a motivator and an elicitor. A stimulus may be used as a motivator and an elicitor depending on the target response desired. A target response in itself may be desirable or undesirable. If an undesirable response competes with an even more undesirable response (as in some treatments of alcoholism), it is desirable; deliberate relaxation is desirable and can also compete with undesirable responses, as in reciprocal inhibition (Thomas, 1974). These relationships are somewhat complicated and are not often treated in BM texts, but it is useful for a practitioner to have them sorted out.

Negative stimuli, by definition, can count on the presence of an appropriate motivation in the subject, viz. self-preservation, dislike, etc. and therefore are often used and misused as punishers. Positive stimuli, on the other hand, require the presence of relevant motivation to be reinforcing. Their effectiveness, therefore has to be first ascertained or established: e.g., a client must be hungry or "taste-hungry" to respond to the offer of food or a delicacy, unemployed to respond to an offer of a job, etc. If motivation can be assumed, e.g., in an employee or an athlete, he will be likely to respond positively to modeling of skills by a supervisor or coach. It is important to recall that in elicitation methods both negative and positive techniques are used to establish the elicited target response. In selection methods, negative techniques are used to eliminate undesirable responses, positive techniques to reinforce acceptable responses.

Momentary treatment-related motivation as discussed above is only a means to an end and should be distinguished from a client's long-term BM-goal-related motivation, on which recovery depends. Recovery is perhaps not the best word, because rather than being an illness treatment, BM is basically a problem-solving proposition.

BM Procedures

A distinction should be made between learning and BM. Learning and memory are the results of BM (although BM of course is not the only reason for them to occur). Historically, BM developed not from learning or cognitive changes studied experimentally, but from scientific method and laboratory procedures used by experimenters in bringing about behavioural and cognitive changes in animals and men (Ladoucer et al., 1977). This interaction was described as a procedure-process relation between the practitioner and client respectively (Cervin & Lasker, 1979).

In the laboratory, a maximum of precision and control can be achieved in setting up the conditions, defining the manipulations, and measuring the responses of subjects. BM on the other hand, is attempting to use similar or analogous methods under real-life or clinical conditions that introduce a great deal of uncertainty and fuzziness into manipulation, instructions, and response measurement. The laboratory environment is strictly controlled; stimuli are well defined (e.g., syllable-pairs, lists of single words, lights and buttons, etc.), and responses are exactly counted or measured (number of words recalled on every trial, number of buttons pressed, etc.), which allows the use of non-fuzzy probabilities as response measures. In the clinic or in real life this is not possible: everyday conditions in homes or work places change, responses and motivation are only approximately known and even with the best of instructions and client-therapist contracts, the goals are only roughly formulated and manipulations and interventions approximately executed. Moreover, variability and fuzziness originate not only in the clients and the environment, but also in the therapists, supervisors, teachers, etc., who may give the "same" instructions in different ways on different occasions.

In order to increase the practical effectiveness of BM interventions, reduce avoidable errors and "precisiate" uncertainty in explicit terms, it seems desirable to pinpoint the essential general features and rationale of BM procedures and specify as precisely and explicitly as possible even their fuzzy aspects. The treatment of procedures is intended to be both general and practical. The basic principles of the two classes of BM methods have been stated before (Cervin & Jain, 1982). What follows are explicit descriptions of the structure of the two BM procedures in the form of fuzzy algorithms. They embody the essentials of the main procedures and some of their important subsidiary aspects. But they do not describe particular BM techniques, such as systematic desensitization, (e.g., Cautela & Kastenbaum, 1967, Rathus & Nevid, 1977), or general BM methods, diagnoses, and treatments recommended for specific problems (Thomas, 1974). These are described in the relevant literature (Bandura, 1977; Kolodny et al., 1979; Plax & Lacks, 1976; Wolpe & Lang, 1964; Wolpe & Lazarus, 1966). Although general, fuzzy algorithms of BM offer procedural precision and describe in the explicit order of handling of fuzzy variables. One can be precise in the presence of fuzziness.

REFERENCE CASE

Before stating the algorithms, a reference example of BM will scribed in order to distil its important features.

A coach is trying to teach his motivated pupil how to strike a forehand in tennis.

0. The coach is satisfied that the learner (a) wants to learn and (b) will respond to his coaching, i.e., skill modeling and reinforcements. After explanations about the grip on the racquet, etc., the relevant stimuli are given:
1. A ball is thrown,
2. The coach hits it himself in front of the learner, accompanying his movements with suitable comments about the important features of the flight of the ball, the racquet movements involved, distance from the ball, the feedback that should be experienced, etc.
3. Then the learner tries to make the stroke himself.
2. The demonstration usually has to be repeated several times.
3. The trials by the learner, including his evaluation of the stimulus, i.e., the oncoming ball, its speed, rotation, etc., are interpolated between demonstrations. The learner gradually begins to hit the ball more and more correctly, and the ball lands in the right court.
4. The coach estimates the various more or less correct aspects of the strokes, as trials are repeated by the learner.
5. The coach praises acceptable strokes.
4. Gradually the more elementary aspects of strokes are no longer commented upon and the definition of good strokes is tightened up.

5. Only the better and better strokes are verbally rewarded and visually confirmed. The coach is using the method of approximation.
6. Eventually the learner develops a good forehand. Thereupon this part of coaching is stopped. The learner then needs only periodic checkups (praise-correction) to maintain a good level of play.

Analysis: In the above example, elicitation (2) and selection (5) of responses are used in tandem: the occurrence of the correct stroke is not left to chance but is elicited by demonstration (modeling). "Correct" responses (4) are estimated liberally at first, then the label is progressively restricted for better strokes to be reinforced (5). This increases the quality of strokes.

Elicitation and selection are the two treatments administered (2 & 5) by the practitioner. The practitioner observes the learner's behaviour and classifies (4) it as more or less "correct" or "incorrect". This definition of "correct" (4) is not only fuzzy but is changing as learning progresses: correct responses are compared with a model on a scale and the more narrowly defined "correct" ones chosen for praise. The definition becomes progressively less fuzzy, until finally "correct" strokes are those that go "in", incorrect those that go "out"*.

The number of correct responses (fuzzy and later non-fuzzy) is noted and the relative frequency of correct forehand responses over all forehands in a game, set, or match, is computed. This determines the level of play.

In order to describe these manipulations in an algorithmic form, they have to be defined in terms of the most elementary actions of the practitioner, such as "observe and classify", "compare and choose", "administer treatment". These actions will be stated in fuzzy algorithmic form.

---

*In "process control" used in industry, the good parts are considered to be those that are "processed" correctly during manufacturing: if processing is correct then the output is good (zero rejects) and is not monitored. By analogy, those strokes that are hit correctly "must" go in. Hence the importance of correct technique.

PRACTITIONER'S OPERATIONS

Elementary Operations

The practitioner will be called upon to use the following elementary operations while dealing with stimuli (subject's responses, events, symbols) and responding to them. It should be noted that the practitioner's responses are stimuli to the subject and vice versa.

Observing, looking, listening, perceiving, recognizing, naming, etc., an event, and responding to it, for example, the coach, seeing a learner respond, might say, overtly or covertly, "He responded", or "She swung her racquet", or "He didn't respond correctly". Observation implies a response from the observer on a nominal scale ; (0 and 1) =, which requires a one-one correspondence between the event observed and its representation, e.g., its name.

Classifying, testing, estimating, recording, etc., a subject's observed response, e.g., as correct, incorrect, or as a good approximation to the correct response. A rating scale between [0 and 10] and subjective judgment are used, for example, 0.0, 0.1, 0.2,...1.0 to evaluate a subject's responses, somewhat similarly to judging performance in figure skating, gymnastics, etc.

Comparing the representations of two (or more) events, for example, the degree of compatibility of a subject's present responses with the ultimately-desired responses, or the present level of responding and that of the modeled response, which may represent the goal of BM.

Choosing, selecting from a list or from a memory store of symbolic representations of events, the coach's action or response to be given to the subject's response, e.g., choosing verbal praise as a reinforcer for a responsive learner, or choosing a live demonstration to satisfy a subject's ambition. In a sense, choice is the inverse of classification: given a class of representations, choose an element with the help of some random or fuzzy device, such as an algorithm.

The Problem and its Solution

The treatment for solution of a behavioural problem used by a practitioner/coach mainly consists of the definition of the situation, evaluation of the present behaviour of the subject/learner (using observation, classification) and, (a) demonstrating, as a prompt to the subject, the desirable behaviour (see Reference Case) followed by subject's/learner's attempt to imitate or execute the demonstrated response - hands-on learning (this part may involve comparing (by the coach) the learner's response with his own); and/or (b) arranging to give the subject an appropriate stimulus after his response (involving choice), for example, given a correct response by the learner (stimulus to the coach), the coach saying "Correct", which is a reinforcing stimulus to the learner. These actions constitute ordered pairs of fuzzy events, that is, the pairs prompt-response or response-reinforcement.

EXPECTED SUBJECT EVENTS

The preceding sections dealt with BM procedures practioners may use. The purpose of applying these procedures is of course to generate certain events in the client. The procedures and subject events and their interrelation have been studied in the laboratory on animals and humans, and in "real life" practice. Thus subject events may be assumed to happen as expected, provided the procedures are carefully applied.

The main result that we expect from applying BM methods is the generating of a learning process. By learning we mean associating responses with stimuli with which they previously had not been associated: hitting a tennis ball correctly, relaxing in the face of previously anxiety-producing stimuli, abstaining from eating in the presence of previously "irresistible" foods, following rules, responding positively to previously frightening social situations, etc.

Elicitation methods do not only generate reflexes, but also any old or new responses that can be elicited by an unconditioned stimulus, a model, a prompt, or instructions in the presence of a to-be-conditioned stimulus that will become a "cue", at the behavioural (i.e., reflex and instrumental responses) and cognitive (e.g., learning a poem) levels. Elicitation has been applied in this way as long ago as the first decades of this century (e.g., Ivanov-Smolensky in the USSR), and is in effect the ancestor of present cognitive methods. Restricting elicitation to reflexes is entirely artificial and arbitrary.

Selection methods imply that their intended function is to "select" and establish responses occurring spontaneously and randomly to a stimulus so that thenceforth they occur regularly in the presence of that stimulus, which then becomes a cue producing that response with a desired frequency.

Often the two methods are combined in tandem. It may be tedious to wait for a desired response to occur spontaneously, as early attempts to train pigeons showed. Elicitation is used to produce the desired responses initially, so that they can be reinforced and taught in short order. Dramatic demonstrations of the superiority of this technique over "pure" operant methods exist in the literature and in practice. Unfortunately, many people, including some psychologists and psychoanalysts, still cling to the idea that the best way to teach is to wait for, and then reinforce, a response; e.g., in order to teach experimental techniques to a student, he should be allowed to potter about in the laboratory on his own without any instruction; to help a mental patient, he should be allowed to "find" his own solution to his problems, while the analyst only listens and does not suggest anything; to teach a student, he should try to discover the solutions himself and then be rewarded, however long it takes. BM eliminates this nonsense and tries to produce results in a much shorter time.

Another important already mentioned aspect of subject events is motivation. Some basic arousal occurs in the subject spontaneously, e.g., hunger, sex, etc. One of the ubiquitous motives is the search for "personal comfort", which includes the basic elements of lifestyle. Other motives are more idiosyncratic, e.g., not everyone interested in making money will take bribes or kickbacks, not everyone wants to devote his life to helping others (altruism motive). But some motivation can be created, e.g., being wined and dined or given "gifts" creates some psychological indebtedness to the giver, which can later be drawn upon.

There is confusion in thinking about the relation between reinforcement/punishment and motivation. The two are not identical but are related by learning. Certain stimulus conditions, including a client's own target-response-produced stimuli, through learning, become signals (cues) to him that reinforcement or punishment is coming. The stimuli that precede the target response and, ipso facto reinforcement or punishment, are also learned as remote cues to reinforcers/punishers. These cues are incentives and disincentives respectively. As mentioned above, secondary motivation created in the presence of incentives and disincentives is called expectation. Thus the cues to reinforcement and/or punishment, i.e., incentives, although learned as cues, can subsequently function as motivators in selection training. In this respect they are different from other learned cues, which simply produce responses. Reinforcers/punishers also produce responses, viz. satisfaction or withdrawal, but these responses per se do not play any important role in selection methods. In elicitation methods, however, conditioned elicitors can function both as incentives for, and as elicitors of, target responses; e.g., a call for dinner may awaken appetite (motivation) and elicit a response to go and get it (instrumental response), for which dinner is the reinforcement. These relationships in a client's behaviour may be part of his problem.

Algorithm PROBLEM

(1) A client comes to a coach with a request to be taught a forehand for a fee.

(2) The coach gathers that: the client is motivated hence will follow instructions and respond to reinforcement.

(3) The coach assesses the fuzzy grade of the quality of the client's stroke-making when thrown a ball, using a subjective scale of the features of good strokes (see Subalgorithm TARGET RESPONSE Figure 3, and Figure 1).

(4) IF assessment shows low quality of strokes (few necessary features)
   THEN the coach either
       recalls and/or

demonstrates
the features of correct high quality strokes which may also be assessed on the above scale.

(5) The demonstrated strokes will serveas a model and a goal for the client.

(6) The recalled model of good strokes will be a basis for the reinforcement strategy of the coach.

(7) The problem is a large difference between the coach's and the learner's quality of stroke-making, as quantified above.

(8) The task for BM is to reduce this difference as much as possible using one or all the BM algorithms; end statement of problem.

(9) Out.

Subalgorithm ELICITATION

(1) The coach demonstrates correct stroke.

(2) The client observes the demonstration and tries to reproduce the stroke when thrown a ball.

(3) The coach evaluates client's stroke-making as in Subalgorithm TARGET RESPONSE Figures 3, 4 and 5.

(4) WHILE d is large at first but decreasing over trials
    DO (repeat (1)-(3)
    END
WHEN d stops decreasing

(5) IF the difference d between coach's and client's stroke quality
    THEN goal achieved (6)
        ELSE
IF d remains large over many trials
    THEN goal elusive (6).

(6). Stop.

Subalgorithm REINFORCEMENT AND PUNISHMENT

(1) The coach observes client hitting the ball on forehand, recognises correct and incorrect aspects of the strokes,

    a.) praises (reinforcement) the correct aspects of responses, and/or

    b.) points out (punishment) the wrong aspects.

(2) The coach evaluates client's stroke-making as in Subalgorithm TARGET RESPONSE Figures 3, 6 and 7.

(3) WHILE d is large at first but
decreasing over trials
    DO repeat (1)-(2)
    END
WHEN d stops decreasing.

(4) IF the difference d between coach's and client's stroke
quality close to 0
    THEN goal achieved (5)
    ELSE
IF d remains large over many trials
    THEN goal elusive (5).

(5) Stop.

Algorithm EXTINCTION 1

(1) The coach observes and recognizes client's bad habits and wrong aspects of his stroke-making when hitting a ball, which interfere with good playing, and.
    a) points them out to the client
    b) demonstrates stroke-making in the
       absence of these habits using
       Algorithm ELICITATION.

(2) The coach evaluates client's stroke-making as in Subalgorithm TARGET RESPONSE Fig. 3, and Fig. 8.

(3) WHILE d is decreasing over trials
    DO repeat (1)-(2)
    END
WHEN d stops decreasing.

(4) IF the difference d between coach's and client's stroke
quality is close to 0
   THEN (goal achieved) (5)
   ELSE
IF d remains large over many trials
   THEN goal elusive (5).

(5) Stop.

Algorithm EXTINCTION 2

(1) The coach observes and recognizes some of the client's bad moves
in stroke-making
   a) shows the consequences of these moves
   b) but ignores the moves themselves.

(2) The coach evaluates client's stroke-making as in Subalgorithm
TARGET RESPONSE Fig. 3.

(3) WHILE d is decreasing over trials
   DO (repeat) (1)-(2)
   END
WHEN d stops decreasing.

(4) IF the difference d between coach's and client's stroke
quality is close to 0
   THEN (goal achieved) (5)
   ELSE
IF d remains large over many trials
   THEN goal elusive (5).

5) Stop.

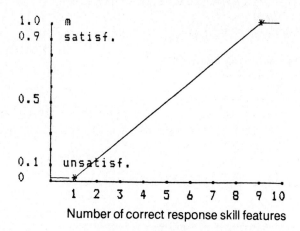

FIGURE 1

BM problem definition: an example

The learner has few skills at the start, the coach many (x-axis).

The problem is to teach the learner many or all of the coach's skills.

The m curve describes the expected gradual improvement of the beginner's initially low, unsatisfactory proficiency as new skills are acquired.

The coach's proficieny is high and satisfactory on account of many skills already acquired.

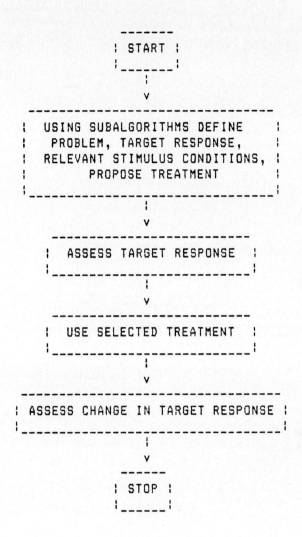

Figure 2
Algorithm FUZZY BEHAVIOUR MANAGEMENT

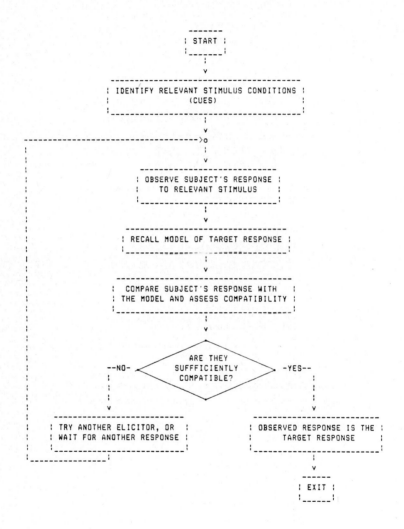

FIGURE 3
Subalgorithm TARGET RESPONSE: recognition and assessment

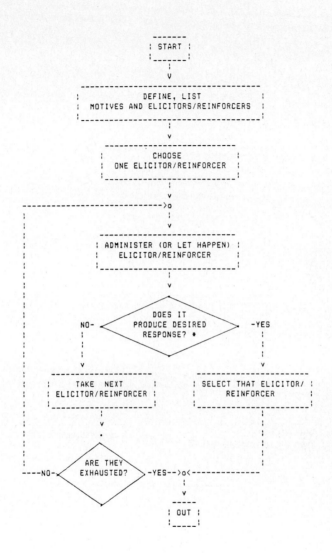

FIGURE 4
Subalgorithm ELICITORS AND REINFORCERS

* Note: Elicitor is to produce target response, reinforcer satisfaction.

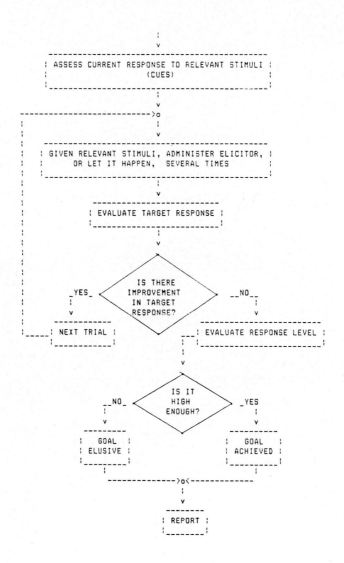

FIGURE 5
Subalgorithm ELICITATION

Towards a Fuzzy Theory of Behaviour Management

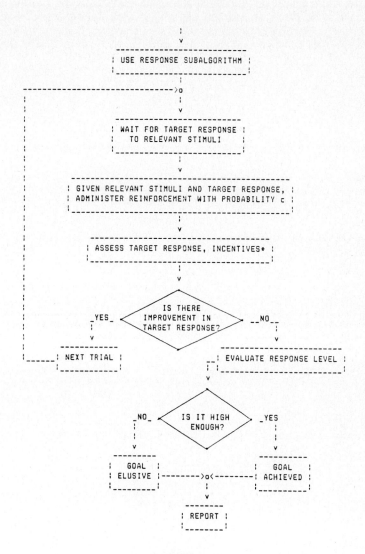

FIGURE 6
Subalgorithm REINFORCEMENT

 * Note: Cues are to be conditioned to target response, incentives to reinforcement.

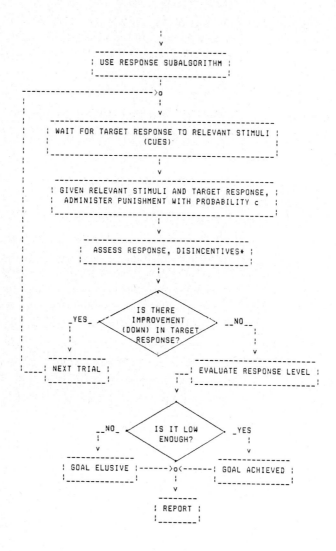

FIGURE 7
Subalgorithm PUNISHMENT

* Note: Cues are to be conditioned to target response, disincentives to punishment.

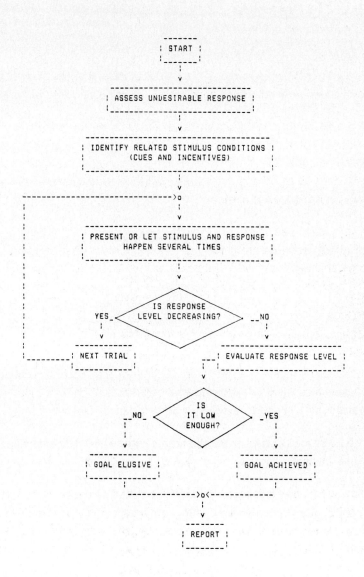

FIGURE 8
Algorithm SIMPLE EXTINCTION

SYMBOLIC STATEMENT OF FUZZY ALGORITHMS

Fuzzy Algorithms (FA), which describe treatment events, use "if-then, else-" statements where client's performance and his goal; else-" statements where

* the "if" part, in selection methods, depends on the observation and classification of the subject's response, for example, "fairly good stroke"; in elicitation methods the "if" part is the coach's demonstration chosen from his repertoire; in the algorithms the "if" part is based on the difference between the client's performance and his goal;

* the "then" part depends, in selection methods, on the choice of an appropriate response from the coach's resources, for example, praise (reinforcement) of the learner's response. In the elicitation case, the "then" part is the observed and classified response of the learner to the (coach's) demonstration (the learner's response, however, underlies the "if" part of the selection method); in the algorithms the "then" part implies "stop" if the goal has been achieved;

* the relation between "if" and "then" depends on the coach's knowledge of, and experience with, for example, what constitutes a "good" stroke, "good" demonstrations, effective methods of teaching and learning;

* the "else" part implies, in the algorithms, "do the next indicated step" if the goal has not been achieved.

* the "while" part means "do the loop", i.e., "repeat treatment", if the goal is not achieved but there has been some improvement in the target response; "else" is not specified beyond "do the next step" in the algorithm, e.g., stop, if the goal is elusive.

It should be noted that the "if-then, else" statements can be extended to include more complex cases (Cervin & Jain, 1982), for example, in the elicitation procedure, the "then" part depends not only on the coach's demonstration, but also on the learner's motivation: "If" the learner is motivated, "and if" he sees a demonstration, "then" he will execute the response in question. In selection procedures, reinforcement depends on the client's response, motivation, etc. Other extended conditional statements can be written as required.

FA have proven useful in summarising and ordering the instructions for practitioners who are trying to solve (behavioural) problems for their clients. FA can be defined as "an ordered set of fuzzy instructions, which, upon execution, yield an approximate solution to a specified problem... The theoretical foundation of /fuzzy theory/ is actually quite precise and rather mathematical in spirit... the source of imprecision is not the underlying theory but the manner in which linguistic variables and fuzzy algorithms are applied to the formulation and solution of real-world problems" Generalizing from the above discussion, verbalized algorithms and flow charts of BM, symbolic statements of the operations entering into these algorithms will be given explicitly with examples.

Observe, classify, assess:

A client's response occurs and the practitioner receives a stimulus s^a produced by the client's response A C B (A is fuzzy subset of client's Behaviour in a BM situation). The practitioner's observation may be a simple "client responded" which may be assessed as (see Notation in the Appendix)

$$m'(s^a \in S) = 1$$

where $m' \in M' = \{0,1\}$ a set of only two elements, S C C set of all the stimuli in a BM situation, and S a fuzzy subset of "correct" response stimuli. This type of rating is known as the rating on the nominal scale $\{0,1\}$ and is little more than labeling.

In a BM situation, the practitioner then proceeds to classify and assess the client's responses "a" generating stimuli $s^a \in S$ he receives, comparing s^a to x on the scale of ratings Sc of skill features Sc = $\{1,2,...10\}$ = and obtaining an associated membership grade $m(s^a \subset S^a)$ for subset $S^a \sim X$, the fuzzy subset of correct features on the scale corresponding to A, X C Sc. This gives the compatibility of the fuzzy subset A of observed responses with the statement "client is proficient". He obtains, e.g., $m(s^a \in S^a) = 0.2$, where now $m \in M = [0,1]$, is the proficiency grade associated with the number of the corresponding satisfactory features on the scale. Classification is thus done by recognizing correct features of responses by comparison with a reference scale. Assessment means determining the number of correct features of S^a and its grade m as a satisfactory response. In response acquisition, the grade is initially low if only a few features of a good response are in evidence, but later increases as the number of correct features goes up. In response elimination the converse is true. Examples of empirical determination of the membership grade m in two fuzzy subsets are shown in Tables 1 and 2. The influence on the ratings of "own behaviour" of the raters is very much in evidence.

More generally, degrees of membership in fuzzy subsets of "good" behaviour or compatibility of observations with the linguistic or numerical description of satisfactory behaviour, in psychology, have been estimated by three main methods:

a) Method of comparative judgment, where the rater is given a standard, say a circle of a certain size, and then is asked to guess if each of subsequently singly presented circles is bigger, smaller, or the same, and how many times bigger or smaller. The standard may be removed during judgments.

b) Ratings: the rater may be given a visual, auditory, or other scale with which to compare the stimulus received. Some scales are multi-dimensional, such as those used in gymnastics, mental health assessment, etc. Some scales are ordinal and some have been refined into interval scales.

c) Method of absolute judgment: the rater is shown a figure or presented with a tone, hue, weight, a speeding object etc., without any standard, and is asked to call the stimuli presented as large, heavy, b-flat, loud, soft, fast, slow, etc. In this case the judging is also done by comparison, as in the above cases, but the judge supplies his own private standard or scale.

The rating's quality, i.e., reliability and validity, depends on the object rated, the rater, and his familiarity with the object rated. A large part of psychology is

TABLE 1
Compatibility Values of Statement:
"X sleeps too much" with actual amount slept.

Av. Hours Slept by "Judge"	6	7	7.5	8	8.5	9
Hours Slept by "X"						
6.0	.5	.3	.3	.3	.1	.2
7.0	.6	.4	.4	.35	.2	.3
7.5	.7	.5	.45	.45	.3	.4
8.0	.9	.55	.5	.5	.4	.5
8.5	1.0	.6	.6	.6	.5	.7
9.0	1.0	.7	.8	.8	.6	.8
10.0	1.0	.9	1.0	1.0	.9	.9
11.0	1.0	1.0	1.0	1.0	1.0	1.0

TABLE 2
Compatibility Values of Statement:
"X drinks too much coffee" with actual amount drunk.

Av. No. Cups of Coffee Drunk by Judge	1	2	3	5	7	11	12
No. of Cups of Coffee Drunk by "X"							
1	.1	0	.4	0	.1	0	0
3	.6	.2	.5	.4	.3	.1	0
5	.8	.6	.6	.5	.5	.2	.5
7	.9	.6	.7	.6	.7	.4	.8
8	1.0	.6	.7	.6	.7	.5	1.0
10	1.0	.8	.8	.9	.8	.7	1.0
12	1.0	.8	1.0	1.0	.9	.9	1.0
15	1.0	1.0	1.0	1.0	.9	1.0	1.0

devoted to the study of measurement procedures under the term Theory of Measurement.

Compare:

This operation first occurs in attempts to recognize correct or desired features of a client's responses. This can be done by taking points from a scale, such as $Sc = \{1,2,3,...n\}$ = given in Figure 1, or recalling a model of the response, and juxtaposing the observed response features with those of the model, much like in comparative judgment. If the response features are more or less independent, comparison of the response and scale can be done in a one-to-one correspondence, feature by feature, and the result quantified by a cardinal number of the subset of desired features obtained (see Figure 1). However, if the response features are not independent but form an integrated pattern or cluster of ordered elements, e.g., holding a racquet and swinging it, concentrating on a mantra and relaxing, then such a pattern must be determined and recognized as a unit. For example, a person might be rated for education level and experience: the two rated traits of the same person form intersecting fuzzy subsets A_1 & A_2. In fuzzy set theory this is described as taking the MINimum of the two sets.

Another comparison occurs when two (or more) people are assigned natural numbers for their comparable traits, or skills, e.g., a coach and his pupil may be rated one 8 and the other 2 on the skill feature scale; the comparison done by subtraction gives the result as 6 additional skill features to be learned by the pupil, if he aspires to reach the coach's level by learning. Assuming a situation similar to that described in Figure 1, the result of such learning will be described by the associated improvement in proficiency $m_s(2) \rightarrow m_s(8)$, i.e., the difference in the two fuzzy subsets $S^p - S^a$ (for coach and pupil respectively), which is (Kaufmann, 1975)

$$S^p - S^a = S^p \,\&\, (-S^a)$$

For example, let the number and ratings of skills be

$$S^a = 0.0/1 + 0.3/2 + 0.1/3 \ (m(x \textbf{C} X) > 3 \text{ being all } 0)$$
$$S^p = 0.9/7 + 1/8 + 0.9/9 \ (m(x \textbf{C} X) < 7 \text{ being all } 1)$$

i.e., fuzzy numbers centered on 2 and 8 for pupil and coach respectively, and other corresponding numbers in the two sets being $1 \rightarrow 7$, $3 \rightarrow 9$. Then subtracting the two sets (taking the minima of sets S^p and $(-S^a)$ where $(-S^a) = (1 - S^a)$), we get

$$S^p \ \& \ (-S_a) = 0.9/7 + 0.7/8 + 0.9/9$$

which can be interpreted as the expected improvement in proficiency the pupil will have to get if he wants to reach the fuzzy level centered on 8 skills (coach's level). A similar result would be obtained (Kaufmann, 1975) by getting a Hamming distance (for the pupil "to go") between the two fuzzy subsets.

Choose

Choice is closely related to decision-making, but does not necessarily require consideration of utility of the alternatives involved. Choice may be made relative to some preset requirements, as shown by Yager (1980), who analyses several methods of making a choice under various scenarios. One in particular is simple and suitable for BM.

Let three attributes (A_1, A_2, A_3) of two people (Y_1 and Y_2) be rated in terms of their compatibility with job requirements:

A_1: $(0.9/Y_1\ 0.5/Y_2)$ (Y_1 is young, Y_2 not so young)
A_2: $(0.2/Y_1\ 0.4/Y_2)$ (Y_1 is not experienced, Y_2 has some experience)
A_3: $(0.6/Y_1\ 0.8/Y_2)$ (Y_1 is O.K., Y_2 is well-spoken)

Suppose you want to decide between the two subjects which is a better candidate for a job you have to offer. You set up a fuzzy decision function (the simplest one among many) as a choice of the person having the largest of the smallest attribute m grades:

$$D_{required} = (A^i_1\ \&\ A^i_2\ \&\ A^i_3), i = 1,2$$

which is read: "The candidate chosen should have all three attributes (& = and) as far as possible". The solution is: "My choice (Decision) will go to the one who has the largest of the smallest m grades on all three attributes." The minima are:

$D'_{min} = (0.2/Y_1\ 0.4/Y_2)$ (both connected with experience, the other two attributes being balanced)

$D_{maximin} = (0.4/Y_2)$ Y_2 is your choice.

Choice is involved in BM when one tries to decide between different methods with restrictions imposed on each of them by the lifestyles or clinical settings in question. There is considerable literature on the topic of fuzzy decision making (Dubois & Prade, 1980; Yager, 1980; Zadeh, 1973).

Inference

Most fuzzy algorithms above have IF-THEN or IF-THEN, ELSE statements. Following Zadeh (1973) one should first distinguish between propositional and set-theoretical interpretation of fuzzy conditional statements IF A THEN B. He opts for the latter, which also seems more appropriate for BM purposes. Second, a distinction should be made between fuzzy conditional statements with and without specification of the operative ELSE. If the alternative is not specified, it can be any fuzzy subset of the universe of discourse; then a fuzzy conditional statement can be defined as "a fuzzy set of ordered pairs (u, v) u $\in$ U, v $\in$ V, with the grade of membership of (u, v) in A x B, given by $u^A(u)$ & $u^B(v)$" (& = $\wedge$ = and = minimum).

In elicitation methods the membership grade of ordered pairs (e^a, s^a), e $\in$ U, s $\in$ C in the fuzzy set E x S is (see Appendix) $m(e^a \in E)$ & $m(s^a \in S)$. This may be read "IF prompt e THEN response a, which produces stimulus $s^A \in$ S". Here the alternative ELSE is of no interest because the operation is under the practitioner's control. If he forgot to give a prompt, perhaps he should not be doing BM.

There are, however, at least two other IF conditons which are important: a to-be-conditioned stimulus, $s^c \in$ C, must always be present in acquisition BM (or conversely, the already conditioned stimulus in elimination BM); and motivation should also be secured. The conditioned stimuli are not necessarily fuzzy, but motivation may be, and it may also be more difficult to ascertain or arrange. For example, in sexual relations a lover would normally observe his/her partner's receptivity to sex, or else would arrange for getting her "in the mood" for sex (rapists bypass this phase).

In the elicitation cases, the expression for compatibility grade of the compound R "motivation and prompt and the client's response and stimulus sa" would be of the form

$$m_R(mot., e, a, s^a) = MIN(m(mot. \in MOT.), m(e^a \in E), m(a \in A), m(s^a \in S))$$

i.e., if prompt is administered and if motivation is present and if the client responds, he will produce a stimulus to the practitioner. This assumes that the conditioned stimulus is present as usual.

The IF-THEN, ELSE compositional rule of inference is used in several algorithms. Algorithm ELICITATION, Step 4, has such an inference: "If the behavioural difference between a model and client is not large, then do 6, ELSE do 5". For example, we have these elements in the situation: if the difference set is D = {1,2,...9} = , then fuzzy subset

"Difference not large", df = 1/1 + 0.9/2 + 0.2/3 + 0/4   (I)

Step "THEN 6", df = 1/6 + 0/5   (II)

i.e., do 6, not 5.

Difference large, df = (1 - not large)

= 0/1 + 0.1/2 + 0.8/3 + 1/4   (III)

Step "ELSE 5", df = 0/6 + 1/5 (IV)

i.e., do 5 not 6.

The conditional rule defines the THEN and the ELSE. The first applies if the difference is no longer large, the second if it is still large. The definition of "large" and "not large" are subjectively rated. The practitioner's rating of the size of difference as large or not large may depend on the ability and motivation of the client. If he/she shows promise, one can aim at a very small difference, if not, one may decide that a fairly large difference is "not large", in order to terminate BM. To make the adjustment without new ratings, Zadeh proposed a simple technique of dealing with such problems: raising the m's to a higher power. Thus in order to make the "not large" into "not at all large", to a desired degree, one can raise a "large" difference to, say, a second (or higher) power, making:

very large = 0/1 + 0.01/2 + 0.64/3 + 1/4   (V)

i.e., the power narrows down the fuzziness.

Zadeh's (1973) conditional expression for the above inference is,

IF A THEN B ELSE C, df. = A x B + (-A x C)   (VI)

In our Algorithm ELICITATION, Step 4, this means

A = *not large* as defined above
B = "THEN do 6" in algorithm
C = "ELSE do 5" " "
-A = 1 - A = *large*

Using numerical definitions we get (taking the minima first)

Algorithmic Steps

	R^1	6	5	
	1	1	0	
Differences	2	0.9	0	IF d "not large" THEN do 6
	3	0.2	0	
	4	0	0	

R^2	6	5
1	0	0
2	0	0.1
3	0	0.8
4	0	1

ELSE: if d "large", do 5

Combining R1 + R2 = R gives

Algorithmic Steps

R	6	5
1	1	0
Differences 2	0.9	0.1
3	0.2	0.8
4	0	1

(VII)

Using the definition of the difference "not large" (NL) and Zadeh's (1973, sect. 5.16) we get for

IF A THEN B ELSE C for the D as above

Maximin dNL o R =

$$= \begin{matrix} & \text{Minima} & & & & & & \text{Steps} \\ & & & & R & & & 6 \quad 5 \\ & d_{NL} & & & & & & \\ [1 & 0.9 & 0.2 & 0] & \circ & \begin{vmatrix} 1 & 0 \\ 0.9 & 0.1 \\ 0.2 & 0.8 \\ 0 & 1 \end{vmatrix} & = & \begin{vmatrix} 1 & 0 \\ 0.9 & 0.1 \\ 0.2 & 0.2 \\ 0 & 0 \end{vmatrix} \\ & & & & & & \text{Maxima} & [1 \quad 0.2] \end{matrix}$$

which may be read: the grade of doing step 6 is 1 and, because of fuzziness, there is a possibility 0.2 of doing step 5 in spite of small (i.e., not large) distance from the goal. If the difference is large, in other words, the client is far from the goal, then a similar calculation with d_L gives:

Maximin d_L o R = [0.2 1.0]

for steps 6 and 5 respectively, which means that that grade of membership of doing step 5 is 1, but again there is a possibility 0.2 of doing step 6, i.e., terminating BM right there. In other algorithms, the final steps and calculations are the same as above.

In Algorithm ELICITATION, step 5, we have a loop "WHILE-DO". The "ELSE" part in this step is not specified, i.e., in the negative case one simply should do the next step 6 (stop). This inference can be described by fuzzy conditional IF-THEN, without specified ELSE-C, which is simply the A x B part of the above, and is calculated in a similar way.

CONCLUDING REMARKS

In the face of the diversity of applications of behavioural methods in various fields by people with different professional backgrounds, it is natural to try and spell out some psychological elements common to all the procedures used. This paper is an attempt to do this in a fuzzy form, while keeping a vigilant eye on applications.

Within the scope of this paper it was possible to analyze to a degree only the main features of some of the algorithms, such as determination of the membership functions (which is an old concept in psychology that was in search of precisiation), if-then inference, loops, linguistic variables, and structure of the algorithms. Some of the algorithmic features remain unanalyzed, for example, the "decreasing" response function. Other forms of the if-then analysis that exist have not been tried out in our behavioural algorithms. These analyses will have to await another opportunity.

However, what is offered in this paper should help in clarifying the distinction between fuzzy data and sloppy thinking. As Zadeh pointed out, the contribution of the fuzzy set theory is to make it possible to think precisely about vague or uncertain variables. One can be clear about what can or should be done, even in the absence of precise measurements and instructions.

An instruction "do this exercise five times at 6 o'clock" is to be executed exactly and does not say anything about deviations. An instruction "do it four to six times between 5 and 6 o'clock" allows a certain amount of leeway, but does not say anything about how to decide what actually should be done, and at what time. A fuzzy instruction "do the exercise preferably five times, as close to 6 o'clock as possible", given with compatibility functions, not only specifies the quantities but also the deviations from the prescription. It is neither rigid nor indeterminate, and not only instructs the client what to do and when, but also allows for weighted deviations from, and subjective interpretations of, the instructions. In that sense, such an instruction is both more realistic and more precise than the other two.

In addition, such instructions usually contain fuzzy conditions under which they are to be executed: e.g., the fuzzy statement "if it is not too hot", with a compatibility function, which can be varied individually, to help the client decide when it is "too hot". Thus we have a complete, flexible, and realistic instruction; unlike the other two types, of which one is unrealistic and other indeterminate: both generate uncertainty rather than help overcome it, and we all know what uncertainty may lead to from the psychological point of view. The effect of fuzzy instructions is much sounder, in the sense that they take from the client the burden of dealing with uncertainty. From the point of view of applicability, instructions in fuzzy form, with their approximate nature but precise and explicit structure, should allow a much easier "portability" from field to field. They are more robust, in the sense of being less vulnerabe to distortion in attempts to adapt them to different environments. In this sense, fuzzy behavioural instructions are fairly foolproof, and may allow people who have less preparation for behaviour management than professional practitioners to apply them to others and especially to themselves. The fuzzy form of behaviour management may contribute to making it a part of everyone's daily life.

APPENDIX

Notation.

BM is usually applied in some given situation, such as the family, sexual scene, home, office, or market place. There is usually a certain aspect of the situation that is associated with a particular behaviour that needs changing or strengthening. The practitioner, after an analysis of the situation and the client's behaviour in it, tries to change the situation physically (e.g., "take a trip", "avoid John Brown"), cognitively ("imagine that you are divorced - is it the end of the line?"), or behaviourally (e.g., systematic desensitization of the client to the particular situation, stopping smoking, etc.). The tools the practitioner uses are measurement scales, elicitors, reinforcers, punishers, the client's motives, and previous learning, i.e., existing stimulus-response associations and cognitive structures. The notation to be used is an attempt to approximate the structure and contemplated restructuring of the situation at hand, the client's behaviour as viewed by the practitioner.

Thus we have:

C = conditions describing the situation at hand (universe of discourse);
S = fuzzy subset of stimuli relevant to the client's target behaviour $S \subset C$;
s = generic element of the conditions (stimuli), $s \in C$;
S^c = subset of conditioned stimuli to client's responses (behaviour triggers, not explicitly used in this paper);

m = number in the membership compatibility function (u in Zadeh (1973) characterizing C: $m_s : C \to [0,1]$, which attaches to each element s in C a number $m_s(s) = m(s \in S)$ in the interval $[0,1]$;

S = $E_i\, m_i / s_i$ where + denotes the union, not the sum;

B = client's behaviour in C;

A = fuzzy subset of responses under analysis A C B;

a = generic element (response) of behaviour $a \in B$;

s^a = stimulus to the practitioner produced by client's response, e.g., his observation of client's target response $s^a \in S$;

$m_s(s^a) = m(s^a \in S)$ from the practitioner's point of view where a is the client's target response to some aspect of C;

U = practitioner's resources for intervention;

E = fuzzy subset of resources used in the situation E C U;

e = generic element $e \in U$ of resouces used: reinforcers, elicitors, punishers, models, prompts, which depend on the client's motivation for effectiveness, to be ascertained by trial and choice.

Sc = scale of items used to ascertain client's target responses in terms of compatibility with a model or criterion; assessment is done by comparative judgment, or (informally) by absolute judgment;

X = fuzzy subset of the correct scale items x, corresponding to the subset S^a of observed response features, with the associated number m to be assigned to the client's response: $m_x(X \sim a)$;

X C Sc scale items, $x \in Sc$;

Sa X = corresponding subsets of client's responses (in terms of stimuli) and scale items;

& = ampersand = = and = minimum (set intersection);

+ = V = or = maximum (set union);

(-Y) = complement of Y; if $m(x\, X) = 1 - m(y\, Y)$, (-Y) = X;

MOT = fuzzy subset of client's motivational states in a situation, used only for illustration in this paper;

mot = generic state of arousal.

REFERENCES

Adam, E. E. Behaviour modification in quality control. *Academy of Management Journal. 18*, 4, 663-679.

Bandura, A. L. *Principles of behaviour modification.* New York: Holt Rinehart and Winston, 1969.

Bandura, A. L. *Social learning theory.* Englewood Cliffs, N.J.: Prentice-Hall, 1977.

Breland, K., & Breland, M. *Animal behaviour,* New York: Macmillan, 1966.

Cautela, J. R., & Kastenbaum, R. A reinforcement survey schedule for use in therapy, training and research. *Pyschological Reports,* 1967, *20,* 1115-1130.

Cervin, V. B., Bonner, M., Rae, R., & Kozeny, J. The learning of cooperation in some feedback and communication systems. *Czechoslovak Psychology,* 1971, *15,* 219-230.

Cervin, V. B. Boredom on the assembly line. *New Democrat, March,* 1973.

Cervin, V. B., & Lasker, G. E. *Algorithmic model of generalized classical conditioning procedures.* Proceedings of 10th Annual Modeling and Simulation Conference. Pittsburgh, PA., 1979.

Cervin, V. B., & Jain, R. Towards fuzzy algorithms in behaviour modification. In M. M. Gupta and E. Sanchez (Eds.), *Approximate Reasoning in Decision Analysis.* Amsterdam: North-Holland, 1982.

Cervin, V. B. *Rationale and procedures in recovery of retrograde amnesia material.* Unpublished, Toronto, 1987.

Dubois, D., & Prade, H. *Fuzzy sets and systems theory and applications.* New York: Academic, 1980.

Hinton, B. L, & Barrow, J. C. The superiors' reinforcing behaviour as a function of reinforcement received. *Organizational Behaviour and Human Performance*, 1975, *14*, 123-143.

Kolodny, R. C., Masters, W. H., & Johnson, V. E. *Textbook of sexual medicine*. Boston: Little Brown, 1979.

Ladouceur, R., Bouchard, M. A., & Granger, L. *Principes et applications des therapies behaviourales*. Paris: Maloine, 1977.

Luthans, F., & Kreitner, R. *Organizational behaviour modification*. Glenview, Ill.: Scott Foresman, 1975.

Plax, K. A, & Lacks, P. B. Behaviour modification: an alternative to traditional vocational rehabilitation techniques. *Human Resource Management*. 1976, *Spring*, 28-31.

Pritchard, R. D., Leonard, D. W., Von Bergen, C. W., & Kirk, R. J. The effects of varying schedules of reinforcement on human task performance. *Organizational Behaviour and Human Performance*, 1976, *16*, 207-230.

Rathus, S. A., & Nevid, J. S. *Behaviour Therapy*. New York: Doubleday, 1977.

Skinner, B. F. *The technology of teaching*. New York: Appleton Century Crofts, 1968.

Suinn, R. M. Behavioural applications of psychology to US world class competitors. In P. Klavora and J. V. Daniels (Eds.), *Coach, Athlete, and Sport Psychologist*. Ottawa: Canadian Coaching Association, 1979.

Thomas, E. J. (Ed.), *Behavior modification procedure: a sourcebook*. Chicago: Aldine, 1974.

Walters, G. C., & Grusec, J. A. *Punishment*. San Francisco, CA: Freeman, 1977.

Watson, D. L., & Tharp, R. G. *Self-directed behaviour*. Monterey, Ca.: Brooks/Cole, 1972.

Wolf, J., et al. *The art of living and growing old*. Praha: Svoboda Press, 1982.

Wolpe, J., & Lang, P. J. A fear survey schedule for use in behavior therapy. *Behaviour Research and Therapy,* 1964, *2,* 27-30.

Wolpe, J., & Lazarus, A. A. *Behaviour therapy techniques: a guide to the treatment of neuroses.* London: Pergamon, 1966.

Yager, R. R., Satisfaction and fuzzy decision functions. In P. P. Wang and S. R. Chang (Eds.), *Fuzzy Sets.* New York: Plenum, 1980.

Zadeh, L. A. Outline of a new approach to the analysis of complex systems and decision processes. *IEEE Transactions on Systems, Man, and Cybernetics,* 1973, *SMC-3,* 513-519

PRACTICAL APPLICATIONS AND PSYCHOMETRIC EVALUATION OF A COMPUTERISED FUZZY GRAPHIC RATING SCALE

Beryl HESKETH[1], Robert PRYOR[2], Melanie GLEITZMAN[3] and Tim HESKETH[3]

School of Psychology, University of New South Wales, Australia [1];
Central Planning and Research Unit, N.S.W. Department of Industrial Relations and Employment[2];
School of Electrical Engineering and Computer Science, University of New South Wales[3]

Although fuzzy set theory is becoming a well recognized field of mathematical enquiry it has "no well established method for assigning degrees of membership, either on a mathematical or empirical basis" (Smithson, 1987, p. 15). Hisdal (1986a) points out that this has inhibited the development and application of fuzzy sets. The lack of a measurement basis for fuzzy set theory may account in part for its limited application in psychology. By adapting traditional psychological measurement to deal with fuzzy concepts, new possibilities are open to both fields of enquiry. The first part of this chapter outlines the need for fuzzy measurement in many fields of psychology. A Fuzzy Graphic Rating Scale and its validation, is then described, and ways in which the scale can be used in various fields are outlined.

THE NEED FOR FUZZY MEASUREMENT IN PSYCHOLOGY

Fuzzy logic defines concepts and techniques which provide a mathematical method able to deal with thought processes which are too imprecise to be dealt with by classical mathematical techniques (Zadeh, 1973). There are many areas of psychology where traditional mathematical approaches have forced a higher degree of precision than is perhaps warranted.

Individual differences.

A perennial issue in psychological assessment has been the extent to which differences in psychological test scores are a function of genuine individual differences rather than differences imposed (or obscured) by the constraints of the measurement procedures. Typically, normative psychological tests restrict the test-taker's responses both in content and process; in content, by specifying a set of dimensions to be measured; in process, by limiting responses to a single choice along a given continuum. A variety of psychologists (e.g. Bannister & Fransella, 1985; Guilford, 1975; Tyler, 1978; Viney, 1986) have suggested that these restraints may deleteriously affect the assessment of human individuality. The imprecision of human cognition, according to Neisser (1967), results from the fact that although the central meaning of concepts remains comparatively stable and is shared across individuals, the boundaries of these concepts are typically ill-defined or fuzzy.

Polkinghorne(1984), writing for counselling psychologists, describes the problem for psychological measurement in the following way,

"When we counseling psychologists impose a matrix of operationally closed sets over human experience, we often reproduce the 'precision fallacy'. The precision of the measuring instrument is more exact than the precision of the experience itself." (p.425).

Fuzzy set theory provides a possible solution to this methodological problem. It takes account of the reality of imprecision of human thought by allowing ranges of scores to be measured and translated into a single score. Although further development work is needed, it is possible that fuzzy variables will be able to be used in standard statistical analyses in the traditional manner. This means that accepted psychometric standards of validity and reliability can be used to evaluate the potential of this application.

State-trait debate in personality theory.

Within psychology a debate has raged for at least 50 years about whether, or to what extent, it is possible to generalise about behaviour across particular contexts (Bowers, 1973; Ekehammer, 1974; Epstein & O'Brien, 1985; Magnusson & Endler, 1977; Valentine, 1982). "Situationists" such as Mischel (1968) have maintained that, because the correlations between self-report measures and behaviour in specific contexts rarely exceed a modest 0.30, situational rather than personal influences are paramount in accounting for human action.

Epstein and O'Brien (1985) note that Mischel's (1968) book occasioned a 'paradigm crisis' in personality theory and assessment. While the debate continues unabated (e.g. Staw & Ross, 1985), one factor that has not been mentioned is Polkinhorne's (1984) "precisions fallacy." That is, no-one has considered that the apparent contextual instability of many assessed psychological traits in the non-cognitive domain, may be a consequence of imposing upon them a level of precision that they do not, in fact, have. Certainly worth consideration is the possibility that many of these attributes may be stable across contexts but only within particular ranges of values. Current assessment methods would interpret such variability as either measurement error or situational effects. Fuzzified measurements of these ranges of scores may promote a methodological resolution of this longstanding debate, by providing a way to validly and reliably measure the range within which an attribute is stable attribute across different situations.

Theories of career development and choice.

In the field of career development most theoretical approaches presuppose a match between the individual's self-view and the perceived characteristics of occupations (Zytowski & Borgen, 1983; Dawis & Lofquist, 1984). Super (1963), for example, presented a self-concept theory which suggested that satisfaction and work adjustment involve being able to implement one's self-concept in an occupation. Testing such a proposition requires measuring concepts of self and occupational stereotypes, and assessing whether satisfaction relates to the degree of match between the two. Most often this is achieved through the use of standardised tests or questionnaires which, as indicated above, specify both the content and the process (rating procedure) of participant responses. Whether the investigator's specifications correspond closely with the participants' own concepts of self remains a moot point. The repertory grid approach (Bannister & Fransella, 1985) does allow individuality in the content of measurement, but the analysis of grouped data with this technique is difficult (Brook, 1986).

Nurius (1986) suggests that advances in cognitive psychology provide a basis for improved measures of the self-concept, but again it would seem that the suggestions offered are more in terms of the content rather than the process of measurement. The graphic rating scale outlined in this chapter, overcomes these difficulties as it includes individual flexibility in the measurement process itself. Such an approach will greatly facilitate testing vocational theories.

Measurement of attitudes in social psychology

Perhaps the strongest case for the need for fuzzy measurement arises from the arguments put forward by earlier social psychologists in relation to the measurement of attitudes. Attitudes are a central concept in social psychology, and, according to Sherif (1976) they are the "stuff of which the self-system is formed" (p. 239).

Self-concepts referred to above, are in fact attitudes towards oneself. While the issue of self-esteem and other evaluative attitudes applied to the self have traditionally been studied by psychologists in the field of personality theory, Sherif reminds us that "the psychological principles governing the formation of attitudes and their incorporation in the self-system cannot be altogether different merely because some psychologists call themselves social psychologists and others call themselves personality psychologists" (p. 240). Hence comments made about the measurement of attitudes are fundamental to the assessment of individual differences and self-concepts.

Traditional approaches to the measurement of attitudes have involved methods such as the semantic differential (Osgood, Suci, & Tannenbaum (1957). The semantic differential approach to measurement requires an individual to describe a concept (e.g. men) in terms of where it falls between bi-polar adjective descriptions such as the one given below:

Strong --:--:--:--:--:--:--:-- Weak

Also widely used is the Likert scale which requires respondents to indicate the extent to which they agree or disagree with statements. For example, respondents may be asked to respond to a statement: "I want a job which allows me to work at my own pace." using the categories "Strongly Agree, Agree, Indifferent, Disagree, Strongly Disagree".

The adequacy of these and other attitude measures depend upon two important features:

1) the adequacy of the belief statements or bi-polar anchors.

2) how well responses reflect the attitudes of those being questioned.

In all of these approaches, responses to several items in a particular domain are summed to provide a single score for the attitude in question. However, individuals may obtain the same score for very different reasons. One individual may simply respond with the midpoint for all items, while another individual may provide a wide diversity of ratings on either side of the midpoint, but which when summed, result in a similar total. Individual difference information is often lost, in traditional attitude measurement as was highlighted by Thurstone (1927) and Sherif & Sherif (1970).

As early as 1927 Thurstone, in a theoretical discussion on the meaning of attitudes, portrayed individual differences graphically. Figure 1 illustrates how he did this for attitudes in relation to pacificism and militarism.

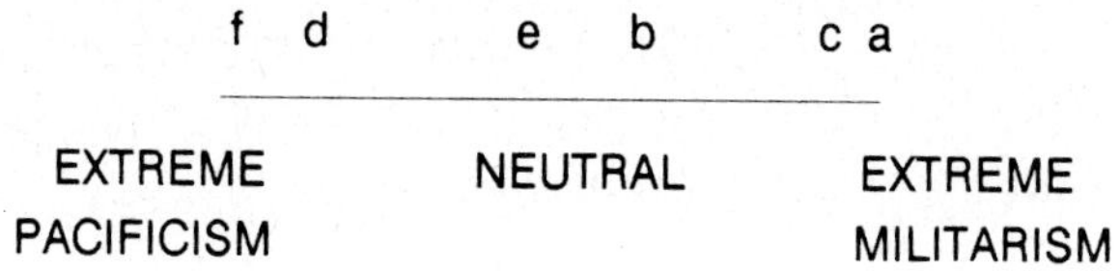

```
        f   d           e    b         c  a
       _____

       EXTREME        NEUTRAL        EXTREME
      PACIFICISM                    MILITARISM
```

FIGURE 1
Attitude variable, militarism - pacificism

One individual (e.g. a) might hold a very militaristic attitude, other individuals (e.g. b and c) might be slightly less extreme. More importantly, from the point of view of our current interest in fuzzy ratings, Thurstone suggested that the range of opinions which a particular person is willing to endorse could also be represented graphically (say from d to e).

Using this graphic representation, Thurstone demonstrated that an individual's opinion could be characterised in terms of three different measures, the range of opinions the individual is willing to endorse, their mean position on the scale, and the one opinion selected which best represented an attitude.

Although these insights into the complex nature of attitudes were identified early in the development of attitude measurement, the subsequent methods used failed to capture this complexity, no doubt because traditional mathematics was not well developed for dealing with ranges and assymetries.

More recently Sherif and Sherif (1970) discussed the notion of latitudes of acceptance and rejection in attitude measurement. They define the latitude of acceptance as the "most acceptable position plus other positions the individual also finds acceptable". The latitude of rejection is defined as "the position most objectionable to the individual, the thing he most detests in a particular domain, plus other positions also objectionable" (p. 300). Conceptually, these ideas demand fuzzy measurement, yet no such developments took place.

The fuzzy graphic rating scale described below offers a direct way of addressing these notions of ranges within the domain of attitude measurement. In particular, the widely used semantic differential (Osgood, Suci, & Tannenbaum, 1957; Heise, 1969; Hesketh & Roche, 1986) has been adapted to provide a fuzzy rating scale which may be represented graphically.

This rating scale allows for assymetries, and overcomes the problem, identified by Smithson (1987, p. 19), of researchers arbitrarily deciding on a most representative value in ranges of scores. This rating scale, which allows respondents to identify a preferred point and to extend the rating in either direction, does not require such assumptions on the part of the researcher. However, it achieves simplicity of presentation to a respondent at the price of an assumption about the shape of the fuzzy function generated (see Hesketh, Pryor, & Hesketh, 1987 for further discussion on this point.)

FUZZY GRAPHIC RATING SCALE.

The fuzzy graphic rating scale presents respondents with the option of indicating a preferred point (indicated by the V pointer in the figure below) and then asks them to extend the rating to the left or right if they wish (indicated by the lines drawn to the left and right of the pointer). Although a fuzzy rating can be elicited using a simple pencil and paper technique, analysis is greatly facilitated through obtaining the rating in a computerised form.

A program FUZRATE, written in Turbo Pascal (Borland International Inc, 1985) for an IBM PC or compatible, elicits interactively a fuzzy graphic rating. After preliminary instructions and tuition in the use of the computerised fuzzy graphic rating scale, the program presents on the screen a modified semantic differential. This is done as follows:

```
        GENERALLY         -V-              GENERALLY
        CONSIDERED |---------------------------------| CONSIDERED
        MEN'S WORK                         WOMEN'S WORK
```

The respondent is first asked to rate the particular concept (e.g. whether they think the job 'psychiatrist' is generally considered men's or women's work) by moving the V pointer using the left and right arrow keys on the key-board (or a mouse if available), and to press the return key when the V pointer reaches the point which best represents their rating. The first phase may result in a rating such as is illustrated below:

```
        GENERALLY      -V-            GENERALLY
        CONSIDERED |----------------------------| CONSIDERED
        MEN'S WORK                     WOMEN'S WORK
```

Next, the respondent is asked to indicate how far to the left the rating could possibly go by extending or removing the left extension using the left and right arrow keys.

```
    GENERALLY      ----V-         GENERALLY
    CONSIDERED |----------------------------| CONSIDERED
    MEN'S WORK                    WOMEN'S WORK
```

And finally, the respondent is asked how far to the right the extension could possibly go.

```
    GENERALLY      ---V--------   GENERALLY
    CONSIDERED |----------------------------| CONSIDERED
    MEN'S WORK                    WOMEN'S WORK
```

The V pointer can be seen to represent a rater's perception of how others generally view a particular occupation, such as psychiatrist. In terms of Hisdal's TEE model, it can be likened to a respondent's best estimate of the probability of the outcome of a poll in which others are required to sex-type the occupation. The fuzzy extensions on either side of the V pointer represent uncertainty inherent in this estimate (Hesketh, Pryor, & Hesketh, 1987).

By making certain simplifying assumptions, not uncommon within Fuzzy Set Theory, this graphic rating can be viewed as a fuzzy variable, hence making possible the use of fuzzy set theoretic operations. The simplifying assumptions are:
1) The global set is represented along the horizontal axis.
2) The fuzzy membership function takes its maximum value (1) at the point on the fuzzy support represented by the V pointer.
3) The extent of the fuzzy support is represented by the horizontal lines to either side of the V pointer.

4) The fuzzy membership function tapers uniformly from its value of 1 at the V pointer to a value of 0 beyond the fuzzy support or the left and right extensions. 5) The union of two or more fuzzy ratings will have a convex membership function. A uniformly tapering membership function may not be appropriate for some applications. Hisdal (1986a) suggests that suitable membership functions might have continuous derivatives (i.e. be smooth).

Union of two fuzzy variables

An example can be used to illustrate the union between a pair of fuzzy variables. The combination of fuzzy variables was performed as follows:

$C^{fuzzy} = A^{fuzzy}$ OR B^{fuzzy} such that

$$f_c(C_i) = \max(\max(f_a(A_i), f_b(B_i)), \min(f_c(C_{i-1}), f_c(C_{i+1}))) \quad (1)$$

Evaluated recursively, this ensures that the resulting membership function is convex. In some applications a concave function may be appropriate but in the data reported in this chapter the convex function was more appropriate (Hesketh, Pryor, & Hesketh, 1987). The way in which two or more fuzzy ratings can be combined via the fuzzy set theoretic Union procedure is illustrated graphically in Figure 2.

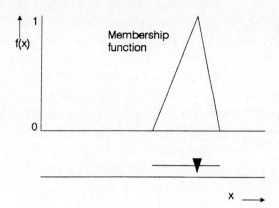

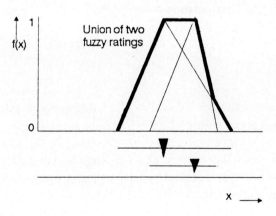

FIGURE 2
Membership function and their union

Expected Value

In calculating an expected value the "average weighting" procedure introduced by Baas and Kwakernaak (1977) was used. Justification for this lies in the link between fuzzy set theory and probability theory (Hisdal, 1986b). To illustrate this let us assume that Fuzzy-Z is a fuzzy rating of some attribute, and that u^{ex} is a "correct" measure of the same attribute elicited under exact conditions of observation. Hisdal (1986b) interprets possibilities as $P(\text{Fuzzy\_Z} | u^{ex})$, (the probability that Fuzzy_Z would be obtained when the exact measure is a priori known to be u^{ex}), and probabilities as $P(u^{ex} | \text{Fuzzy\_Z})$ (the probability that u^{ex} is the exact score given an a priori Fuzzy_Z). The position of the subject's response on the x-axis corresponds to the u^{ex} of the TEE model. As an estimate of u^{ex} is required, given Fuzzy_Z, a reasonable approach is to rescale each ordinate of the rating function (the $f_x(x_i)$ values) so that all the ordinates sum to 1 ($\Sigma f_x = 1$), as required of a probability distribution), and to calculate the expected value of the resulting distribution. This assumes a uniform prior $p(u^{ex})$, a reasonable application of Bayes' postulate. The expected value was calculated as follows:

$$X = \Sigma_U X_i \ (f_x(X_i) / \Sigma_U f_x(X_i)) \tag{2}$$

where $f_x(X_i) / \Sigma_U f_x(X_i)$ is the rescaled ordinate of each point X_i, and the outer summation of all X_i multiplied by their respective probabilities gives the expected value of the distribution. The procedure used to calculate the expected value is illustrated graphically with an example in Figure 3.

The calculations required to form the union of fuzzy ratings, and the expected value are performed within the program FUZRATE which is described more fully in Hesketh, Pryor, Gleitzman, and Hesketh (1987). The graphic scale discussed above allows a fuzzy rating to be elicited from respondents who have no knowledge of the mathematics underpinning fuzzy variables.

EXPECTED VALUE OF A FUZZY VARIABLE

Take a fuzzy variable represented by the following membership function:

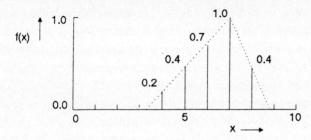

$\sum f(x) = 0.2 + 0.4 + 0.7 + 1.0 + 0.4 = 2.7$

Rescale so that $\sum f_{new}(x) = 1.0$

(Divide each f(x) value by 2.7).

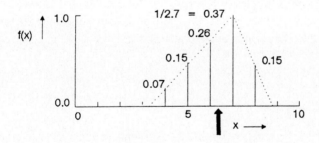

Expected x value = $\sum f_{new}(x) \cdot x$

= 0.07 x 4 + 0.15 x 5 + 0.26 x 6 + 0.36 x 7
  + 0.15 x 8

= 6.38

FIGURE 3
Example illustrating calculation of expected value

EVALUATION OF THE FUZZY GRAPHIC RATING SCALE

Although the fuzzy graphic rating scale can be adapted for use in many areas we chose to validate it in relation to Gottfredson's (1981) map of occupations. We were particularly interested in this application because of the need to develop a measure of social space to test Gottfredson's (1981) theory of career circumscription and compromise. Gottfredson introduces for the first time, a spatial account of "the person-occupation matching paradigm." The theory states that an individual's "social space" (the zone of acceptable occupational alternatives) is successively restricted in terms of sex-type, perceived prestige level of occupations and occupational interests. Compromise in career choice follows a reverse order, with interests compromised most easily, followed by prestige while preference for sex-type of occupations is most resistant to change. The concept of measuring a space fits more easily with a fuzzy rating scale than with traditional approaches to psychological measurement.

Two studies have been undertaken within the context of Gottfredson's occupational map. As these studies provide reliability and validity data for the fuzzy rating scale, they are reviewed briefly below.

Study One

The subjects for the first study were 10 males and 10 females aged 18 to 45 with an average age of 29 years. These respondents used the fuzzy graphic rating scale with five prestige and three sex-type anchors to rate nine occupations. The nine occupations, carefully chosen to represent three levels of prestige based on Daniel's (1983) scale and three levels of sex-type based on Shinar's (1975) scale are given in Table 1. Also given in Table 1 is an example of a sex-type and a prestige anchor used with the fuzzy graphic rating scale described earlier. Although no numbers appeared on the screen, for analysis the responses were converted to the scale [0,99] with each arrow key press representing one point. Ratings can also be elicited with a mouse. The presentation of the eight scales was randomized as were the nine occupational stimuli.

TABLE 1
Occupations and sample anchors used in Study One.

Occupation	Daniel (1983) (Prestige)	Shinar (1975) (Sex-type)
Engineer	2.6	1.9
Motor Mechanic	4.7	1.5
Groundskeeper	5.2	2.0
Psychiatrist	1.9	3.7
Real Estate Agent	4.3	3.4
Cook	4.6	4.2
Primary Teacher	3.8	5.6
Registered Nurse	3.8	6.6
Receptionist	4.6	6.3

Prestige ratings from 1 (high) to 7 (low).
Sex-type ratings from 1 (masculine) to 7 (feminine).

Sample Prestige Anchor:

NOT VERY   VERY
WELL PAID |---------------------------------------| WELL PAID

Sample Sex-type Anchor:

[0.99] scale used for analysis only!

The computerized graphic rating procedure was explained to respondents using a trial occupation not included in the nine stimuli. Various possible manipulations were demonstrated, including ratings with no spread on either side, one with spread only on the one side, a rating with one end hard against the upper or lower end of the scale and other variations. Instructions on the screen also explained to respondents how to use the arrow keys for rating the occupations.

RESULTS

For the purposes of comparison with the a priori ratings obtained by Daniel (1983) for the prestige of the nine occupations and by Shinar (1975) for the sex-type of the occupations, it was necessary to calculate the expected value for individual scales and for various combination of prestige and sex-type scales.

Intercorrelations between the graphic ratings of the nine occupations on the eight scales and the a priori Shinar and Daniel scale values for the same nine occupations were obtained for each of the 20 subjects. These correlations were then averaged across the 20 subjects. Analyses were undertaken using Gauss Version 1.49B (Edlefsen & Jones, 1986). Although Fisher's z-transformation is often used before averaging, Hunter, Schmidt & Jackson (1982, p. 42) have recently noted that the use of this procedure may bias the averaging of correlations, with larger r values, receiving larger z scores. Since the effect of using Fisher's z-transformation on the present data was found to be relatively small, only untransformed correlation averages which probably represent conservative estimates are reported.

Validities for the five prestige scales ranged from -0.74 to -0.88, while the three sex-type scales had validities of 0.86, 0.91 and 0.93. Of particular interest were the low correlations between the prestige ratings of the nine occupations using the Fuzzy Rating scales and Shinar's sex-type ratings (0.02 to 0.25), and also between the sex-type ratings using the fuzzy scales and Daniel's prestige ratings (0.03 to -.1). This provides a measure of construct validity for the scales.

Combining Scales.

The expected values, calculated after various composite combinations of fuzzy ratings, were correlated with the a priori scale values for the nine occupations obtained by Daniel and Shinar, and the correlations averaged across the 20 subjects. These data provided a basis for eliminating from the union any scale(s) which diminished the match between the composite fuzzy ratings and the a priori values for the occupations. Optimum criteria correlations were found by combining three of the fuzzy ratings for prestige ($r = 0.90$) and two of the fuzzy ratings for sex-type ($r = 0.94$).

The results from the first study were encouraging, and provided adequate justification for extending the fuzzy graphic rating scale to obtain measures of preferences on dimensions of prestige and sex-type.

Study Two

Study two aimed to examine both the reliability and the validity of four of the prestige scales and two of the sex-type scales from study one, but applied to the rating of a preferred position on the scale. As the data were collected with a view to testing Gottfredson's (1981) theory, respondents also provided fuzzy ratings of their occupational interests. Participants were 15 males and 15 females aged between 21 and 57 years, with a mean age of 34 years and 9 months. Follow up data were not available for 2 participants. Based on the findings from Study One, four anchors were used to measure prestige (money, education, status and power), and two to measure sex-type (women's and men's work, and jobs women and men choose). As well, six interest scales based on the Holland categories (Realistic, Investigative, Artistic, Social, Enterprising and Conventional) were used with the anchors "extremely unattractive" and "extremely attractive". An example of a prestige, sex-type and interest anchor is given in Figure 4.

In addition to using the fuzzy rating scale with the interest, sex-type and prestige anchors, respondents completed a computerised paired comparisons expercise using the nine occupations from Study One. The Vocational Preference Inventory, (Holland, 1978) was also completed.

The computerised paired comparisons exercise required respondents to rate their preference for the nine occupations in Table 1 presented two at a time. The output from this exercise ranked the occupations in order of preference. The most preferred occupation for each subject was then scored in terms of its a priori prestige score (Daniel, 1983) and sex-type score (Shinar, 1975).

The VPI was scored to provide information on the six interest categories in Holland's theory (Realistic, Investigative, Artistic, Social, Enterprising and Conventional.)

Subjects were interviewed twice. At the first interview they completed the fuzzy ratings, the paired comparisons exercise and the VPI. At the follow-up interview two weeks later respondents again provided fuzzy ratings of the same

FUZZY GRAPHIC RATING SCALE

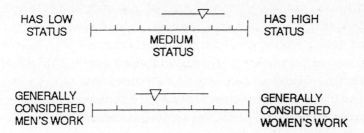

How attractive do you find REALISTIC type occupations which involve manipulating objects, tools, machines or animals?

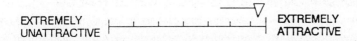

QUESTIONS

1) What type of occupation would you prefer? One which is ....

2) How far would you be willing to move to the left?

3) How far would you be willing to move to the right?

FIGURE 4
Example of prestige, sex-type and interest scales

prestige, sex-type and interest scales. After giving their ratings two forms of feedback were provided. First, the fuzzy rating provided on each individual scale was fed-back to the respondent and they were asked to describe what they meant by the rating they gave. Following this, the union was formed for the four prestige scales and the two sex-type scales, and this was reproduced in the form of a two dimensional map representing Gottfredson's social space concept. An example is given in Figure 4.

The expected value was calculated for combined fuzzy prestige scales and sex-type scales and these were correlated with the Daniel (1983) and Shinar (1975) ratings for prestige and sex-type of the most preferred occupation from the paired comparisons exercise. The correlations were 0.415 for prestige and 0.579 for sex-type, both of which were significant ($p < .01$).

The expected value for the Fuzzy ratings of interests were correlated with the VPI score for each of the six Holland dimensions. The obtained correlations of each Holland graphic rating scale with its corresponding Vocational Preference Inventory scale for the sample were all significant ($p < .05$). The correlations ranged from 0.37 for Social to 0.63 for Investigative. Considering the differences in the context presented to the participants by these two procedures and the differences in scoring the participants' responses, these results suggest that the fuzzy graphic rating scales are quite adequate single item representations of the Holland vocational scales. In addition, the scales provide a measure of flexibility on each of the dimensions, a feature not available with more traditional approaches to measurement.

The test-retest reliability data for the left extension, right extension, and most preferred position, and expected values derived from the fuzzy ratings on individual prestige, sex-type and interest scales are given in Table 2. The expected value reliabilities vary from 0.62 to 0.87.

Test-retest reliabilities for various combinations of the prestige scales yielded the highest reliability when three of the four status anchors were combined (0.78). The two sex-type scales give a high test retest reliability of 0.92. Extensive reliability and validity data from both studies are reported in Hesketh, Pryor, Gleitzman & Hesketh (1987).

TABLE 2
Test-Retest Reliability coefficients for single prestige, sex-type and interest scales.

SCALES	Left Extension	Most Preferred Point	Right Extension	Expected Values
Prestige:				
Pay	.74	.54	.76	.71
Education	.67	.62	.77	.74
Status	.76	.57	.68	.62
Power	.71	.59	.77	.70
Sex-type:				
Choose	.41	.73	.57	.82
Work	.66	.78	.62	.85
Interest:				
Realistic	.77	.72	.77	.77
Investigative	.76	.61	.51	.67
Artistic	.73	.90	.86	.87
Social	.72	.75	.70	.76
Enterprising	.85	.61	.69	.74
Conventional	.63	.66	.53	.74

POTENTIAL APPLICATIONS

The computerised fuzzy graphic rating scale proved to be an unexpected aid to vocational counselling. The program FUZRATE feeds back to participants their fuzzy ratings on individual prestige, sex-type and interest scales, and then presents the results of forming the union for the prestige and sex-type scales. As part of the validation process, respondents in the second study were asked what they meant by their fuzzy ratings. The resultant discussion about the factors influencing preferences allowed the exploration of issues in greater depth than is usually possible without such a graphic aid or stimulus. Pryor, Gleitzman & Hesketh (1987) summarise several examples of the descriptions given by respondents of what they meant by their ratings.

Within the vocational psychology literature there is considerable concern about sex-typing of preferences, and the resistance of these preferences to attempts to encourage both males and females to consider a wider range of occupations. It is important that young people are helped to become aware of the underlying influences on their occupational preferences. Completing a program such as FUZRATE in discussion with a counsellor, helps clients to make more concrete the influences of prestige and sex-type on preferences. Although research has clearly shown these to be two of the most important factors affecting choice (Gottfredson, 1981), they are seldom discussed directly in career education and guidance interventions. Through the use of the Fuzzy Graphic Rating Scale in the FUZRATE program these latent influences can be made more manifest.

The application of fuzzy graphic rating scales to computerized vocational counselling systems more generally needs to be explored. Watts (1986) warns would-be users of computers in careers guidance of the potential abuse of this technology. According to Watts, computerised guidance may foster a "pigeon-hole" approach to clients by conveying a spurious impression of precision. Since the fuzzy rating procedure takes more cognizance of individual differences than traditional self-estimate measures and stresses score ranges rather than single points, it may offer some safeguard against this technological tendency. In fact, any technique that allows counsellors to more accurately and more wholistically understand the complexity of a person's attitudes deserves to be fully investigated and evaluated. Fuzzy set theory and fuzzy graphic rating scales hold just such promise.

Moreover, fuzzy measures may in the longer term allow the assessment of greater human individuality since fuzzy variables applied to attitude measurement can be interpreted as indicating not only intensity of the attribute but tolerable ranges of the attribute. In traditional affective psychological measures tolerable range has not been measured. That such measurement would represent an important breakthrough for psychometrics can be illustrated by reference to two major empirical problems drawn from career development theory.

First, several writers (Gottfredson, 1981; Tversky, 1972; Zytowski, 1965) have noted the disproportionate effect in decision making of "avoiding the negative" in comparison with "obtaining the positive". That is, there are some alternatives and outcomes from decisions that people will not tolerate and which influence decision making before one will consider choosing among desirable outcomes and alternatives. Until the present time the possibility of accurately measuring these tolerable boundaries and being able to use them for theoretical purposes has proved elusive. Similarly, Hilton's (1962) cognitive dissonance model of decision making draws the distinction between "maximising" and "satisficing" as choice outcomes. The crucial difference is whether people choose the very best option or simply an acceptable one. Hilton and probably most vocational development theorists would see "satisficing" as the more realistic conceptualisation. The concept, however implies the notion of tolerable range as in the Gottfredson formulation, which has proved difficult to operationalise. Fuzzy set theory may provide a mathematical basis on which to establish the delineation of individuals' "satisficing" range of occupational alternatives.

Second, a number of recent theories related to career development have incorporated notions of ranges of attributes, for example, Gottfredson's "social space" (a set of occupational alternatives rather than a single choice) and Dawis and Lofquist's (1984) "flexibility" in work adjustment. To date the measurement of these conceptualisations has remained unsatisfactory. So much so, that Gottfredson (1985) for example, observes that this inability to measures ranges is the single biggest drawback to the proper evaluation of her theory. Research is currently underway using the program FUZRATE to test Gottfredson's (1981) theory of career compromise.

It is possible that the use of fuzzy measurement, particularly in relation to determining areas of intolerance, may help to improve the predictive validity of instruments. It is common to find that the hit rate of interest inventories in predicting occupational choice is usually quite low, in the region of 30% to 40% (Holland, 1978). By examining the zones of intolerance in terms of interests and work adjustment (the areas of the scale outside the person's fuzzy rating of liking or preference), it may be possible to develop a new understanding of predictive validity in terms of forecasting the negative.

Further areas where fuzzy measurement may be applicable can be found outside the immediate context of vocational psychology. Firstly, the evaluation of many psychological interventions could be made more accurate if the evaluation measures were fuzzified. For example, large sums of money have been spent in the last two decades in many different countries on programmes aimed at altering sex-role stereotyping in a wide range of contexts. Fuzzified evaluation measures may be able to provide more sensitive indications of the intervention impact by providing not only an absolute rating of attitude but also a rating of response range. This may be extremely important since a programme's effect might not be on participants' most preferred point on a sex-type continuum but rather on the tolerable range of responses. Thus a primary education teaching programme designed to encourage girls to play with nontraditional objects such as trucks, may not alter the girls' preference for dolls as the most desirable toy but may broaden their willingness to spend at least some time playing with nontraditional objects. Without a measure of range this important outcome could be obscured and the programme evaluated too unfavourably.

In addition fuzzy set theory and fuzzy logic through conceptualisation of grades of membership provides a set of mathematical relations which much more closely approximate recent formulations of human thought in terms of paradigms and deviations (Hunt, 1982; Neisser, 1967). Lohaus (1986) has both observed and provided evidence on the crucial effects of experimenters' predecisions on the validity and ultimate interpretability of any empirical findings. In particular he points to the importance of establishing experimental procedures, including variable measures which are congruent with the nature of the phenomena being studied. In terms of human cognition fuzzy measurement may take us one step closer to fulfilling this experimental ideal.

The state-trait debate in personality theory offers yet another application of the fuzzy graphic rating scale. Mood states are known to have a more stable trait component upon which situational variables have their effect. It might be possible to represent traits in terms of ranges with states being reflected in the most preferred position. A similar analysis can be applied to the measurement of attitudes more generally. For example, Staw & Ross (1985) have demonstrated that job satisfaction has a stable component, related more to individual differences than to features of the job itself. While personality theorists can be accused of neglecting situational influences, the job satisfaction literature over the past half century has failed to acknowledge the obvious contribution of more stable affective orientations in individuals. The prospect of measuring stability through ranges with situational influences reflecting more specific responses offers exciting opportunities for resolving the debate. In traditional affective psychological measures, tolerable range has not been measured. Early social psychologists recognised the importance of this concept in attitude measurement, but the empirical measures capable of supporting the idea were not available. The Fuzzy Rating Scale offers a way of applying earlier ideas about attitude measurement.

Finally, Hisdal (1986b) has highlighted the need to explore further which factors contribute to fuzziness. For example, fuzziness may be inherent in a particular situation, it may arise due to perceptions of an observer, or it may be a function of definition or measurement. In practice, it remains difficult to determine how much of the fuzziness is due to each of these sources, although psychometric reliability analyses may help determine some contributions to fuzziness. Test-retest reliabilities over a two week period, reported in this chapter, were in the order of .61 to .87 for individual scales which shows that a proportion of the fuzziness is due to lack of stability over such a period. Further research examining reliabilities over varying intervals and with variations of the scale presentation is needed to explore the different contributions to lack of reliability in the fuzzy graphic rating scale. Such analysis may help to throw light on sources of fuzziness.

The procedure of asking respondents what they understood by their fuzzy ratings is another important way of increasing our understanding of fuzziness. Respondents found it easy to describe the components that went into making a fuzzy rating after having provided the rating. The graphic representation appears to act as a stimulus to respondents to recreate their thought processes while providing the rating. As such, the scale provides an opportunity for intensive exploration of meanings of fuzziness in a wider range of applications.

Future research is needed to explore the best way in which to present the scale graphically. The two studies reported briefly in this chapter did not use numbers under the graphic scale, but these could easily be included. Other issues which have not yet been addressed relate to the dimensionality of anchors. For example, should "masculine" and "feminine" concepts be opposite ends of the same scale, or do they require two separate scales? Answers to this question would be of considerable theoretical interest in the androgyny literature.

Although the idea of using extensions to a semantic differential rating scale is quite simple, without computerised data collection, analysis is extremely tedious. The combination of computerised data collection and fuzzy set theory offers an opportunity for new and creative approaches to measurement of individual differences.

REFERENCES

Baas, S. M., & Kwakernaak, H. Rating and ranking multiple-aspect alternatives using fuzzy sets. *Automatica*, 1977, *13*, 47-58.

Bannister, D., & Fransella, F. *Inquiring Man*. Beckenham: Croom Helm, 1985.

Borland International Inc. *Turbo Pascal*. Scotts Valley, CA.: Borland, 1985.

Bowers, K. Situationism in psychology: An analysis and critique. *Psychological Review*, 1973, *80*, 307-336.

Brook, J. Research applications of the repertory grid technique. *International Review of Applied Psychology*, 1986, *35*, 489-500.

Daniel, A. *Power, privilege and prestige: Occupations in Australia*. Cheshier: Longman, 1983.

Dawis, R. V., & Lofquist, L. H. *A psychological theory of work adjustment: An individual differences model and its applications*. Minneapolis: University of Minnesota Press, 1984.

Edlefsen, L. E., & Jones, S. D. *Gauss Programming Language Manual: Software Version 1.49B*. Kent, W.A.: Aptech Systems, 1986.

Ekehammer, B. Interactionism in psersonality from a historical perspective. *Psychological Bulletin*, 1974, *81*, 1026-1048.

Epstein, S., & O'Brien, E.J. The person-situation debate in historical and current perspective. *Psychological Bulletin*, 1985, *98*, 513-537.

Gottfredson, L. S. Circumscription and Compromise: A developmental theory of occupational aspirations. *Journal of Counseling Psychology. Monograph*, 1981, *28*, 545-579.

Gottfredson, L. Role of self concept in vocational theory. *Journal of Counseling Psychology*, 1985, *32*, 159-162.

Guilford, J. P. Factors and factors of personality. *Psychological Bulletin*, 1975, *82*, 802-814.

Heise, D. R. The semantic differential and attitude research. In G. F. Summers (Ed.), *Attitude Measurement*. Chicago: Rand McNally, 1969.

Hesketh, B., & Roche, S. Research into classification procedures at a youth prison. *Australia and New Zealand Journal of Criminology*, 1986, *19*, 42-52.

Hesketh, B., Pryor, R., Gleitzman, M. & Hesketh, T. *A New approach to measuring individual differences: Manual for a Computerized Fuzzy Graphic Rating Scale*. Sydney: N.S.W. Department of Industrial Relations & Employment, 1987.

Hesketh, T., Pryor, R. & Hesketh, B. An application of a computerised Fuzzy Graphic Rating Scale to the psychological measurement of individual differences. *International Journal of Man-Machine Studies* (to appear)

Hilton, T. L. Career decision-making. *Journal of Counseling Psychology*, 1962, *9*, 291-298.

Hisdal, E. Infinite-valued logic based on two-valued logic and probability. Part 1.1. Difficulties with present-day fuzzy-set theory and their resolution in the TEE model. *International Journal of Man-Machine Studies,* 1986, *25,* 89-111. (a)

Hisdal, E. Infinite-valued logic based on two-valued logic and probability. Part 1.2. Different sources of fuzziness. *International Journal of Man-Machine Studies,* 1986, *25,* 113-138. (b)

Holland, J. *Manual for the VPI.* Consulting Psychologist Press, 1978.

Hunt, M. *The universe within: A new science explores the human mind.* Brighton: Harvester Press, 1982.

Hunter, J. E., Schmidt, F. L., & Jackson, G. B. *Meta-analysis: cumulating research findings across studies.* Beverly Hills, CA.: Sage, 1982.

Lohaus, A. Standardizations in methods of data collection: Adverse effects on reliability. *Australian Psychologist,* 1986, *21,* 241-252.

Magnusson, D., & Endler, N. Interactional psychology: Present status and future prospects. In D. Magnusson & N. Endler (Eds.), *Personality at the crossroads: Current issues in interactional psychology.* Chicago: Rand McNally, 1977.

Mischel, W. *Personality and assessment.* N.Y.: Wiley, 1968.

Neisser, U. *Cognitive psychology.* New York: Appleton-Century-Crofts, 1967.

Nurius, P. S. Reappraisal of the self-concept and implications for counseling. *Journal of Counseling Psychology,* 1986, *33,* 429-438.

Osgood, C. E., Suci, G. J., & Tannenbaum, P. J. *The measurement of meaning.* Urbana, Ill.: University of Illinois Press, 1957.

Polkinghorne, D. E. Further extensions of methodological diversity for counseling psychology. *Journal of Counseling Psychology,* 1984, *31,* 416-429.

Sherif, M., & Sherif, C. W. Attitudes as the individual's own categories: the social judgement-involvement approach to attitude and attitude change. In G. F. Summers (Ed.), *Attitude Measurement.* Chicago: Rand McNally, 1970.

Sherif, C. W. *Orientation in social psychology.* N.Y.: Harper & Row, 1976.

Shinar, E. H. Sexual stereotypes of occupations. *Journal of Vocational Behavior,* 1975, *7,* 99-111.

Smithson, M. *Fuzzy set analysis for behavioural and social sciences.* New York: Springer Verlag, 1987.

Super, D. E. Self concepts in career development. In D. E. Super, R.Starishevsky, N. Matlin, & J. P. Jordaan (Eds.), *Career Development: Self Concept Theory.* New York: College Entrance Examination Board, 1963.

Staw, B. M., & Ross, J. Stability in the midst of change: A dispositional approach to job attitudes. *Journal of Applied Psychology,* 1985, *70,* 469-480.

Thurstone, L. L. Attitudes can be measured. *American Journal of Sociology,* 1922, *33,* 529-554. Also In G. F. Summers (Ed.), *Attitude Measurement.* Chicago: Rand McNally, 1970.

Tversky, A. Elimination by aspects: A theory of choice. *Psychological Review,* 1972, *79,* 281-299.

Tyler, L. E. *Individuality: Human possibilities and personal choice in psychological development of men and women.* San Francisco: Jossey-Bass, 1978.

Valentine, E. R. *Conceptual issues in psychology.* London: Allen and Unwin, 1982.

Viney, L. Editorial. *Australian Psychologist,* 1986, *21,* 193-194.

Watts, A. G. The role of the computer in careers guidance. *International Journal for the Advancement of Counselling,* 1986, *9,* 145-158.

Zadeh, L. A. Outline of a new approach to the analysis of complex systems and decision processes. *IEEE Transactions on Systems, Man, and Cybernetics,* 1973, *SMC-3,* 28-44.

Zytowski, D. G. Characteristics of male university students with weak occupational similarity on the Strong Vocational Interest Blank. *Journal of Counseling Psychology,* 1965, *15,* 182-185.

Zytowski, D. G., & Borgen, F. H. Assessment. In W. B. Walsh & S. H. Osipow (Eds.), *Handbook of Vocational Psychology* (Vol.2) Hillsdale, N.J.: Erlbaum, 1983.

AUTHOR INDEX

A

Abelson, R.P., 266,292
Adam, E.E., 384,421
Ajdukiewicz, K., 280,293
Anderson, J.A., 205,231
Anderson, J.R., 61,62,69
Anderson, N.H., 99,100,121,122
Armstrong, S.L., 159,197
Arrow, K.J., 267,293
Arzumanov, Y., 302,316
Asmonov, N.M., 346,380
Aunon, J.I., 300,315
Axelrod, S., 92,123

B

Baas, S.M., 436,450
Ballard, D.H., 163,197,205,231
Bandura, A.L., 384,385,388,421
Bannister, D., 426,428,450
Bar-Hillel, Y., 163,197
Barr, A., 315
Barrow, J.C., 384,422
Barsi, F., 180,197
Barwise, J., 81,87
Bass, B.M., 58,69
Bauer, F.L., 266, 293
Begg, I., 55,64,69
Begleiter, B., 300,315
Bellman, R.E., 95,122,241,246, 260,365,380
Benesova, E., 298,319
Bergen, Von C.W., 384,422
Bergman, G., 346,380
Bertozzi, P.E., 302,318

Beyth-Marom, R., 91,92,122
Bienkowski, M., 128,153
Blin, J.M., 250,251,255,260
Bloom, P., 129,152
Bonner, M., 384,421
Borg, I., 266,293,294
Borgen, F.H., 428,454
Borges, M.A., 59,69
Bouchard, M.A., 384,422
Bowers, K., 427,428,450
Braune, R., 322,342
Breland, K., 384,421
Brook, J., 427,450
Brown, W., 302,313,315,316
Brownell, H.H., 97,123
Bruner, J.S., 266,293
Budescu, D.V., 17,49,61,62,67,69, 72,81,88,92,122,124,125

C

Caramazza, A., 97,98,120,123, 368,380
Carlstein, E., 92,125
Carroll, K.K., 7,47
Cascio, W.F., 58,69
Cautela, J.R., 388,421
Cervin, V.B., 384,385,388,406,421
Chapman, R.H., 55,66
Chase, C.I., 60,65,69
Cheeseman, P., 5,47
Chen, C.H., 171,197
Chesney, G.L., 302,316
Chlewinski, Z., 267,293
Cholewa, W., 92,122

Chon, L.C., 300,315
Chwedorowicz, J., 267,268,271, 278,293
Civanlar, M.R., 18,47
Clarke, D.D., 5,47
Cohen, B., 156,157,159,160,190, 195,197
Cohen, J., 26,47
Cohen, L.J., 6,47,191,192,195,197
Coleman, J.S., 4,5,47
Collingridge, D., 41,47
Collins, R., 6,47
Collis, J., 58,59,60,64,71
Coombs, C.H., 266,293,331,342
Cooper, R., 81,88
Courchesne, E., 300,316
Cox, J.A., 58,60,65,72,92,98,124
Curry, S.H., 300,317
Czogala, E., 93,95,122

D

Daniel, A., 438,439,440,442,444,450
Dawes, R.M., 331,342
Dawis, R.V., 428,447,450
Debrue, G., 331,342
Deng, D., 219,232
Dickstein, L.S., 53,69
Donchin, E., 300,316
Douglas, M., 44,47
Dubois, D., 78,88,92,93,95,97,120, 122,191,192,197,237,260,412,421
Dyckhoff, H., 96,97,122

E

Ebanks, B.R., 183,197
Edlefsen, L.E., 440,451
Eisen, P., 322,343

Ekberg, P.H.S., 62,70
Ekehammer, B., 427,451
Elster, J., 9,10,13,47
Endler, N., 427,452
Englander, T., 62,70
Epstein, S., 427,451
Ercegovac, D.V., 300,316
Evans, J.St.B.T., 52,70

F

Fansella, F., 426,428,450
Farkas, A., 62,70
Farris, D.R., 360,380
Fedrizzi, M., 259,261,260
Feigenbaun, E.A., 315
Feldman, J.A., 163,197,205,231
Ferrell, W.R., 370,374,381
Fillenbaum, S., 58,60,65,72,
Fischler, I., 129,152
Fisher, D.L., 53,70
Fodor. J.A., 157,193,196,197
Ford, J.M., 300,301,316
Forster, K., 128,129,131,152
Forsyth, B., 17,49,62,72,81,88, 92,124
Fraenkel, A.A., 163,197
Francozo, E., 298,303,318,
Frank, M.J., 101,123
Franke, C.H., 271,289,295
Fraser, L., 45,48
Friedman, D., 300,317
Friedman, M., 211,231
Fu, K.S., 175,184,197
Fuhrmann, Gy., 160,166,170,176, 178,180,184,187,192,193,197,198

G

Gabrielian, A., 172,196,198
Gackowski, Z., 266,293
Gaines, B.R., 37,48
Galanter, E., 4,49
George, F.H., 346,380
Giertz, M., 365,380
Ginsburg, G.P., 4,48
Giorno, F., 298,318
Glass, A.L., 61,62,70
Gleitman, H., 159,167,189,192,196, 197,198
Gleitman, L.R., 159,167,189,192, 196,197,198
Gleitzman, M., 436,444,445,451
Glucksberg, S., 160,200
Goguen, J.A., 160,198,368,380
Gold, C., 129,154
Goocher, B.E., 58,65,70
Goodglass, H., 302,317
Goos, G., 266,293
Gopher, D., 322,342
Gottfredson, L.S., 438,445
Granger, L., 384,422
Greco, G., 298,299,305,306,307, 314,316,318
Grice, H.P., 55,65
Griggs, R.A., 55,56,60,64,65,70,71
Grusec, J.A., 384,422
Guilford, J.P., 426,451
Guttman, L., 266,293

H

Hacking, I., 10,45,48
Hajicova, E., 298,319
Hall, E.W., 346,380
Hall, L.O., 207,231
Halmos, P., 160,162,183,198
Hampton, J.A., 170,195,199
Hansen, J.C., 301,316
Harris, G., 55,64,69
Hebb, D.O., 157,175,176,193,196, 199,208,231
Heckerman, D.E., 30,48
Heise, D.R., 431,451
Hempolinski, M., 267,293
Herning, R., 300,319
Hersh, H.M., 97,98,120,123,368,380
Hesketh, B., 431,433,434,436,444, 451,445
Hesketh, T., 433,434,436,444
Hickey, T., 25,26,47
Hildebrand, D.K., 33,36,48
Hillsdal, E., 97,123
Hillyard, S.A., 301-303,317,316
Hilton, T.L., 448,451
Hinton, B.L., 384,422
Hisdal, E., 425,436,449,452
Holland, J., 442,447,452
Holyoak, K.J., 61,62,70
Hörmann, H., 59,63,65,70,84,85,88
Horvath, M.J., 370,374,381
Horvitz, E.J., 30,48
Huchings, D.W., 349,381
Hunt, M., 448,452
Hunter, J.E., 440,452

I

Iversen, L.L., 177,199
Iwanski, C., 249,260

J

Jackson, G.B., 440,452
Jain, R., 388,406,421

Johnson, V.E., 384,422
Johnson-Laird, P.N., 156,160,189, 190,195,199,200
Johnston, V.S., 302,316
Jones, S.D., 440,451
Josiassen, R.C., 300,301,316
Just, M., 61,70

K

Kacprzyk, J., 235-242,246,248,249, 251,258-262
Kahneman, D., 41,49
Kaiser, C.J., 350,351,381
Kandel, A., 158,199,208,211,231
Kapell, B., 300,316
Karczmar, A.G., 176,199
Kass, C.E., 368-370,374,381
Kastenbaum, R., 388,421
Kaufmann, A., 164,199,283,293, 365,381
Keele, S.W., 167,200
Kelly, G.A., 10,48
Kim, J., 19,48
King, B., 322,343
Kintsch, W., 268,293
Kirk, R.J., 384,422
Kirk, S.D., 361,381
Kirk, W.D., 361,381
Kline, M., 74,88
Knibb, K., 37,49
Koch, C., 177,199
Kohonen, T., 206-208,221,231
Kolodny, R.C., 384,388,422
Kopell, B.S., 302,318
Körndle, H., 82,85,88
Korner, S., 365,381
Kostandov, E., 302,316

Kozeny, J., 384,421
Kozielecki, J., 266,294
Krantz, D.H., 98,123,327,331,342
Kreitner, R., 384,422
Kruskal, J.B., 266,294
Kulka, J., 120,123
Kullback, S., 37,48
Kutas, M., 291,301,302,303,317
Kwakernaak, H., 436,450

L

Lacks, P.B., 384,388,422
Ladouceur, R., 384,385,388,422
Laing, J.D., 33,48
Lang, P.J., 388,423
Langlotz, C.P., 30,48
Lasker, G.E., 388,421
Layder, D., 42,48
Lazarus, A., 384,388,423
Leao, B., 298,318
Lee, E.T., 177,199,206,231,232
Lee, S.C., 177,199,206,231,232
Lees, R.B., 192,196,199
Lehrer, A., 268,294
Leiman, J.M., 128,153
Leonard, D.W., 384,422
Lerner, J.W., 358,381
Levy, A., 163,197
Lewis, E.R., 177,199
Lichtenstein, S., 353,381
Lindsay, P.H., 266,294
Lingoes, J.C., 266,293
Lloyd, B.B., 167,201
Lofquist, L.H., 428,447,450,
Loftus, E.F., 168,199
Lohaus, A., 448,452
Lombroso, C.T., 302,317

Loo, S.G., 183,199
Lopes, L.L., 62,70
Lorge, I., 51,72
Luce, R.D., 98,123,327,342
Luchandjula, M.K., 96,123
Lukes, S., 10,42,48
Luria, A.R. 303,304,313,317
Luthans, F., 384,422

M

Macht, M., 301,317
Maestrini, P., 180,197
Magnusson, D., 427,452
Makridakis, S., 92,125
March, J.G., 41,48,360,381
Marciszewski, W., 267,269,278, 279,294
Mars, G., 44,48
Marsh, J.T., 302,313,315,316
Marslen-Wilson, W., 128,145,152, 302,317
Massaro, D.W., 133,146,152,153
Masters, W.H., 384,422
Matsumyia, Y., 302,317
Maurer, H.A., 172,196,200
McCallum, W.C., 300-302,316,317
McCarthy, G., 300,317,319
McClelland, J.L, 128,152,204, 208,232
McCloskey, M.E., 160,200
Medin, D.L., 159,195,200
Meinong, A., 267,294
Menger, K., 93,123
Mervis, C.B., 158,159,167,168,171, 189-191,195
Miller, G.A., 4,49,156,189,200
Miller, R.S., 167,201

Mischel, W., 433,452,
Molfese, D.L., 300,317
Money, L., 322,343
Moray, N., 322,342,343
Morton, J., 128,129,145,152
Mountcastle, V.B. 157,176,193, 196,200
Muller, E.N., 7,49
Mumford, L., 346,381
Munson, R., 300,319
Murhy, G.L., 156,157,159,160,190, 195,197

N

Nakao, M.A., 92,123
Neimark, E.D., 55,71
Neisser, U., 426,448,452
Nevid, J.S., 384,388,422
Neville, H.J., 302,313,317
Newell, A., 266,294
Newstead, S.E., 55,56,58-60,64,65, 67,71
Niez, J., 206,232
Norman, D., 266,294
Norwich, A.M., 98,123,330,343
Nosal, C., 266,294
Novak, V., 120,123
Novick, M.R., 351,382
Nowakowska, M., 266,294,365,382
Nurius, P.S., 432,452
Nurmi, H., 250-252,256,260

O

O'Brien, E.J., 427,451
O'Connor, E.J., 58,69

Oden, G.C., 98,99,120,124,132, 133,141,146,147,149,152,153
Oleron, P., 281,304,317
Olsen, J.P., 41,48
Olson, D.R., 298,304,317
Ore, O., 177,200
Osgood, C.E., 429,431,452
Osherson, D.N., 156-160,168, 188-192,195

P

Pap, A., 278-280,294
Papakostopoulos, D., 300,317
Pedrycz, W., 96,97,122
Pepper, S., 58,65,67,71,98,124
Peretto, P., 206,232
Peterson, P., 81,88
Pfefferbaum, A., 300,316
Piaget, J., 175,200
Plax, K.A., 384,388,423
Pocock, P.V., 300,317
Poggio, T., 177,199
Pohl, N.F., 58,71
Polkinghorne, D.E., 426,427,452
Pollard, P., 60,71
Porjesz, B., 300,315
Posner, M.I., 167,200
Prade, H., 78,88,92,93,95,97,120, 122,191,192,197,237,260, 412,421
Pribram, K.L., 4,49
Pritchard, R.D., 384,422
Pryor, R., 431,433,434,436,444, 445,451
Prytulak, L.S., 58,65,71

R

Rae, R., 384,421
Ramsey, E.P., 278,294
Rapaport, A., 17,49,62,72,81,88,92, 98,101,124
Rasmussen, J., 322,343
Rathus, S.A., 384,421
Reed, S.K., 167,191,200,201
Reschner, N., 161,201
Riezebos, D., 60,71
Ritter, W., 300-302,314,316,317,318
Roberts, F.S., 266,289,295
Rocha, A.F., 298,299,301-308, 314-316,318
Rocha, M.T., 298,301-304,307,308, 314,316,318
Roche, S., 431,451
Roemer, R.A., 316,400
Rosch, E., 159,167,168,171,190,195
Rosenfield, L.C., 346,382
Rosenthal, H., 33,48
Ross, J., 427,448,453
Roth, E.M., 159,189,191,195,201
Roth, W.T., 302,318
Rubin, D.C., 97,124
Ruchkin, D.A., 291,319
Rueckl, J., 132,141,149,153
Rumelhart, D.E., 128,145,152,153, 204,208,232

S

Sage, A.P., 360,380
Salomaa, A., 172,174,196,200,201
Santos, M.R., 298,319
Sawyers, B.K., 59,69
Schmidt, A., 302,317
Schmidt, F.L., 441,452

Schneider, M., 211,231
Schriesheim, C., 58,71
Schwartz, B., 41,49
Seidenberg, M.S., 128,153
Sgall, E.P., 298,304,319
Shackle, G.L.S., 30,49
Shafer, G., 6,49,267,280,295
Shagass, C., 300,316
Shaw, M.L.G., 37,48
Sherif, C.W., 430,431,452
Sherif, M., 429-431,452
Shinar, E.H., 438-440,442,444,452
Shortliffe, E.H., 315,319
Simon, H.A., 266,294,360,381
Simpson, C., 161,201
Simpson, G.G., 167,201
Simson, R., 300,301,317,318
Skinner, B.F., 46,49,384,422
Smith, E.E., 156-160,168, 188-192,195
Smith, J.C., 302,313,315,316
Smith, N.V., 76,88
Smithson, M., 12,17,30,32,36-38,49, 78,88,93,95,96,124,425,431,453
Sperry, R.W., 176,201
Squires, K.C., 300,301,319
Stachowski, R., 266,295
Stanovich, K.E., 128-130,153
Staw, B.M., 427,448,453
Stevens, S.S., 58,71
Stiles, G.S., 219,232
Stone, G., 219,221,232
Straumai, J.J., 300,316
Suci, G.J., 429,431,452
Suinn, R.M., 384,422
Super, D.E., 428,453
Suppes, P., 98,123,279,295,327,342

Sutton, S., 300,319
Szabo, N., 180,201
Szentágothai, J., 157,176,193,196, 201,202

T

Tagliasco, V., 302,317
Tanaka, R., 180,201
Tannenbaum, P.J., 429,431,452
Tannenhaus, M.K., 128,153
Taylor, S.E., 43,49
Teasuro, G., 223,232
Tharp, R.G., 384,385
Theil, H., 37,49
Theoto, M.T., 298,299,308,314,319
Thöhle, U., 98,99,120,124
Thomas, E.J., 387,388,422
Thorndike, E.L., 51,72
Thornton, C., 323,343
Thurstone, L.L., 430,453
Torre, V., 177,199
Toulmin, S.E., 74,76,88
Trussell, H.J., 18,47
Tukey, J.W., 266,295
Tulving, E., 129,154
Turksen, I., 98,123,322,330,343
Tversky, A., 41,49,98,123,327,331, 342,447,453
Tyler, L.E., 426,452

U

Uchiyama, N., 298,319

V

Valentine, E.R., 427,453
Vaughan, H.G., 300,301,317,318
Viney, L., 426,453

Vizi, E.S., 177,202

W

Waller, R.J., 361,362,364,382
Wallsten, T.S., 17,49,58,60-65,67,
 69,72,81,82,85,88,89,92,98,
 101,122,124
Walters, G.C., 384,422
Warfield, J.N., 360,362,382
Warner, S.A., 56,70
Washabaugh, W., 298,304,319
Waterton, K., 322,343
Watson, D.L., 384,385,422
Watts, A.G., 445,453
Weber, S., 93,125
Welsh, A., 128,145,152
West, R.F., 128,129,153
Whinston, A.P., 250,251,255,260
Wiener, N., 266,295
Wierwille, W., 322,343
Wilkinson, A.C., 149,154
Wilkinson, S.J., 10,50
Williams, R.J., 206,209,232
Williges, R., 322,343
Winkler, R.L., 92,125
Witwicki, W., 267,295
Wolf, J., 384,422
Wolpe, J., 384,388,423
Wood, D., 172,196,200

Y

Yager, R.R., 81,84,88,92,125,236,
 239-242,246,260,263,315,319,
 411,412,423

Z

Zadeh, L.A., 6,9,10,11,14,30,50,77,
 78,81,82,84,88,93,95,122,125,
 156-159,202,206,232,238,241,
 246,260,263,268,271,282,289,
 295,345,348,349,365-367,382,
 412,413,415,423,425,454
Zimmer, A., 74,78,81,82,84,88,89
Zimmermann, H.-J., 93,95,96,
 98-100,120,122,124,125,237,
 241,263
Ziolkowski, A., 259,262
Zwick, R., 17,49,50,62,72,81,88,89,
 92,98,124,125
Zysno, P., 96,98,99,100,120,124,
 322,343
Zytowski, D.G., 428,447,454

SUBJECT INDEX

A

ability 2,205,222
acceptance 270-272,278,284,288, 290,291
activation 208,209
aesthetic 193
agency 1,6
aggregartion 95,215,313,315
algorithm 206,227,235,367
altruism 393
analogue theory 61,62
animal 7,8
animal training 384
ANOVA 335-337
anxiety 386,392
approximate 78
Archimedian 330,332
Aristotelian 159,162,190,192
arithmetic 79,87,376,377
arousal 393
artificial intelligence 266
association 5,34,35,
    auditory 361
    visual 361
associativity 94
assumption 330
asynchronous 223
attention 301,324,347,375
attitude 58,429
auto-associative 216,223,226
automatic 350
autonomous processing framework
    128-150
aversive 385
Ayres Tactile Discrimination Test 376

B

back-propagation 228
Bacon 349,350
baseline 131
BASIC 378
Bayesian 75,83,350
belief 265-292,297-315,430
Benton Finger Agnosia Test 376
Bernoulli 10
BESTFIT 149
binary operator 92-94
biological 224
bipolar 429
BM (behavoiur management)
    383-420
Boolean 92,207
Borel 10
Bose-Einstein statistics 20
bottom-up 128,129
brain 175,204-206,228,304,309, 313,346,348
brain damage 205

C

cancer 7
career 438
Cartesian 162,346
category 155,157
category membership 155-196
cell 178-181,204,205,223
cell-assemblies 157,196,176
censorship 276
central area 309-311,313

cerebral hemisphere 297
Chi-square 35
choosing 391,392,407
classical conditioning 223
classical statistics 351
classifying 391,392,407
client 388
closure
    auditory 361
    grammatic 361
    visual 361
co-norm 94,120
co-t-norm 94,96,100,108,111,120
cognitive 204
    function 206
    paradigm 265
    psychology 127
    science 156,158,159,177,
        189,193,196
communication 178,224,234,
    281,372
comparing 391,392
complexity 207
comprehension 53,128,377
computation 205
computer 51,83,102,105,107,175,
    224,289,306,346,349,350,364,
    367,378
    science 156,204-206,251
concatenation 78,79
concept 155-196
    core 156
conditional proposition 73
confidence 207,297,299,300
congruent 131,132
conjunction 110,275
connectedness 328,330,331

connection machine 206,223
conscious 324
consciousness 346,347
constraint 17,19,44,248
context effect 130-150
continuous scale 69
control 323,324
conversion error 53,55,56
correlation 297,299,300,309,313
cortex 228
cybernetics 346

D
DDS (decision support system)
    234,235
de dicto 10,11,41,43
de re 10,11,41,43,44,
deaf 302
decision 2,4,6,42,43,53,128
    making 233-259
    multiobjective 235-259
deduction 75
default reasoning 75
delta rule 221
density 222
Descartes 347
desensitization 388
deterministic 280,281,330
diet 375
discriminant analysis 2
disjunction 110
dissonance 276
distance 282,288,289
dysfunction 358
dyssymbolia 370

E

EEG 298,302,303,305-307,314
elicitation 385,390,393,395,415
elimination 408
emotion 64
empiricism 346
engineering psychology 384
epistemic 274
ergonomics 325
error correction 180
    detection 180
Euler diagram 53,55
event related 300,302
excessive crying 378
excluded middle 75,157,275
exogenous 300
expectancy 304
expected frequency 58,65
    value 436,437
expert system 81,82,314
extinction 398

F

factor analysis 2
featural theory 61,62
feminine 439,448
FEV (fuzzy expected value) 203, 210,211-215,217,218,228
FG (group freedom) 21-32
filler sentence 135-146
finite 155,181
    automata 206
forecats 91
formal language 174
FORTRAN 378
frame 298
freedom 17,22,29,41,42,46
Frege's axiom 161
frontal area 303,309-311,313
FS (finite subset) 30
functional layer 185
FUZRATE 432,436,445-447
fuzzy grafic rating scale 425-450

G

game theory 158
gender 44
Gestalt psychology 269
gestures 372
glb (greatest lower bound) 18
Goodman & Kruskal lambda 35
goodness-of-fit 101
graceful degradation 225
Grelling's antinomy 191,195

H

Hamming distance 216,219,221,226,272,283,411
handicap 369,377
haptic 372,376
hearing 302
hedges 366
hemisphere 306,308-314
hesitation 306
hetero-associative 216,226
homomorphism 274,289
Hovercraft Simulator 323
humanistic 345,346,349,367
hypercube 205,223
hypnotic 384
hypothesis 266

I

ill-defined 69,156

imagery 377,384
immaturity 377
implementation 175
implication 275,276
inclusion 35,36,40,193
incomplete representation 53
incongruent 130,132
individual differences 430
   lexicon 82
induction 75
inference 83
information integration 100
   processing 266
informationally encapsulated 194
inhibition 407
inhibitory 130
input 216,222,225
instrumental responses 393,394
intelligence 204,371,374
interactive processing framework
   128-150
interest scale 442,445
interpresonal 377
interpretation error 53
intersection 96,100,102,104-109,
   111-113,116,336
   bold 337
   drastic 337

J
judgment 17,43,265-292
just noticable difference 332,333

K
Kendal tau 111
knowledge 5,219
knowledge based 321-342

L
LaMettrie 346
Laplace 10
learning 157,196,228,229,267,387
learning disability (LD) 345-378
Leibniz 10
lethargy 378
linguistic scale 270
listening 307-309
log-linear 2

M
m-fuzzy grammar 155-196
   set 155-196
masculine 439,448
massively parallel 205,206
matrix 208,211
max operator 341
McCulloch-Pitts model 177
measurement theory 271
mechanistic physiology 346
membership 63,64,92,97,98,101
memory 53,224,226,372,387,397
mental 236,266
   image 3
   patient 393
   representation 183,196
   workload 321-342
message 266,267
metabelief 269
militarism 430
min-max operator 17,85,98,100,
   101,341
mind/brain 157
Misses von- 10,75
modal logic 10
modular organization 196

monistic 176,196
monotonic 97
monotonicity 94,100,110-113,321,
   330,333,334
Monte Carlo algorithm 25
motivation 387,393
motor movement 375
multidimensional 370

N
natural language 84,234,273
necessity 11
network 299
Neumann machine 204
neural model 205
   network 155,203,204,207,
   227,228
neurological disorder 372,373
neuron 175,177,178,181,204
module 157,176,179,180,183,184
neuronal architecture 194
neurophysiological 204
neurophysiology 206
neuroscience 204
node 299
non-linear 209,210
non-monotonic reasoning 75
non-numerical 102,104
non-orthogonal 219

O
observing 391,392,407
occupation 438-440
opinion 193,352,431
option 33-35
orthogonal 183,221,226
   quasy - 203

orthogonality 221-223
orthographic 132
output 206
overlap 35,36

P
P300 300
pacificism 430
paradigm 131
parallel 227
parietal area 309-311,313
pattern 222
pattern recognition 155,206,368
PDP-parallel distributed processing
   203-208,210,215,219,223,
   227,228
perception 236,302
permutation 20
perseverance 375
phonological 303,314
Popper, Karl 75
possibility theory 1-46
PRE-proportional reduction of error
   35,37
precision fallacy 427
predicate calculus 75
preference 12,20
premature birth 372
premise 53
prestige 439,445
Prisoner's Dilemma 43
probability theory 1-44,158
procesor 224
processing unit 206-208
proposition 76,146
propositional theory 61,62

prototype theory 159,167,182, 190,195
PRU-proportional reduction of uncertainty 37
psychoanalist 393
psycholinguistic 53,61,62,68, 146,350
psychometrics 447
punishment 387,394,398

Q
quantifier (universal, colloquial) 51-90
  linguistic 233-259

R
r-option 23,24
random dots 81
Raven Coloured Progressive Matrices Test 375
reading 370
real number 217
reasoning 51,53,63,68
recall 225,226,309
reception 179
  auditory 361
  visual 361
recognition 157,174,187,196, 302,303
recollection 223
recursive 170
reductionistic 196
redundancy 180
regression 336,337
rehearsal 372
reinforcement 267,269,392,394,398
reliability 445
remembering 371
representation 221
representative 210
residue number 155,179,180,187
response latency 130
  time 130
reward 386
rheme 304-306,313,314
robotics 206
rule based 321-342
Russel's antinomy 156,163,191,195

S
scale analogue 66
  interval 327,408
  numeric 266
  ordinal 327,408
  rating 53,63
scaling 58
  multidimensional 266
school 353
screeching 378
script 298
selection 385
self-concept 427,429
semantic differential 429,431
sensory 300,371,374
sentence-picture verification 61
sex-type 440,444,445,448
short-term memory 372
sigle-layer 210
single processor 204
skill-based 321-342
sleep disorder 384
slow wave 307
sociobiology 44
speech shadowing 302

spelling 376,377
state array 219,220
state-trait 427
stimulus 146-148
    conditioned 393
    unconditioned 393
stochastic 1,207,248,321,328, 330-333
stuttering 306
subalgorithm 395,398
subjective lexicon 196
subjectivism 346
subordinate 174,192,194,196
sum-product 98,97
superordinate 174,192,194,196
swallowing 378
syllogism 53,57,63,68
syllogistic 83
symmetry 94
synapses 177
synaptic 223,224

T
t-conorm 336
t-norm 94,100,108,111,120,336
t-test 108
theme 298,304-306,313,314
thermodynamic 178,180,346
think-aloud protocol 81
threshold 145,146,219,223,251, 271-273
tipicality 159
tolerance 270,272
top-down 128,129
transitivity 56,283,321,328,332-334
Turbo PASCAL 432
typical 210

U
uncertainty 41
unconscious 40
union 96,100,105,106,108,109, 111-113,116
    bold 337
    drastic 337

V
vague 177
vagueness 91
value 266
variation coefficient 307,309
vector 208
    binary 216,219
Venn diagram 53
verbal absurdities test 375
    analogies test 375
    opposites test 375
    uncertainty 82
visual-motor coordination 361
vocabulary 375
vocational 428
Vocational Preference Inventory (VPI) 442,444
VOT (voice onset time) 303

W
weak association 34
    entailment (WEM) 34,37,38
    order 327,328
    prediction 33
well-defined 176,187
Wernicke 313
WFEV weighted fuzzy expected value 203,210,217,218, 226-228

WISC-R Mazes subtest 376
word identification 130-150
word list (random) 130,131

Y
Youden Square 326

Z
Z-transformation 440